T0305687

The Study of Behavior

Organization, Methods, and Principles

Behavior studies now span a variety of sub-disciplines, including behavioral ecology, neuroscience, cognitive psychology, and evolutionary developmental biology. While the fields' rapid growth has led to startling new insights into animal behavior, it has brought increasingly fragmented approaches to the subject.

Integrating ideas and findings from a range of disciplines, this book provides a common framework for understanding diverse issues in behavior studies. The framework is derived from classical ethology, incorporating concepts and data from research in experimental psychology, neurophysiology, and evolutionary biology. Hogan outlines the origin and development of major ideas and issues in the field, drawing on examples throughout to highlight connections across sub-disciplines. Demonstrating how results in one area can directly inform work in others, the book ultimately proposes concepts to facilitate new discussions that will open the way for improved dialog between researchers across behavior studies.

JERRY A. HOGAN is Emeritus Professor of Psychology at the University of Toronto. A researcher of animal behavior for more than 50 years, his work focuses on understanding the structure, motivation, and development of behavior, with tropical fish and junglefowl as primary models. A past president of the International Society for Comparative Psychology (2002–2006), Hogan is also a co-author of *Causal Mechanisms of Behavioural Development* (1994) and a leading contributor to *Tinbergen's Legacy: Function and Mechanism in Behavioral Biology* (2009), also published by Cambridge University Press.

The Study of Behavior

Organization, Methods, and Principles

JERRY A. HOGAN
University of Toronto

CAMBRIDGE
UNIVERSITY PRESS

University Printing House, Cambridge CB2 8BS, United Kingdom

One Liberty Plaza, 20th Floor, New York, NY 10006, USA

477 Williamstown Road, Port Melbourne, VIC 3207, Australia

314-321, 3rd Floor, Plot 3, Splendor Forum, Jasola District Centre, New Delhi - 110025, India

79 Anson Road, #06-04/06, Singapore 079906

Cambridge University Press is part of the University of Cambridge.

It furthers the University's mission by disseminating knowledge in the pursuit of education, learning and research at the highest international levels of excellence.

www.cambridge.org
Information on this title: www.cambridge.org/9781107191976
DOI: 10.1017/9781108123792

First published 2017

A catalogue record for this publication is available from the British Library

ISBN 978-1-107-19197-6 Hardback
ISBN 978-1-316-64219-1 Paperback

Cambridge University Press has no responsibility for the persistence or accuracy of URLs for external or third-party internet websites referred to in this publication, and does not guarantee that any content on such websites is, or will remain, accurate or appropriate.

Contents

Preface

The idea for this book germinated more than 50 years ago when I was a postdoc in ethology at the *Rijksuniversiteit* in Groningen, Netherlands. I had recently received my Ph.D. in experimental psychology from Harvard and realized that ethology and psychology had much in common. At the time, most people felt that the two disciplines were conceptually very far apart, which was definitely not the case. A book pointing out the similarities and differences between the fields seemed a worthwhile endeavor. In the event, I was somewhat lazy and Robert Hinde beat me to the punch with his book, *Animal Behaviour*, in 1966, and the second edition in 1970. It was actually a good thing because Robert knew much more than I did, and also knew all the players. It would have been a very uneven competition had I written a book then. But I continued to harbor the idea of a book. Getting married, having children, getting a real job, and moving to a new country, however, made writing a book out of the question for quite a while, and time went by. Things changed, and experimental psychology and ethology almost merged for a short time, but then ethology largely transmuted into behavioral ecology and experimental psychology became either cognitive or neurophysiological; behavior genetics, neuroethology, and evolutionary psychology also emerged as new fields. Each field was asking its own questions and developing its own concepts and vocabulary. And there was very little communication among fields. A kind of intellectual isolation set in, and the reasons for writing a book became even more compelling than they had been originally. But by then, I had retired, and writing a book was not high on the priority list of this retired professor.

Some years later, however, while visiting Brazil, I was having lunch with my old friend César Ades, a professor of experimental psychology at the University of São Paulo. César knew I liked Brazil and he suggested that, if I had a project, I could apply for a visiting professorship at the Institute for Advanced Studies at the University. Of course, he knew that I had a book project, and after some thought, I did apply. Less than a month later, César was dead, tragically struck by a car while jogging. But he had already spoken to others at the Institute. My application was considered and I was offered a fellowship. That was the motivation I needed, and thus began the book.

There are several features of the book I should mention. First, it is not meant to be a normal textbook. It is a monograph that presents my ideas about various aspects of behavior. For the topics I cover, I make no attempt to review all the relevant literature. I usually discuss only one or two examples, though I often suggest other sources where further information can be found. Second, it is meant to be semi-historical. For most topics, I provide some background to the ideas I discuss, and if there is a choice, I generally pick the oldest good example that supports the point I wish to make. In many cases, I also refer to a relevant recent example. Third, I make extensive use of quotations from original sources. If an author makes a point clearly, I prefer using the author's own words rather than trying to paraphrase. In a few cases, I do both. Occasionally, this method interrupts the flow of the text, but I am convinced that the educational advantages far outweigh the minor disadvantages. Finally, Chapter 1 is a rather condensed statement of my worldview with respect to the study of behavior, and provides the rationale for the organization and content of the book. Some readers may have difficulty appreciating these points. I suggest that such readers quickly move on to the rest of the book that is written in what I hope is a more accessible style. All the issues raised in the first chapter are discussed again in various contexts throughout the book. Further, although it is possible to read each chapter independently, I generally assume the reader has already read what has gone before. I also assume the reader knows a bit of basic biology and psychology.

I have borrowed liberally from material I have previously published. Most of Chapter 1 was published as Hogan (2015), and many sections of Chapters 3, 5, and 6 can be found in Hogan (2001, 2005). However, all sections have been revised and updated and thoroughly edited. The finished product reflects my current thinking.

And now, some acknowledgments. I am deeply grateful to the *Instituto de Estudos Avançados da Universidade de São Paulo* (IEA) for taking me into their midst and providing me with the community and facilities that made this book possible. I would especially like to thank the director, Martin Grossmann, and his assistants Richard Meckien and Rafael Borsanelli for their help in many ways throughout my stay in Brazil. I am also grateful to Camila Nihei and Altay de Sousa for helping me adjust to life in Brazil, and for their continuing help in my times of need. I would also like to acknowledge the fantastic resources of the libraries of the University of Toronto, the second pillar of support for the book.

Many people read and commented on parts of the book as it was being written, or helped in other ways. They include (in alphabetical order by first name: the Brazilian way): Allert Bijleveld, Geoffrey Hogan, Hendrik van Kampen, Johan Bolhuis, Luc-Alain Giraldeau, Marco Varella, Martin Daly, Michael Ashton, Murillo Pagnotta, Paul Frankland, Peter Slater, Robert Lockhart, Roland Maurer, Stefan Köhler, Stephen Hutchings, Theunis Piersma, and Tom Morgan. I thank them all, especially Hendrik who read and commented on almost the entire manuscript.

1

The Framework

The purpose of this book is to discuss a broad range of behavioral phenomena using a common framework. In the mid-twentieth century, several general frameworks for studying behavior were proposed that became widely used (e.g. behavioristic psychology: Skinner, 1938; ethology: Tinbergen, 1951). But as many studies of behavior became more molecular (behavioral neurophysiology, behavioral genetics), more cognitive, or more molar (behavioral ecology, evolutionary psychology), new concepts evolved and many old concepts were discarded. Further, in both the older and the newer fields, new questions about behavior were being asked. A major result of this expansion was that scientists in one field were often unaware of or did not understand the work being done in other fields; in many cases, they also found such work to be irrelevant or even misguided. I hope to show that, insofar as one is actually interested in behavior, a common set of concepts can be used to understand and discuss issues in all these fields. In this chapter, I first discuss my definition of behavior, its units and levels of analysis. I then analyze the questions that can be asked about behavior, and finally consider problems related to teleology, cause, and function. I end with some comments on the functional study of behavior.

WHAT IS BEHAVIOR?

One definition of behavior that is used by some behavioral scientists, and that corresponds in many ways to common sense, is that which an animal does; what it does consists of muscular contractions and glandular secretions. This definition, however, does not include many phenomena, such as perceptions and feelings, that intuitively belong in the concept. Another definition brings in the concept of mind, because mind does include all the phenomena one expects. Although I am sympathetic to this solution, I prefer a more corporeal concept. I define behavior as the expression of the activity of the nervous system, which may be manifested as activity in neurons, muscles, and glands (Hogan 1984; 1994a).

An important point to be made about this definition is that it does not imply that the study of behavior involves neurophysiology. The study of behavior is the study of the *functioning* of the nervous system and must be carried out at the behavioral level, using behavioral concepts (cf. Skinner, 1938; von Holst & von St. Paul, 1960). Physiology and neurophysiology in particular may provide useful insights into the functioning of the nervous system, but the major concern of behavioral science is the *output* of the nervous system, manifested as perceptions, thoughts, feelings, and actions. I discuss aspects of this definition throughout the book.

Units of Behavior

No two occurrences of behavior are ever identical, and it is therefore necessary to sort behavior into categories in order to make scientific generalizations. What are these categories or units of behavior, and how are they organized? Categories can be defined in different ways (e.g. structurally, causally, functionally, historically: Hinde, 1970, Ch. 2) and at different levels of complexity (e.g. individual muscle movements, limb movements, acts: Gallistel, 1980). The answer to the question 'what are the units of behavior?' depends on one's definition of behavior, as well as on one's level of analysis. It also depends on the questions one hopes to answer. My behavior units (*perceptual, central, motor mechanisms*) are defined structurally, from a causal perspective, ultimately, in terms of neural circuits that produce the perceptions, thoughts, and motor patterns that are relevant to the basic functional categories of life (feeding, defense, reproduction). A similar approach is taken in the book by Gallistel (1980): *Organization of Action.* Although Gallistel's units (*reflexes, oscillators, servomechanisms*) are more complex than mine, both proposals define units structurally from a causal perspective.

However, as mentioned above, Hinde (1970) has discussed how behavioral units can also be defined functionally. Skinner's (1938) distinction between operants and respondents is an example of functional categorization. These units are distinguished by the operations that affect them, and not by their structure. The 'bar press' becomes the unit of analysis; and it does not matter (in most cases) how the bar comes to be pressed. Functional units are also used in the field of behavioral ecology (see Giraldeau, 2005, in press). Investigators in this field are interested in foraging or migration or optimality, and it does not matter which behavior patterns are used to accomplish these ends. As Hinde points out, different kinds of classification are appropriate for answering different kinds of questions. This issue is discussed further below.

Units of Behavior: Behavior Systems

I will develop the concept of a behavior system using structural categories at a level of complexity indicated by terms such as feeding behavior, aggressive behavior, play behavior, and so on. These terms can be considered names for behavior systems as a whole, but my analysis begins with a consideration of the parts of which these systems are constructed.

Three kinds of parts can be distinguished: motor parts, perceptual parts, and central parts. All of these parts are viewed as corresponding to structures within the central nervous system, and I will refer to them as *mechanisms*. The word *mechanism* usually connotes analysis at a molecular level, and might seem to imply that behavioral analysis should occur at a neural level. The *American Heritage Dictionary* (Morris, 1969), however, defines a mechanism as "the arrangement of connected parts in a machine" or "any system of parts that operate or interact like those of a machine". It defines a machine as "any system ... formed and connected to alter, transmit, and direct applied forces in a predetermined manner to accomplish a specific objective". This definition is agnostic with respect to the level of analysis. I use the word mechanism to emphasize the fact that the perceptual, motor, and central units of behavior systems are structural concepts arrived at by causal analysis at the behavioral level, as I discuss below.

Each motor mechanism, perceptual mechanism, or central mechanism is conceived of as consisting of some arrangement of neurons (not necessarily localized) that is able to act independently of other such mechanisms. I call these parts *behavior mechanisms* for two reasons. First, the actual neural connections, their location, and their neurophysiology are not of direct interest in the study of behavior. Second, the activation of a behavior mechanism results in an event of behavioral interest: a particular perception, a specific motor pattern, or an identifiable internal state. This conception can also include entities such as ideas, thoughts, and memories, which are cognitive structures proposed by many psychologists. Behavior mechanisms can be connected with one another, and the organization of these connections determines the nature of the behavior system. In order to make the discussion more concrete, I shall use the feeding system of a chicken as my example, but the concepts apply to all behavior systems in all species, including human language (Hogan, 2001).

Motor Mechanisms

We say a chicken is feeding when it walks about looking at the ground, when it scratches at the substrate, and when it pecks and swallows small objects. Walking, scratching, pecking, and swallowing are all easily recognizable motor patterns and can be viewed as reflecting the motor mechanisms of the feeding system. Three points here are worthy of mention.

First, although the behavior patterns of walking and so on are easily recognizable, there is considerable variation among different instances of the 'same' pattern. In a practical sense, this variation does not usually interfere with the identification of a pattern, and that is sufficient for most purposes. The second point is essential. What we observe is only a reflection or manifestation of the motor mechanisms of the system. The motor mechanisms themselves are groups of neurons located inside the central nervous system of the animal; activation of a motor mechanism is responsible for coordinating the muscle movements and glandular secretions that we actually observe. Finally, the concept of a *motor mechanism* is clearly related to the ethological concept *Erbkoordination* (Lorenz, 1937) or *fixed action pattern* (Tinbergen, 1951; Hinde, 1970), but is meant to be much broader in scope and to encompass all types of coordinated movements.

Perceptual Mechanisms

Corresponding to the motor mechanisms on the efferent side of a behavior system are perceptual mechanisms on the afferent side. Perceptual mechanisms solve the problem of stimulus recognition and are often associated with particular motor mechanisms. In the feeding system of a chicken, there must be perceptual mechanisms for recognizing the objects at which the bird pecks, for what it swallows, and for the type of environment in which the bird scratches. There must also be perceptual mechanisms that are sensitive to changes in the chick's internal state consequent to its behavior. Particular perceptual mechanisms may be restricted to a single sensory modality, but they frequently integrate information from several modalities.

Perceptual mechanisms are inherently more difficult to study than motor mechanisms, because the output of a perceptual mechanism can generally only be 'seen' after it has activated some motor mechanism. However, in some cases, modern imaging technology allows more direct observation of perceptual mechanisms, and it will undoubtedly become even more useful as the technology becomes more precise. In any case, the general method used to study perceptual mechanisms is to present stimuli that vary along different dimensions and to ascertain which combination of characteristics is most effective in bringing about certain responses.

The concept *perceptual mechanism* is clearly related to concepts such as *releasing mechanism* (Lorenz, 1937; Tinbergen, 1951; Baerends & Kruijt, 1973); *Sollwert*, or *comparator mechanism* (von Holst, 1954; Hinde, 1970); *cell assembly* (Hebb, 1949); and *analyzer* (Sutherland, 1964). However, as with the term motor mechanism, perceptual mechanism is meant to encompass all types of stimulus recognition mechanisms.

Central Mechanisms

Central mechanisms do not differ in any basic way from motor or perceptual mechanisms; they are distinguished separately because of their function, and come in several varieties. I find it useful to distinguish among representations (thoughts, memories, ideas), comparator mechanisms, oscillators, and central coordinating mechanisms. These distinctions are discussed in detail in Chapter 2.

Here, I want to emphasize the central coordinating mechanism, which is the part of the behavior system that is responsible for integrating the input from various perceptual mechanisms and coordinating the output to the various motor mechanisms associated with it. In many cases, it is also responsible for the timing and activation of the whole behavior system. It is the central coordinating mechanism that usually corresponds to the name we give to a behavior system: a hunger mechanism, an aggression mechanism, a sexual mechanism, and so on. The concept *central coordinating mechanism* is clearly related to the neurophysiological concepts *central excitatory mechanism* (Beach, 1942), *central motive state* (Stellar,

1960), or *neural center* (Doty, 1976), but it is used here in a still more general sense that would also include the *modules* posited by many cognitive psychologists (Barrett & Kurzban, 2006).

Behavior Systems

We can now return to the concept *behavior system* and define it as an organization of perceptual, central, and motor mechanisms that acts as a unit in some situations. A pictorial representation of this definition is shown in Figure 1.1. The first part of the definition is structural and is basically similar to Tinbergen's (1951, p. 112) definition of an instinct, but it is not meant to have connotations of 'innate' that were implied in Tinbergen's definition. It is also similar to the *functional organization* of von Holst & von St. Paul (1960). Hierarchical organization of control is also implied in this part of the definition (see below in section on levels), and it is thus related to conceptions of Tinbergen (1951), Baerends (1976), and Gallistel (1980); see also Hogan (1981). As well, these concepts are clearly compatible with current models of neural networking (e.g. Enquist & Ghirlanda, 2005). Further, as we shall see, there are various levels of perceptual and motor mechanisms, and the connections among them can become very complex. A diagram such as Figure 1.1, if expanded to encompass all the facts that are known, would soon become unmanageable. In the extreme, it would become congruent with a wiring diagram of the brain. The main function of such a diagram, and of the concept of a behavior system, is to direct our thinking into particular pathways. The second part of the definition of a behavior system is causal. Ideally, it might be desirable to have a purely structural definition, but at present, the only method for determining behavioral structure is through causal (or motivational) analysis.

Finally, it should be obvious that the concepts I am proposing are derived from the concepts of classical ethology. Many of these concepts have been derided and discarded as being old-fashioned and out of date. It is my opinion that the core of many of these classical concepts still provides the best vocabulary for describing and understanding behavior. It is my hope that my generalizations of the original concepts will prove useful for encouraging constructive dialog among the various fields of behavioral study. I also hope that my emphasis on the idea of behavioral structure will lead to improved ways of conceptualizing many aspects of current studies of behavior, as I plan to show in later chapters.

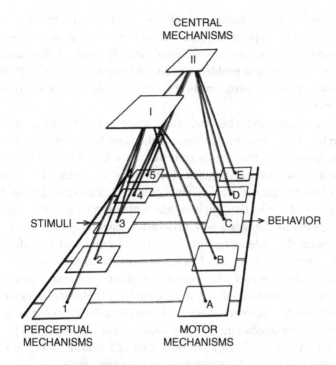

Figure 1.1 Conception of behavior systems. Stimuli from the external
world are analyzed by perceptual mechanisms. Output from the
perceptual mechanisms can be integrated by central mechanisms and/or
channeled directly to motor mechanisms. The output of the motor
mechanisms results in behavior. In this diagram, central mechanism I,
perceptual mechanisms 1, 2, and 3, and motor mechanisms A, B, and
C form one behavior system; central mechanism II, perceptual
mechanisms 3, 4, and 5, and motor mechanisms C, D, and E form a second
behavior system. Systems 1-A, 2-B, and so on can be considered less
complex behavior systems.
From Hogan, 1988, with permission

Levels of Analysis

The phrase 'levels of analysis' has been used in several ways in the
study of behavior. For example, Tinbergen (1963), Hinde (1970), and
Hogan (1994a) use this phrase to distinguish among analyses on a scale
that goes from genes to biochemistry to physiology to behavior.
The word 'level' has also been used by Schneirla (1949) in his discussion
of levels of organization in both evolutionary and developmental contexts. In all these cases, the scale is one of complexity. Although the

idea of levels of complexity is usually straightforward, problems can
arise because concepts used to analyze behavior at the various levels
can be different and are often incompatible. Schneirla & Rosenblatt
(1961) emphasize these problems, as do von Holst & von St. Paul (1960).
I discuss several such problems in the section on the relations of cause
and function.

Tinbergen (1951) also talks about levels in his discussion of the
hierarchy of instincts. In this case, higher and lower levels refer to
which elements of the system control each other. In discussions of
hierarchy, one can distinguish hierarchies of connection from
hierarchies of classification or embedment (R. Dawkins, 1976a). In
a hierarchy of connection, superior elements exert control over inferior elements in the same way in which a captain gives orders to
a lieutenant. In a hierarchy of classification or embedment, inferior
elements are actually a part of superior elements in the same way in
which a platoon is part of a company. Platoons and companies are
different classifications of the same elements: soldiers. Tinbergen's
hierarchy is one of control, whereas 'levels of analysis' as used here is
a hierarchy of embedment. (See also Gallistel, 1981; Hogan, 1981.)

Yet another use of the phrase is that of Pagnotta (2014). In his
discussion of different interpretations of the word 'culture' by anthropologists and biologists, he distinguishes among levels of description,
explanation, theory, and worldview (*Weltanschauung*). Somewhat similar distinctions were also made earlier by Vygotsky (1934). These distinctions can be very useful for focusing on the actual points of
disagreement when scientific controversies arise. This is also
a hierarchy of embedment.

Finally, the phrase 'levels of analysis' has also been used in
a completely different way by many evolutionary biologists and psychologists. They use the phrase as a synonym for Tinbergen's four
questions. Problems with this usage are discussed in the next section.

WHAT DO WE WANT TO KNOW ABOUT BEHAVIOR?

In his well-known article, "On Aims and Methods of Ethology",
Tinbergen (1963) stated that there are four kinds of questions one can
ask about any biological phenomenon: causation, survival value, ontogeny, and evolution. He felt that understanding any example of behavior required answers to all of these questions. These distinctions are
important, and have had, and continue to have, a great influence on
students of animal behavior (e.g. Dewsbury, 1992; Hogan, 1994a;

Hogan & Bolhuis, 2009; Bateson & Laland, 2013). However, as I have pointed out elsewhere (Hogan, 1984; 1994a), these questions do not cover all the important aspects of behavior. Tinbergen (1963, p. 426), himself, stated: "in speaking of the 'four problems of Biology' we apply a classification of problems which is pragmatic rather than logical." A more general classification of questions can be derived from some distinctions made originally by Aristotle, and I will propose such a classification in the next section.

A somewhat different classification of problems in biology has been important among evolutionary biologists. Baker (1938) introduced the terms 'proximate causation' and 'ultimate causation', and Mayr (1961, p. 1503) used these terms to distinguish between the problems of interest for the functional biologist (proximate causes: "the immediate set of causes") and for the evolutionary biologist (ultimate causes: "causes that have a history and that have been incorporated into the system through many thousands of generations of natural selection"). Mayr's use of the phrase 'proximate cause' corresponds almost exactly to Tinbergen's use of the word 'causation'; but the ultimate causes of Mayr correspond only partially to Tinbergen's problems of survival value and evolution, even though most current authors treat them as the same (cf. Cuthill, 2009).

It is Mayr's terminology that has become the standard in the fields of behavioral ecology and evolutionary psychology, but as with Tinbergen's terms, the words 'proximate' and 'ultimate' have many meanings, and various interpretations of these terms soon appeared. Francis (1990) published a critique of the distinction, especially of the term 'ultimate'. Francis argued that explanations referring to ultimate causes typically emerge from functional analyses, but that functional analyses do not identify causes of any kind resembling those of proximate explanations. He also pointed out that the term 'ultimate cause' implies that functional analyses are somehow superordinate or superior to those involving proximate causes. He concluded that ultimate causes are neither ultimate nor causes.

Mayr (1993) responded to Francis's critique quite vehemently. I will not discuss all the relevant issues here, but one general point is worth making. Mayr chided Francis for going to the dictionary for the definition of the word ultimate: "It has always seemed to me a dubious procedure to search in a dictionary for authoritative information on a scientific term" (p. 93). I agree that the dictionary is not the place to look for an authoritative scientific definition of a term, but the problem is that most people who read a word will understand that word in its usual meaning, whether an author intends it that way or not.

Mayr (1993, p. 94) provides a scientific definition of 'ultimate causations': the laws "which cause changes in the DNA of genotypes". This formulation is an improvement on his earlier definition, cited above, because it focuses on the causes of the variation that lead to natural selection. Unfortunately, this meaning is not what most authors of papers in behavioral ecology and evolutionary psychology have in mind when they use the term. A typical interpretation, which is still widely held, is: "Ultimate causation concerns adaptive signifi-cance" (Daly & Wilson, 1978, p. 9), a meaning Mayr specifically eschewed. These issues have continued to be discussed and debated in both psychological and biological circles (e.g. Dewsbury, 1999; Bolhuis, 2009). Laland *et al.* (2011) have even suggested that the prox-imate-ultimate dichotomy is no longer useful, and may actually ham-per progress in the biological sciences.

A further complication concerns the use of the term 'levels'. Daly & Wilson (1978, p. 11), for example, in a section entitled 'levels of behavioral explanation', mention Tinbergen's "four great problems ... cause, function, development, and evolution. They correspond to our proximate causation, ultimate causation, ontogeny, and phylogeny." Similarly, Sherman (1988, p. 616) says: "In summary, there are four different levels of analysis: evolutionary origins, functional conse-quences, ontogenetic processes and mechanisms." This terminology continues to be used in the present (e.g. Barnard, 2004; Ryan, 2009). Although all behavioral scientists would agree that there are different *kinds* of explanations, the use of the word 'level' is unfortunate. The AHD defines 'level' as "relative position or rank on a scale". It is not possible to rank Tinbergen's questions on any reasonable scale, although the word 'ultimate' may suggest to some that ultimate explanations are somehow higher or more important than other kinds of explanations (cf. Francis, 1990; Dewsbury, 1994; 1999).

Lehrman (1970, pp. 18–19) addressed similar issues with a very wise statement that is as relevant today as it was when it was written:

> When opposing groups of intelligent, highly educated, competent scientists continue over many years to disagree, and even to wrangle bitterly, about an issue which they regard as important, it must sooner or later become obvious that the disagreement is not a factual one, [...].
> If this is, as I believe, the case, we ought to consider the roles played in this disagreement by semantic difficulties arising from concealed differences

in the way different people use the same words, or in the way the same people use the same words at different times; by differences in the concepts used by different workers (i.e., in the ways in which they divide up facts into categories); and by differences in their conception of what is an important problem and what is a trivial one, or rather what is an interesting problem and what is an uninteresting one.

I hope the following discussion will clarify some of the current misunderstandings.

A Classification of Questions

In discussing physics and metaphysics, Aristotle pointed out that one and the same thing could be described or explained in four different ways. These types of explanation have been called 'causes' and are usually listed as: 1) material, 2) efficient, 3) formal, and 4) final cause. A standard example is the description of a chair: It may be made of wood (cause 1), have been manufactured in such and such a way by Mr. Smith (cause 2), have four legs, a seat, and a back arranged in a particular way (cause 3), and be used to sit in (cause 4). Just as Tinbergen argued that our understanding of any biological phenom-enon is increased by asking all four questions, so Aristotle argued that our understanding of any phenomenon is increased by considering all four of its causes.

If we apply Aristotle's classification to behavior, we discover that it can subsume Tinbergen's four questions, and that additional ques-tions arise as well. It will be helpful first to translate Aristotelian terminology into terms that are more suitable for the analysis of beha-vior. I shall talk about 1) the matter, 2) the causation, 3) the structure, and 4) the consequences of behavior. For most purposes, the matter of behavior is the same regardless of the kind of behavior being consid-ered, so this question is primarily relevant to the definition of behavior itself. As stated above, I define behavior as the expression of the activity of the nervous system, and I will discuss a number of issues arising from this definition later. I will first consider some of the implications of Aristotle's other three distinctions for the analysis of behavior, and also indicate some of the relationships among the questions.

Causation of Behavior ('Efficient Cause')

Aristotle defines 'efficient cause' as "the primary source of the change or coming to rest; e.g. the man who gave advice is a cause, the father is cause of the child, and generally what makes of what is made and what causes change of what is changed" (McKeon, 1947, p. 122). In the remainder of this chapter, I shall use the terms cause and causal, without quotation marks, to refer to Aristotle's 'efficient cause', but retain quotation marks for Aristotelian usage. The causes of behavior include stimuli, the internal state of the animal, various types of experience the animal has had during its development, as well as the genes with which it is endowed. Understanding the mode of action of these causal factors, including their interaction with each other, is the primary goal of a causal analysis of behavior. I find it useful to distinguish among three types of causation in terms of the changes that are of interest.

The first type of causation is *motivational*, in which the immediate effects that causal factors have on behavior are of primary interest. This use of the word corresponds to Tinbergen's causal question; it is also similar to the phrase 'proximate causation' as used by many biologists. An electric shock causes the rat to jump, hunger causes the chicken to peck; advice causes the man to spend his money in a certain way. Note that the term motivation is used here in a much broader way than is often the case, in that it refers to external as well as internal causes of behavior. In fact, this use of the term corresponds to its original sense: that which causes motion.

The second type of causation is *developmental* (Tinbergen's ontogenetic question), in which changes in behavior during the course of an individual's lifetime are of primary interest. What effects does the presentation of a small red cube to a young chick have on its behavior when it grows up, and how do those effects come about? Will the adult chicken prefer to mate with the red cube rather than with a conspecific? The causal factors in development are the same stimuli, internal states, and so on that are important for motivation, but one is primarily interested in their longer term effects. The red cube certainly affects the behavior of the young chick when it is presented, as does the father of the child affect the behavior of the mother prior to conception, but it is the origins and course of development of adult sexual behavior or of the child that one wishes to explain.

It should be noted that distinguishing between motivation and development on the basis of time scale (e.g. Tinbergen, 1972) is not fundamental, because a more basic distinction is whether the changes brought about by the causal factors are reversible or permanent (i.e. result in a change in behavioral structure – see below). For example, the yearly cycle of gonad growth and regression in some birds and fish as a result of changes in day length is generally considered to be a problem in motivation, whereas one-trial learning to avoid electric shock in a rat is a problem in development. The former changes occur over days or weeks, whereas learning occurs in a few seconds. On the other hand, reversible and permanent are themselves relative concepts and in any particular example, the two can fade into each other (cf. Frankland *et al.*, 2013). Thus, it is probably best to consider that motivational and developmental changes often reflect a different time scale, but more basically reflect underlying causal mechanisms that are more or less reversible.

A third type of causation is *phylogenetic* (Tinbergen's evolutionary question), in which changes in behavior over generations are of primary interest. Here, the concern is what effect causal factors have on genetic structure, because in most species, only genes can be passed on from one generation to the next. It is generally thought that these are the factors that are responsible for genetic recombination and mutation, but recent evidence suggests that other factors in an animal's lifetime can also cause genetic change, and this evidence will be discussed later. Further, in species in which specialized modes of conspecific communication have developed, as in humans, there is no reason to exclude culturally transmitted information as a causal factor in the phylogeny of behavior. And finally, it may be surprising to some readers that I consider phylogenetic changes to be a causal problem. I also discuss this issue later in the section on the relationship between cause and function.

Structure of Behavior ('Formal Cause')

Aristotle's 'formal cause' refers to the parts of an object and the relations among them. A chair consists of four legs, a seat, and a back arranged in a particular way with respect to each other. By analogy, one can also describe the form or *structure* of behavior. What are the units of behavior and how are they organized? This question has already been discussed extensively. The units are the perceptual, central, and motor mechanisms described above and arranged as depicted in Figure 1.1. These units are all described in more detail in Chapter 2.

Consequences of Behavior ('Final Cause')

Aristotle's fourth question, 'final cause' introduces the notion of purpose. As I argue below, using the purpose or function of behavior as a cause of behavior is unscientific and should be avoided. However, asking about the consequences, or effects, of behavior raises many questions that can be investigated. For example, the consequences of a behavior such as pecking may include closing a microswitch in a Skinner box, enjoying the performance of pecking itself, reducing hunger, providing nutrients necessary for egg production, or increasing fitness. Closing a microswitch and enjoying the activity are immediate consequences of performing the behavior, hunger reduction takes many minutes, producing eggs takes hours or days, and increasing fitness can only be measured over generations.

These various consequences introduce the problem of time scale (cf. Zeiler, 1992), which we have already met when discussing the causes of behavior. Time itself is no more satisfactory a criterion for distinguishing among consequences of behavior than it was for distinguishing among causes of behavior. A better criterion might be the type of changes being considered. Thus, one could distinguish motivational, developmental, and phylogenetic consequences, depending on the question of interest. Eating reduces hunger, reinforces a stimulus-response association, and allows an animal to pass its genes to future generations.

Alternative terms could also be used to describe the different effects of behavior. For example, feedback is a concept that includes many motivational consequences of behavior, and reinforcement and learning would be a primary developmental consequence of behavior. Most biologists would use the word function for the phylogenetic consequences of behavior (e.g. Hinde, 1975). In this context, function is synonymous with 'survival value', which is Tinbergen's remaining question.

The Concept of Function

The term 'function' is actually used in various ways even among biologists. This is not the place to discuss all these uses (see Bock & von Wahlert, 1965; Hinde, 1975; Wouters, 2003; Cuthill, 2009), but there are four uses that are especially relevant in the study of behavior (Wouters, 2003, p. 633):

> ... there are at least four different ways in which the term 'function' is used in connection with the study of living organisms, namely: (1) function as (mere) activity, (2) function as biological role, (3) function as biological

advantage, and (4) function as selected effect. Notion (1) refers to what an item does by itself; (2) refers to the contribution of an item or activity to a complex activity or capacity of an organism; (3) refers to the value for the organism of an item having a certain character rather than another; (4) refers to the way in which a trait acquired and has maintained its current share in the population.

The first two uses are basically the everyday meanings of the word. Notion (1) asks 'how does it work'? It is a structural concept more or less synonymous with the concept 'mechanism'. When an animal is deprived of food, it eats; that is how the hunger system works if it is functioning properly. Notion (2) asks 'what is it for'? What function does this behavior serve in the life of the animal? Eating results in ingestion and provides the animal with nutrients and energy. These uses of the word function are essentially descriptions of current behavior, and are what biologists often call 'functional biology' (e.g. Mayr, 1961; Wouters, 2003). They make no claims about the fitness consequences of eating (survival value), or of its evolutionary history. Behavior systems are generally organized (function) in such a way (sense 1) that, when activated, they usually produce a functional outcome (sense 2). The other uses of the word 'function' will be discussed in the chapters on phylogeny.

A Revised Classification of Questions

The considerations presented above suggest a classification of questions one can ask about behavior. One can ask questions about the cause, structure, and consequences of behavior with respect to current behavior, ontogeny, and phylogeny. This classification is different from Tinbergen's classification, but includes all his questions. It also provides a framework in which one can see the relations among different fields of investigation, because all these questions can be investigated at different levels of analysis – from genetics and physiology to individual behavior, social behavior, and populations. It also provides the basic framework for the structure of this book. Chapters 2, 3, and 4 consider the structure, cause, and function of current behavior. Chapters 5, 6, and 7 discuss the structure, cause, and function of behavioral ontogeny, and Chapters 8, 9, and 10 discuss the structure, cause, and function of behavioral phylogeny.

THE RELATION OF CAUSE AND FUNCTION

'Cause' and 'function' are two concepts central to the content of this book. They are concepts that are defined independently of each other. With respect to behavior, cause is what instigates or modulates some behavior, and function is a special category of the consequences of performing that behavior. In spite of their logical independence, cause and function seem to co-mingle with each other in ways that are often confusing and sometimes just wrong. I will discuss four such problems.

Aristotle's 'Final Cause', and the Problem of Teleology

Aristotle defines 'final cause' as "that for the sake of which", and uses the following example to illustrate his meaning: "health is the cause of walking about. ('Why is he walking about?' we say. 'To be healthy ...')" (McKeon, 1947, p. 123). Another example is provided by various answers to the question: Why did the chicken cross the road? To get to the other side; to avoid being run over; to get food; to increase the likelihood of passing on its genes to subsequent generations. These answers all specify the goal or purpose of the behavior, and all imply that some outcome that has not yet occurred is the cause of the behavior. That this approach is fallacious can be seen by considering, for example, that the chicken might be hit by a car when crossing the road, or that there might be no food on the other side when it gets there; the man might drop dead of a heart attack because of walking about too strenuously. *The outcome of behavior can never determine its occurrence.* At best, the outcome of behavior can be one of the determining (causal) factors of future occurrences of similar behavior.

All the examples given above are examples of teleology, "the belief that natural processes are not determined by mechanism but rather by their utility in an overall natural design" (*American Heritage Dictionary*, Morris, 1969). In this sense, teleology does not lend itself to scientific investigation because it implies that the goal of behavior (what it is designed for) has become its 'efficient cause'. Unfortunately, Aristotle's 'final cause' is basically a teleological concept, and the biologists' concept of 'ultimate causation' (Baker, 1938; Mayr, 1961) is often used teleologically as well. Although it is unlikely that most scientists believe in intelligent design, the language they use in describing their results can often be interpreted that way. A common example is the statement that "natural selection has designed some aspect of an animal's behavior". Natural selection designs nothing! Natural selection is the *consequence* or outcome of the fact that some individuals are more successful than others.

It is possible to rephrase both the questions and answers about behavior so that teleological implications are avoided, and one can then talk quite scientifically about the outcomes or consequences of behavior (cf. Pittendrigh, 1958). For example, we can say the man is walking about because he believes that walking about promotes health, and he wants to be healthy. This belief and desire reflect the activation of specific cognitive mechanisms that exist before the behavior occurs. In people, such beliefs and desires reflect cognitive structures that have been built up by the specific experience of the individual. If we say only that the chicken crossed the road because it intended to get to the other side, we have merely restated the fact that it was seen to cross the road. A causal analysis would inquire into the origins of the intention (was it a product of phylogenetic experience or individual experience; and what were the selective forces or the particular experiences presumed to be necessary) and the causal factors activating the intention (what specific stimuli or internal state led to activation of that specific intention). It should be clear that without such an analysis, the use of cognitive language is no advance on the behavioristic language that has been used in the past.

This conception of cognitive structures also allows a way to resolve some of the current controversies in the field of animal cognition (e.g. Barrett & Kurzban, 2006; Shettleworth, 2010; Bolhuis *et al.*, 2011). We can begin with the notion that cognition only implies knowledge, and that knowledge is another way of saying that cognitive structures exist. I have already argued that activation of a cognitive structure is the cause of behavior. Insofar as ideas, beliefs, intentions, purposes, and the like are considered to be cognitive structures, the activation of the structures representing these entities can be a cause of behavior. In this framework, the concept of consciousness, for example, becomes an epiphenomenon, an outcome of the activation of a cognitive mechanism, and definitely not a cause of behavior. Interestingly, this approach was already espoused many years ago by T. Huxley (1893), who wrote: "The consciousness of brutes would appear to be related to the mechanism of their body simply as a collateral product of its working, and to be as completely without power of modifying that working as the steamwhistle which accompanies the work of a locomotive engine is without influence upon its machinery" (p. 240). Needless to say, I include human beings with the brutes in this context.

Finally, some authors use the word teleology in a somewhat different sense from the definition given above (e.g. Mayr, 1988; Hopkins & Butterworth, 1990). Such authors consider that teleological concepts can be used to explain the occurrence of some behavior. It is my opinion that it is better not to use the word teleology when discussing causes of behavior for the reasons given by Lehrman above.

The Problem of Units

As mentioned above, behavioral units can be described both structurally (as inferred from causal analysis) and by consequence. Any behavioral unit can be described both ways. A dog wags its tail in a certain way or it gives a friendly greeting to its master; both statements describe the same event. Similarly, a squirrel moves its front legs in a certain way for a specific length of time or it buries a nut. In both examples, the first description is structural and the second description is by consequence. As Hinde (1970) discusses, sometimes one kind of description is more useful than the other kind, and in many cases, one mixes both kinds of description. Except for sometimes not realizing that these descriptions are alternatives and that both are correct, this mingling of cause and function seldom causes any problems.

A more serious confusion stems from "our habit to coin terms for major functional units such as nest building, fighting, or sexual behavior and treat them as units of mechanism" (Tinbergen, 1963, p. 414). For example, sexual behavior, defined functionally, would be all behavior that serves a reproductive function. Defined structurally (as inferred from causal analysis), sexual behavior would be all behavior controlled by a specific sexual mechanism. The structural (causal) definition would exclude many displays seen during courtship; these displays may be necessary for successful reproduction, but they are believed, in many cases, to be controlled by aggression and escape mechanisms (Tinbergen, 1952; Baerends, 1975; Groothuis, 1994). The behavior systems of Timberlake (1994) are an example of mixtures of causal and functional classification.

The Problem of Levels of Analysis

A more serious problem is the co-mingling of cause and function with different levels of analysis. For example, in his concluding remarks to the section on causation, Tinbergen (1963, p. 416) states: "As far as the study of causation of behaviour is concerned the boundaries between these fields are disappearing, and we are moving fast towards one Physiology of Behaviour, ranging from behaviour of the individual and even of supra-individual societies all the way down to Molecular Biology." Although Tinbergen is here recommending that we abolish the distinctions between levels, another of his examples shows what confusion can result if this is done. He points out (p. 414) that "... the more sophisticated ethologist is fully aware of the fact that the term 'Innate Releasing Mechanism' refers to a type of function, of achievement, found in many different animals; and that different animal types may well have convergently achieved [this] by entirely different mechanisms ..."

The confusion is that, at the neurophysiological level, 'releasing mechanism' is a functional concept as Tinbergen states: it is postulated that somewhere in the nervous system, there must be an organization of neurons that is responsible for the properties attributed to releasing mechanisms. These properties could come about as a result of any of several different neural organizations. The actual neural organization and its location in the brain are not of special interest when analyzing behavior; what is of interest are the consequences of the activation of that mechanism. It is in this sense that von Holst & von St. Paul (1960) and Baerends (1976) refer to the functional organization of behavior. At the behavioral level, however, releasing mechanism is a causal concept; it is postulated that activation of a particular releasing mechanism is one of the controlling (causal) factors for the occurrence of a certain behavior. As von Holst & von St. Paul (1960) noted, it is necessary to use level-adequate concepts when studying complex behavior.

The Interaction of Cause and Function

By far the most serious problem concerning the relation of cause and function is the interaction attributed to them. The biological definition of function (sense 4, p. 15) as 'consequences that have been selected' and current biological usage of the phrase 'ultimate cause' both imply some sort of causal relationship between cause and function, a relationship that I maintain is teleological (and un-Darwinian). Several authors have argued specifically, however, that such usage is not teleological (e.g. Daly & Wilson, 1983, p.15; Alcock, 1989, Ch. 1). Their argument is that natural selection is a mechanism, and therefore a cause of evolution. But here is where the confusion of levels of analysis arises. From the point of view of the individual (or the gene itself), the cause of evolution is the mutations or other factors that bring about variability among individuals. Natural selection is the consequence or outcome of the fact that some individuals are more successful than others. Natural selection is only a mechanism of evolution at the level of the group or species, where it is a mechanism for changing the frequency of genes in the population pool. That is, in some particular environment, individuals with a certain genotype will be more successful than individuals with a different genotype. However, that environment does not cause the successful genotype to appear; it only maintains that genotype once it has been caused by other (causal) factors. The situation is similar to that in operant psychology. Reinforcement is an outcome when a response occurs; the response must occur first 'for other reasons' (Skinner, 1938).

One way to attack many of the controversies that currently surround the concept of causation is by always asking the simple question: what is causing what, and what are the consequences? I will take the case of niche construction as an example. Laland *et al.* (2011, p. 1514) state:

> earthworms change the structure and chemistry of the soils in which they live and, by constructing their environment, modify selection acting back on themselves, for instance, influencing their water-balance organs. Here again there is reciprocal causation. The ultimate explanation of the earthworm soil-processing behavior is selection stemming from a soil environment, but a substantial cause of the soil environment is the niche-constructing activity of ancestral earthworms.

Let's assume for the moment that the changing environment has no substantial effect on the reproductive output of the earthworm whose activities are changing the soil. However, the descendants of that earthworm will be growing up in an environment different from the one of their ancestors. Insofar as there is genetic variation among these descendants, some may be better adapted to the new environment than others (e.g. have better water-balance organs to deal with the new environment), and those genotypes will come to predominate in the population (i.e. the outcome of selection). The earthworm has *caused* a change in soil composition. Selection has *caused* a change in the population frequency of certain genes. So far, no problem.

A problem does arise, however, with the term 'reciprocal causation'. Does a change of gene frequency in a population imply any causal changes to the genes themselves? I would say no. *Selection is not a cause of genetic change.* Traditionally (i.e. in the framework of the Modern Synthesis – see Chapter 8), genetic change is thought to be exclusively due to random processes (mutation, recombination, drift). But, more generally, following Mayr's (1993) definition of 'ultimate causations' (the laws which cause changes in the DNA of genotypes), genetic change theoretically can be brought about by non-random forces. Beginning with Waddington's (1942, 1953) work on 'genetic assimilation', there are now many cases that show that the 'state' of particular genes can be inherited, and further that the 'state' of a gene can depend on the environment it has encountered in an individual organism's lifetime (Gilbert, 2012). The precise mechanism of inheritance still has to be worked out, but it is clear that environmental factors can cause heritable changes (e.g. Bohacek & Mansuy, 2015). In these cases, 'reciprocal causation' is a warranted description of one of the mechanisms of evolution. Interestingly, these cases are examples of 'ultimate causations' occurring within the lifetime of an individual organism!

THE FUNCTIONAL STUDY OF BEHAVIOR

Finally, I would like to make a few comments about the study of behavior from a functional point of view, in both psychology and biology. In psychology, the study of consequences was introduced by Skinner (1938) in his formulation of operant behavior and reinforcement. Operant behavior could be 'controlled' by its consequences, and reinforcement assumed the role of selection. As Timberlake (2004) points out, Skinner has commented: "operant behavior is essentially the field of purpose". It was assumed that any operant in any species using any reinforcer would obey the same laws of contingency. Any behavior that was reinforced with a certain contingency would show a pattern of responding that was characteristic for that specific contingency. The actual behavioral mechanisms involved in producing these effects were not considered to be relevant. In an extensive review of the properties of various reinforcers in various species, Hogan & Roper (1978) showed that these assumptions underlying the study of operant behavior needed to be modified, though this demonstration had very little effect on the field. More recently, Timberlake (2004) has also argued that analysis of the mechanisms of operant contingencies would lead to a more accurate picture of the control of purposive behavior. Further aspects of operant psychology are considered in Chapter 7.

In biology, the functional perspective underlies the field of behavioral ecology. McNamara & Houston (2009, p. 670) state: "Behavioural ecology often makes the assumption that animals can respond flexibly by adopting the optimal behaviour for each circumstance." These authors review many studies that demonstrate that behavior "is determined by mechanisms that are not optimal in every circumstance". They suggest that it is necessary to integrate the study of functions and mechanisms; they specifically propose that the field should consider the evolution of mechanisms. Fawcett *et al.* (2013, p. 2) take a similar stance: "Behavioral ecologists have long been comfortable assuming that genetic architecture does not constrain which phenotypes can evolve ... To understand the functional basis of behavior, we would do better by considering the underlying mechanisms, rather than the behavioral outcomes they produce as the target of selection." They also review studies that contradict the basic functional hypothesis, and suggest that one can only understand the functional basis of behavior by considering the evolution of learning and decision rules. This issue returns in Chapters 9 and 10.

In these parallel developments, practitioners in both psychology and biology are calling for an explicit recognition of the role mechanisms play in determining the behavior that we observe, even if their main interest remains the function of that behavior. However, cause and function (consequences) remain separate questions. One or the other or preferably both questions need to be asked about any behavior, and at different levels of analysis. How far understanding the cause of some behavior is helpful in understanding its function, and vice versa, is actually an empirical question. Sherry (2009), Bolhuis (2009), Bolhuis & Wynne (2009), Sherry & Strang (2015), and Lefebvre (2015) provide good discussions of this issue, and various examples are given throughout this book.

2

Structure of Behavior
Actions, Perceptions, Representations,
Behavior Systems

Gerard Baerends (1916–1999). Photo by Eddy de Jongh.

In Chapter 1, I described the units of behavior that I will be using in this book: perceptual, central, and motor units that can be combined into behavior systems. These are the structural units of behavior that I will be examining in more detail in this chapter.

MOTOR MECHANISMS: ACTIONS

If one looks at the behavior of any animal, it is obvious that behavior does not consist of random movements of the limbs and body. Various patterns of movements can be seen to recur, and in general, one has the impression that behavior is highly organized. As pointed out in Chapter 1, the way one describes and analyzes these movements depends on the question being asked. In this chapter, I will be discussing actions from a structural point of view, which means that behavior can be considered to be a series of spatial-temporal patterns of muscle contractions. For some purposes, contractions of one or a few individual muscle fibers may be of behavioral interest, but for questions about the behavior of the animal as a whole, more complex units are usually necessary. Lorenz (1937) proposed such a unit that he called an *Erbkoordination* (inherited coordination). Tinbergen (1951) translated this concept into English as *fixed action pattern*. Many aspects of this proposed unit of behavior will be discussed below, but I will begin by describing the example that gave rise to this proposal.

Lorenz & Tinbergen (1939), in a series of experiments, analyzed the egg-retrieval behavior of the greylag goose (*Anser anser*). When a brooding goose sees an egg just outside the nest, she is likely to rise, orient toward the egg, and then stretch her neck toward it. If she can reach the egg, she places her bill behind it and pulls it back to the nest in a very characteristic and stereotyped manner (see Figure 2.1). An observation of special significance in these experiments was that the egg-retrieval movement itself continued quite normally to completion even if the egg rolled away. The incongruous sight of a goose retrieving an imaginary egg led Lorenz to suggest that the form of the movement was not determined by stimuli from the environment, but instead was programmed in the central nervous system. The idea of central nervous system programming was derived from work of von Holst (1937) on the control of fin movements in fish. Von Holst had shown that the rhythmic movements of the fins were controlled by oscillators (rhythmic signal generators) in the brain in the absence of any sensory input, and Lorenz suggested that the egg-retrieval movement might also be controlled in a similar way.

Another important observation was made when the egg rolled away. During retrieval with an egg present, the goose can be seen to make small lateral movements with her bill that serve to keep the egg balanced. When the egg rolls away, however, the retrieval movement continues, but the balancing movements of the bill stop. This result suggested to Lorenz that the egg-retrieval movement as a whole could be analyzed into a fixed component (the *Erbkoordination*) and a variable or orientation component that he called a *taxis*, which is a series of simple reflexes. The form of the fixed component is determined

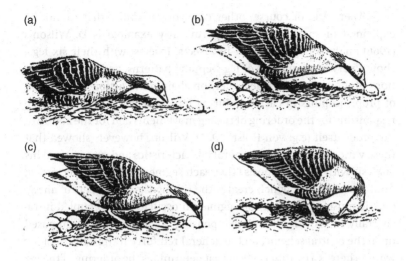

Figure 2.1 Greylag goose retrieving an egg. (a) The goose sees the egg outside the nest. (b) She gets up off the nest and approaches the egg with outstretched neck. (c) She places the underside of her bill over the egg, at which point the behavior pattern begins and continues until (d), when the egg is on the nest rim.

After Lorenz & Tinbergen, 1939, with permission.

directly by the central nervous system and is not affected by environmental stimuli, whereas the form of the taxis is continuously determined by stimuli from the external environment.

Lorenz assumed that these coordinations were stereotyped, inherited, and species-specific. He also proposed that they only are expressed in specific situations and that they had certain motivational properties; these requirements are discussed later. With respect to stereotypy, many examples of species-specific behaviors were soon shown to be often quite variable, and Barlow (1968), for example, suggested deleting the word 'fixed' and replacing it with 'modal'. However, variability in most of the cases that have been studied can easily be accounted for by the taxis components of the behavior. In terms of the concepts developed in Chapter 1, the behavior observed would be the outcome of the simultaneous activation of the motor mechanism responsible for the coordination of the fixed component together with the simpler motor mechanisms underlying the taxis component. (For a recent discussion of the concept of stereotypy and its relation to other concepts of variability, see Japyassú & Malange, 2014.)

There are, of course, other cases of variability that cannot be explained in such a simple way. An early example is D. Wilson's (1966) analysis of locomotion in insects. Insects, with their six legs, show a very large number of leg-stepping patterns, and early explanations of these patterns relied solely on chain-reflex action: stimulation of proprioceptors in each leg during locomotion was assumed to be responsible for the ordering of the leg movements and for the nature of the order itself (e.g. von Holst, 1935). Wilson, however, showed that these various patterns had general characteristics and summarized his analysis with the hypothesis "that each segment contains a reciprocal inhibition network which creates the timing signals for sets of antagonistic muscles and that the segmental oscillators are coupled intracentrally to provide the relative phasing. Reflexes are superimposed upon the central scheme, and in general reinforce it" (p. 118). In other words, there is a central rhythm that determines the ordering of the leg movements in conjunction with separate oscillators for each leg, but the movements are also subject to superimposed reflexes. I discuss oscillators later in this chapter. (See also Gallistel, 1980.)

The control of movement is a subject properly treated in a separate book, but in general, the analysis of movements into centrally controlled coordinations plus superimposed reflexes is sufficient for most behavioral studies. Nonetheless, because the original concept proposed by Lorenz has so many controversial facets, and because there are many actions that are relatively stereotyped but are not inherited or species-specific, I prefer to use the neutral term *behavior pattern*: any sequence of muscle movements that can be reliably recognized as the same on different occasions by a competent observer. An important aspect of this definition is that it makes no mention of the consequences or usefulness of the movement. We still speak of the egg-retrieval pattern even when the egg rolls away. We recognize it by its form and not by its consequences. One other aspect of this definition is that it is agnostic with respect to development: learned behavior patterns are ubiquitous, especially in humans.

Using this broad definition, we can recognize many behavior patterns in our own species. Some of these even conform to Lorenz' narrow definition. For example, studies of the reactions of newborn babies to various taste stimuli show that very specific facial expressions are associated with bitter, sour, and sweet substances: distilled water evokes a 'neutral' expression, whereas the sweet taste evokes a 'smile', the sour taste a 'pucker', and the bitter taste a 'disgust' expression. Each expression can be reliably recognized. The same

(a)   (b)

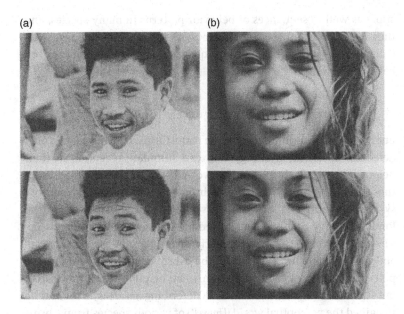

Figure 2.2 The eyebrow flash filmed in different cultures. This behavior pattern is used as a greeting at a distance. In both sets of pictures, the upper photograph shows the lowered eyebrows at the beginning of the greeting, and the lower photograph shows the maximally raised eyebrow during the greeting. (a) Balinese (b) Samoan.
From Eibl-Eibesfeldt, 1972.

expressions are also seen in older children and adults, even in those that have been blind from birth (Steiner, 1979). Another example, seen in most societies, is the eyebrow flash: raising the eyebrows for about one-sixth of a second when giving a friendly greeting from a distance (see Figure 2.2). Other behavior patterns are unique to specific individuals. The various movements used in the course of cigarette smoking, for example, may be highly stereotyped in one individual, but often differ considerably from person to person. Still other behavior patterns, such as walking and running, probably never strike most people as being especially stereotyped, but if you compare the gait of any human with that of any chimpanzee, the similarity of these behavior patterns within the two species becomes apparent.

A recent paper by Gadbois *et al.* (2015) gives a brief history of conceptualizations of behavior patterns in modern ethology. They make a number of points similar to those presented above. They also point out that the general properties of behavior patterns seem to apply as well to sequences of behavior patterns in many species, and they present examples from their work on several canid species. I discuss the issue of behavior sequences below in the section on behavior systems.

PERCEPTUAL MECHANISMS: PERCEPTIONS

Energy in the environment comes in many forms, but animals can only detect those forms of energy that can stimulate receptors in their nervous system. Each species has evolved sensory systems that detect those forms of energy that it uses in its everyday life. These sensory systems can differ greatly among species, both with respect to the kinds of energy that can be detected and the sensitivity of the receptors. Von Uexküll (1921) was an early investigator of the sensory capabilities of animals. He coined the word *Umwelt* to describe the perceptual world of each species. In 1934, he published a now dated, but still informative, illustrated book describing his ideas about how he imagined the perceptual world (*Umwelt*) of various species from worms and ticks to birds, dogs, and humans might look (von Uexküll, 1934/ 1957). Reviews of some of the early studies on these topics, especially with respect to behavior, can be found in Tinbergen (1951, Ch. 2) and Hinde (1970, Ch. 4). More recent information can be found in Stevens (2013) and Burghagen & Ewert (in press).

Once we know what forms of energy an animal can detect, a more interesting question is what use can the animal make of that information: how is sensory information analyzed? In spite of the very wide differences in what stimuli a species can detect, the basic organization of sensory systems is generally very similar. The receptor cell transforms the incoming energy into a neural impulse, which can then activate a perceptual mechanism. At the neurophysiological level of analysis, the concept of perceptual mechanism is a functional concept: the analysis of incoming (or imagined) stimuli can occur at the level of the receptor or at higher levels in the brain or both; it can involve only one or a few nerve cells; it can involve complicated brain circuits, etc.; but the outcome of activating a perceptual mechanism is always a perception of some kind in humans (and probably in most other animals as well) and/or a signal to higher brain centers or directly to some motor mechanism. I will mention some neurophysiological examples below, but I will first discuss the concept of perceptual mechanism at a behavioral level, where it is a causal concept: it is one component of behavioral structure.

Unlike behavior patterns that can be directly observed, perceptions occur within the brain and can only be studied indirectly. One way of studying the perceptual world of animals has been very successful. The first step is to observe the objects or events that are associated with particular behavior patterns in a relatively natural environment. For example, with which fish does a Siamese fighting fish (*Betta splendens*) fight? What kinds of prey does a frog eat? From what situations does a chick flee? On the basis of these observations, we can infer what the natural stimulus is for a particular reaction. For example, the natural stimulus for egg retrieval in the greylag goose is a goose egg. The second step in the analysis is to determine what it is about the stimulus that makes it effective. What makes an object an egg for a goose? Its size, color, shape, or smell? Using models, we can study an animal's responses to selected aspects of the natural stimulus.

One of the first insights gained from such experiments is that many animals pay very little attention to much of the information available to them, and what they respond to often depends on their motivational state. Tinbergen (1951) has discussed several examples. Nesting greylag geese will retrieve a cardboard cube (of the right size and color) almost as readily as a real egg. Male three-spined sticklebacks (a small fish, *Gasterosteus aculeatus*) in breeding condition will attack a crude, cigar-shaped model painted red underneath, but will ignore a detailed, accurate model of another male that is uniformly grey. Male European robins (*Erithacus rubecula*) will furiously attack a few orange feathers stuck to a twig in their territory, but will ignore a real stuffed male robin whose breast has been painted brown. These examples are extreme, but it can be shown that all animals react to only some of the stimuli in their environment. This selectivity by animals led to the concept of a sign stimulus. A *sign stimulus* is that part of the total stimulus situation that is relatively most effective for releasing a response. Some authors refer to this concept as *key stimulus*.

An Experimental Analysis of Animal Perception

The stimuli that release the egg-retrieval response of the herring gull (*Larus argentatus*) have been studied in great detail by the Dutch ethologist Gerard Baerends and his colleagues (see Baerends & Kruijt, 1973; Baerends & Drent, 1982). Herring gulls nest in colonies and normally brood a clutch of three eggs. If the gulls are alarmed while brooding, they fly off the nest, but return as soon as the potential danger has disappeared. In Baerends' experiments, two experimenters walked into the colony, which caused the birds to leave their nests. One nest was chosen for study and two of its three eggs were removed. These eggs were replaced by two model eggs. The model eggs were not placed in the nest cup, but were placed instead on the rim of the nest. One of the experimenters then hid in a small portable tent that was set up near the chosen nest, and the other experimenter left the colony. The nest owner soon returned, and the experimenter in the tent could easily observe and record the behavior of the returning bird. Typically, the gull would enter the nest and sit on the one egg. It would then rise, look at the model eggs on the nest rim, retrieve one egg and then retrieve the other. (The retrieval movement itself is similar to the egg-retrieval movement of the greylag goose described above.) The egg that was retrieved first was considered to be a better releasing stimulus than the other egg. Thousands of such experiments were performed using pairs of models that differed from each other in shape, size, color, pattern of speckles, and various other attributes; possible side preferences and other experimental details were also controlled for.

The results of these experiments showed that some aspects of the stimulus were not very important for the gull. Shape, for example, had relatively little effect on the releasing value of the egg: round, square, oblong, and egg-shaped models were retrieved with about equal frequency. Color, size, and speckle pattern, on the other hand, were all very important. Green was the most highly preferred color, larger models up to five times normal size were always preferred over smaller ones, and speckled models were always preferred over nonspeckled ones. Thus, green, a large size, and speckles are three of the sign stimuli for egg retrieval in the herring gull. By varying several stimulus attributes of the egg models at the same time, it was possible to show that each of the sign stimuli increased the releasing value of the model independently. This is a demonstration of the law of heterogeneous summation (Tinbergen, 1951). Thus, the most effective stimulus is one that includes all the sign stimuli. The case of the herring gull is especially interesting in this respect. The normal egg is dark beige, speckled, and slightly larger than a large chicken egg, whereas the most effective egg is green, speckled, and almost as large as a football. This egg has been called a supernormal stimulus because it is invariably chosen over the gull's own egg in a choice test (see Figure 2.3).

Figure 2.3 Supernormal stimulus. Herring gull choosing to retrieve a large, green, speckled egg rather than its own smaller, brown egg. Courtesy of Gerard Baerends.

Releasing Mechanisms

The fact that it is possible to determine the most effective stimulus for each response has led to the concept of releasing mechanism. A *releasing mechanism* is a group of neurons in the central nervous system that is responsible for analyzing, evaluating, and summating the different attributes of an object, and that can release a specific response. The word 'release' implies disinhibition of a motor mechanism that has certain motivational properties, and I discuss this implication in Chapter 3. A more neutral term might be 'object recognition mechanism' or 'object detector', but for the time being, I will continue to use the word 'releasing'. Each response must have its own releasing mechanism because the most effective stimulus is different for each response. For example, herring gulls not only retrieve eggs, they also eat them. They do not eat eggs in their own nest (unless they become broken), but they do eat any eggs they can grab from another gull's nest. Experimental tests have shown that small, red eggs are preferred for eating, whereas large, green eggs are preferred for retrieval (Baerends, 1982). I should also mention here that in the older ethological literature, this concept was termed '*das angeborene auslösende*

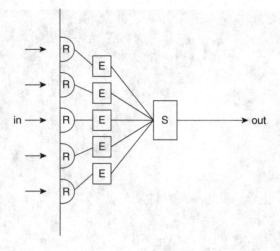

Figure 2.4 Model of a releasing mechanism. R = receptor, E = evaluation unit, and S = summation unit. Further explanation in text. From Baerends & Kruijt, 1973.

Schema' by von Uexküll and by Lorenz, and was translated into English by Tinbergen as 'innate releasing mechanism'. Since some releasing mechanisms and many equivalent neural structures are learned or are at least influenced by learning, as I discuss in Chapters 6 and 7, and because the word '*angeboren*' or 'innate' has dubious scientific meaning, most modern authors now omit the word innate.

Baerends & Kruijt (1973) presented a model of a releasing mechanism that is shown in Figure 2.4. In the figure, the receptor units, R, correspond to the various stimulus features of an object that are relevant to a specific response, such as size, shape, and color for the egg-retrieval response. The output from each receptor unit is evaluated (E) with respect to its associated response and the output of the evaluation unit is directed to the summation unit (S), which is responsible for integrating input from all the evaluation units. So, for example, if a stimulus object is presented to the bird, the receptor unit that detects color may determine that the color of the object is brown; the evaluation unit for color for the egg-retrieval response determines that brown is a moderately good color and sends a moderate positive signal to the summation unit. In a similar way, the other receptor units detect the value of the stimulus for their particular dimension, and the integration of these inputs in the summation unit can be said to be the releasing value of that stimulus for the egg-retrieval response. Whether the response actually occurs depends not only on the releasing value of the stimulus, but also on the entire motivational (causal) context, as will be discussed in Chapter 3. In this model of the releasing mechanism, the receptor units are general-purpose feature detectors. The unit that detects color, for example, can send its output to evaluation units for all responses for which color is important, such as the egg-eating response. The output of the evaluation and summation units, however, is directed to a specific response mechanism.

A question that can be asked about this model is: Do these behavioral units have any correspondence to neurophysiological units? The answer is both yes and no. There are definitely feature detectors in the sensory systems of animals. For example, the retina of the vertebrate eye comprises three layers of neurons: the receptor cells, the interneurons, and the ganglion cells, the axons of which form the optic nerve. The structure of these neurons and the interconnections among them differ among species, but in all cases, the information conveyed by the ganglion cells through the optic nerve has already undergone considerable analysis. One of the pioneer studies was conducted by Lettvin et al. (1959) on the frog's visual system. These investigators recorded activity in the retinal ganglion cells in response to various moving stimuli. They identified five types of ganglion cell that responded selectively to different features of the stimuli: sustained-contrast, net-convexity, moving-edge, net-dimming, and absolute-darkness detectors. Although none of these features corresponds exactly to the size or shape of the stimulus objects, together with further analysis at higher levels of the nervous system, they do provide the basis for the sort of perceptual phenomena envisaged in the releasing mechanism concept. More recent neurophysiological studies on prey selection of toads to various visual stimuli by Ewert (1997) have elucidated a more complete characterization of the releasing mechanism for this response. Ewert's neurophysiological units do not fall neatly into the receptor, evaluator, and summation categories of Baerends; nonetheless, there is a close correspondence between the neurophysiological and ethological units. From a behavioral perspective, activation of the summation unit of Baerends would correspond to perception of the object.

There is one other approach that has been used to model sensory systems with respect to behavior: neural networks (Enquist & Ghirlanda, 2005). This approach is similar in many ways to the ethological approach, but it begins by mimicking the architecture of nervous systems and connects elementary neuron-like units into networks with various inputs and outputs. These units and their connections are mathematically defined, and one can test hypotheses about behavior by making different assumptions about inputs and the mode of operation of the connections between the units. This approach, also called connectionist modeling, has been widely used in cognitive psychology.

Enquist & Arak (1993, p. 446) point out:

> ... even the most simple artificial networks, consisting of a few interconnected cells, exhibit many of the properties shown by animal recognition systems: they are easily trained to classify objects and perform generalizations. Such networks provide convenient tools for uncovering general principles of recognition free from much of the complexity found in the nervous systems of real organisms.

They used such a model to show that the mechanisms concerned with signal recognition have inherent biases in response, and that these biases can act as important agents of selection on signal form. Their demonstration used a simple neural network that was trained to discriminate long-tailed birds (conspecifics) from short-tailed birds. After training, the model was tested with a wide array of stimuli and was found to respond more strongly to a stimulus with a longer tail than to the stimulus with which it was trained (cf. supernormal stimulus). If we assume that the perceptual mechanism possessed by a female has similar properties, and that this mechanism is attached to her sexual central mechanism, we see that she would prefer a male with an exaggerated trait; this could provide the basis for selection of the signal form.

This demonstration was originally met with some skepticism (M. Dawkins & Guilford, 1995), but further work has confirmed the usefulness of these models (Enquist & Ghirlanda, 2005). Ghirlanda & Enquist (2003) have used artificial neural networks to understand the many phenomena associated with stimulus generalization, and Ryan (2009) has used them in his analysis of the evolution of mating calls in frogs (see p. 250). An amusing example of the former is provided by an experiment in which chickens were required to discriminate between human faces. In the test, they were found to prefer the same faces that human subjects rated as more beautiful (Ghirlanda *et al.*, 2002). We can infer from this result that our sense of beauty is closely tied to the structure of our perceptual mechanisms.

Models such as those of Baerends are basically very simple network models. The major difference between such ethological models and most neural network models is that the latter incorporate the possibility of change as a function of experience, whereas most ethological models are static in that respect. I discuss change of behavioral structure in Chapters 6 and 7.

Representations are the neural correlates of objects, ideas, concepts, memories, feelings, and the like. Because these entities are difficult to specify empirically, there have been many different approaches to studying them (see, for example, the discussion by Gallistel, 1990a). And there is definitely no consensus about their characteristics. So in the following, you will be getting my opinion about these matters, including my opinion about the role these entities play in the study of behavior.

The summation unit of a releasing mechanism is the representation of the object appropriate for the associated response. There must theoretically be as many of these summation units as there are responses that are associated with particular objects. I will come back to this point in my discussion of behavior systems later in this chapter, but let me now change my terminology and call the summation unit of a releasing mechanism by the more neutral term, *object recognition mechanism*. As with the term *behavior pattern*, one can immediately include perception of any object in the category, even if it is not associated with a particular response and/or if it depends on learning.

Representations of ideas, memories, meanings, etc. are also central behavior mechanisms. In what ways are they similar to or different from object recognition mechanisms? At the neurophysiological level, they are in most ways all very similar: they all consist of organized connections of neurons that can excite or inhibit one another. But as neural systems, they are also organized in specific ways and respond in particular ways to various inputs.

The neuroscientist David Marr (1982) proposed a tri-partite classification scheme for understanding neural systems that is widely used in cognitive science: computation, algorithm, and hardware. Computation is what the system does (accomplishes), algorithm is how the system is organized, and hardware is the neuronal basis of the system. In a recent paper, Insel & Frankland (2015) point out that these three aspects of neural systems correspond approximately to the function (consequences), structure, and matter of behavior mechanisms as I define them in Chapter 1. In their section on challenges in identifying mechanisms of behavior, Insel and Frankland provide an excellent analysis of the many problems encountered when trying to define appropriate units of analysis for neural systems (behavior mechanisms) and their parts. How to deal with some of these problems will be discussed below in the section on behavior systems. They also give many examples of how these ideas have been used by neuroscientists investigating the behavioral functions of neural circuits in the hippocampus and prefrontal cortex. But the problem of categorizing the neural systems remains. For the time being, I will consider all kinds of representations to be one class of central mechanisms: activation of a representation leads to an event of behavioral interest either within the brain itself or through the motor system. Representations are one component of behavioral structure. I discuss this issue again in Chapter 7 in the section on memory.

CENTRAL MECHANISMS: COMPARATOR MECHANISMS,
OSCILLATORS

There are two other classes of central mechanisms that are important for the study of behavior: comparator mechanisms and oscillators. These mechanisms are also organized connections of neurons, but differ from representations in their functional relation to motor action. At a basic level, comparator mechanisms are central mechanisms that function to distinguish sensory information caused by movements of the animal itself from general sensory information. They are similar to Gallistel's (1980, 1981) *servomechanisms*, in that negative feedback is fundamental to their functioning. At a more complex level, they can be characterized as *expectations* (see Chapter 4).

A very important example of a basic comparator mechanism is the *efference copy* of von Holst & Mittelstaedt (1950; von Holst, 1954). They distinguish between *exafferent stimulation* that is produced by movement in the external world and *reafferent stimulation* that is produced by the animal's own movement. They propose a mechanism that allows an animal to discriminate these two types of stimulation as illustrated in Figure 2.5. When the animal 'intends' to do something,

> a motor impulse, a 'command' C (Fig. 2.5a), from a higher centre HC causes a specific activation in a lower centre LC (Fig. 2.5b), which is the stimulus-situation giving rise to a specific efference E (Fig. 2.5c) to the effector EF (i.e. a muscle, a joint, or the whole organism). *This central*

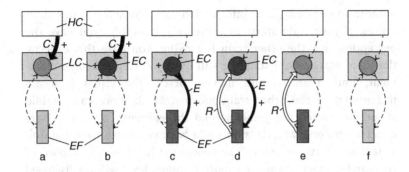

Figure 2.5 Illustration of the reafference principle. Explanation in text. From von Holst, 1954, with permission.

stimulus situation, the "image" of the efference, may be called 'efference copy' EC [my emphasis]. The effector, activated by the efference, produces a reafference R, which returns to the lower centre, nullifying the efference copy by superposition (Fig. 2.5d–f). Because of the complementary action of these two components we can arbitrarily designate the whole efferent part of this process as plus (+, dark coloured) and the afferent part as minus (-, white coloured). When the efference copy and the reafference exactly compensate one another, nothing further happens. When, however, the afference is too small or lacking, then a + difference will remain or when the re-afference is too great, a – difference will remain. This difference will have definite effects, according to the particular organisation of the system. The difference can either influence the movement itself, or for instance, ascend to a higher centre and produce a perception (von Holst, 1954, p. 91).

In other words, if the feedback from the movements of the animal (reafferance) matches the expectation of what that feedback should be (the efference copy), the animal perceives the world as stable, and perceives itself as having moved. If there is no efference copy (i.e. there is no expectation of movement), stimulus input (afferance) is perceived as the world moving. It should be noted that afferance and reafferance refer to all the stimulation the animal receives while moving, in all sensory modalities, not just stimulation from the particular muscles that are contracting.

The 'reafference principle' is essential for understanding how an animal can distinguish between feedback from its own movements (reafference) and stimulation from external sources (exafference). If you are sitting in a train next to another train, the only way you can visually distinguish whether your train is moving or the other train is moving is by information you receive from your other senses. If your train starts to move very smoothly, you have the perception that the other train is moving. You have the 'expectation' that you are still standing still until you can feel some movement; until then, stimuli moving across the visual field are interpreted as the other train moving. Much more can be said about the reafference principle and the efference copy, but I am using it here only as an example of a basic comparator mechanism, a second class of central behavior mechanisms. Examples of comparator mechanisms at a more complex level will be discussed later.

Oscillators or central pattern generators are a third class of central behavior mechanisms. At a neural level, oscillators are neurons or groups of neurons (neural systems) that fire in a rhythmic pattern in the absence of any rhythmic input. Oscillators are also called pacemakers. Oscillators can be found throughout the brain and are generally thought to be responsible for rhythmic activity in many systems. Their rhythms range from milliseconds to days and perhaps even years. It was von Holst (1937) who showed that fin movements in fish are controlled by central pattern generators, and D. Wilson (1966), as we have seen, postulated oscillators as controlling factors in insect locomotion (see Gallistel, 1980, 1981 for a general discussion of the role of oscillators in the organization of action). Lorenz' (1937) concept of *Erbkoordination* also invokes the operation of central pattern generators in many types of actions. Beyond their role in determining the form of behavior patterns, oscillators are also involved in the timing of most behavioral events. They are probably important for our perception of time, and are definitely involved in biological rhythms, as discussed in Chapter 4. As with most basic comparator mechanisms, so also with oscillators, both are central behavior mechanisms that generally operate in the background of consciousness or awareness.

CENTRAL COORDINATING MECHANISMS: BEHAVIOR SYSTEMS

Motor mechanisms, perceptual mechanisms, representations, comparator mechanisms, and oscillators, as discussed above, are relatively 'stand alone' groups of neurons, albeit with important implications for behavior. However, we are often interested in more complex behavioral entities. These can range from simple reflexes to motivational systems such as feeding behavior, aggressive behavior, sexual behavior, fear behavior, etc. For that purpose, it is necessary to consider how these units are integrated into more complex units that I call *behavior systems* – any organization of perceptual, central, and motor mechanisms that acts as a unit in some situations.

Examples of simple behavior systems would include the sweet taste stimulus-smile response and the friendly face-eyebrow flash in humans and the egg recognition-egg retrieval response in the greylag goose and herring gull. Lorenz (1937) originally thought that analysis of such simple behavior systems would be sufficient for understanding much of an animal's behavior. But early work by Baerends (1941) on the digger wasp (*Ammophila campestris*) and by Tinbergen (1942) on the stickleback indicated the need for considering superimposed mechanisms. Such mechanisms include the central coordinating mechanisms discussed in Chapter 1. These mechanisms have a coordinating function that is different from that of the central mechanisms already discussed.

Each animal has a large number of motivational behavior systems. The number and kind of systems, as well as their precise structure, depend on the species, the sex, and the developmental history of the animal. Hunger and sexual systems are probably ubiquitous. Even so, the structure of the sexual system is usually different in males and females of the same species. The stimuli to which males and females react are often different (which means males and females must possess different object recognition mechanisms), as are many of their sexual behavior patterns. Other systems such as fear and aggression are very common, though their relative importance varies widely in different species. Parental systems are totally lacking in many species of amphibians, reptiles, and fish that deposit fertilized eggs or bury them, and then desert them and leave the newly hatched young to fend for themselves. Parasitic species of birds such as cuckoos and cowbirds, which lay their eggs in the nests of other species, have unconventional parental behavior systems. Large differences between males and females are also characteristic of parental systems. The precise structure of a behavior system in an individual also depends on its developmental history. The behavior pattern of lever pressing, for example, will be part of the hunger system of a rat that has been trained in a Skinner box, but may be totally absent in a rat that has not been so trained (see Chapter 7). Here, I will discuss the hunger system of a young chick and the incubation system of the herring gull as examples. Further examples are discussed in Chapter 3.

The Hunger System of a Young Chick

Figure 2.6 depicts the hunger system of a young chick of the red junglefowl (*Gallus gallus spadiceus*) and includes a number of features that characterize most behavior systems. This figure was developed on the basis of the results of experiments published in Hogan (1971) and later experiments summarized in Hogan (1984). In the figure, boxes represent putative cognitive (neural) mechanisms. Perceptual mechanisms include various feature recognition mechanisms (such as of color, shape, size, and movement), object recognition mechanisms (such as of grain-like objects [G], worm-like objects [Wo], and possibly others) and a function recognition mechanism [Food]. Motor mechanisms include those underlying specific behavior patterns (such as pecking [P], ground scratching [S], walking [Wa], and possibly others), and a motor pattern coordinating mechanism that could be called foraging [For]. There is also a central hunger coordinating mechanism [H]. Solid lines indicate mechanisms and connections among them that develop prefunctionally; dashed lines indicate mechanisms and connections that develop as a result of specific functional experience.

Feature recognition, object recognition, central coordinating, and motor mechanisms have been discussed above, but the function recognition and the motor pattern coordinating mechanisms are new. Which objects are food is something a young chick must learn, and the development of such learned behavior mechanisms is discussed in

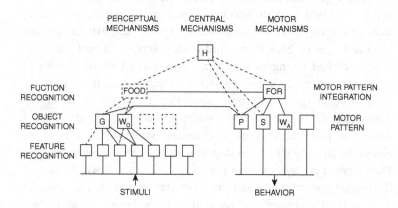

Figure 2.6 Model of the hunger behavior system of a young chick. Explanation in text.

From Hogan, 1988, with permission.

Chapter 5. Once developed, they would be similar in structure to the summation unit of a releasing mechanism and would filter input from various object recognition mechanisms.

Like the muscle contractions in a motor pattern, the occurrence of individual behavior patterns is not random. In the young chick, the movements of walking, scratching, looking, and pecking occur in predictable sequences under the influence of the foraging mechanism. Characteristics of such sequences are discussed below; in the young chick, the foraging behavior mechanism coordinates the movements necessary for discovering, uncovering, and ingesting food. It can be seen in the figure that perception of grain or of a mealworm can lead directly to pecking (and swallowing), but that the sequences of foraging require perception (or imagination) of food. The influence of the central hunger mechanism also requires functional experience to develop (dashed lines in the figure), and this aspect of the hunger system is discussed in detail in Chapter 5.

The Incubation System of the Herring Gull

The top portion of Figure 2.7 shows Baerends' (1970) depiction of the incubation (or nesting) system (N) of the herring gull. The lower portion of the figure shows interactions of the incubation system with the escape (E) and preening (P) systems. These interactions are discussed in Chapter 3. This figure depicts primarily the motor side of the incubation system. The boxes represent putative cognitive (neural) mechanisms. N is the central coordinating mechanism for the incubation system as a whole. It controls the building, settling, and incubation mechanisms, which in turn control all the behavior patterns (mechanisms) as shown. At the top of the figure, the egg outside the nest can activate the 'retrieving' behavior mechanism (dashed line from the egg), and a comparator unit (CU) compares the efference copy (EC), which is the expected feedback from eggs in the nest, with the actual feedback from the eggs in the nest (input, IP). If the feedback matches the efference copy, the bird continues 'incubating' the eggs. If the feedback is insufficient (too few eggs), the comparator mechanism can activate 'looking around' and can also activate the escape system and such behaviors as 'rising' from the nest and 'walking' or 'flying' away. In this diagram, the settling and building mechanisms can be considered motor coordinating mechanisms, comparable to the foraging mechanism in the hunger system of the chick.

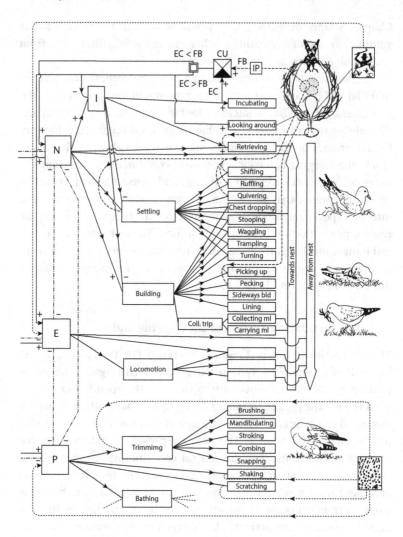

Figure 2.7 Model of the incubation system of the herring gull.
Explanation in text.
From Baerends, 1976, with permission.

Behavior Sequences

Most behavior systems have a number of motor mechanisms associated with them. The hunger system in chicks and the incubation
system in herring gulls are two examples. Grooming systems in many
species of birds and mammals also contain multiple motor mechanisms (van Iersel & Bol, 1958; Berridge, 1994). Food caching systems in

canids provide other examples (Gadbois *et al.*, 2014). In many of these cases, when the system as a whole is activated, the associated actions occur in recognizable sequences. For example, the dustbathing system in chicks contains motor mechanisms for the behavior patterns bill rake, scratch, vertical wing shake, head rub, side rub, and body shake (see Figure 3.5, p. 77). During the 30-minute bout of dustbathing in adult chickens, these behavior patterns recur regularly in easily recognized sequences. Most sequences are the same, but there are also many deviations in which some patterns are skipped or repeated (Larsen *et al.*, 2000). A similar organization of dustbathing motor patterns has been described in bobwhite quail (*Colinus virginianus*) by Borchelt (1975), although quail appear to have a more rigid dustbathing bout structure than the chickens.

Another example is the sequence of behaviors shown by rats during a grooming bout (see Figure 2.8). Similar sequences are seen in other rodents including mice, hamsters, gerbils, squirrels, and guinea pigs (Berridge, 1990, 1994). Because the serial order of the discrete

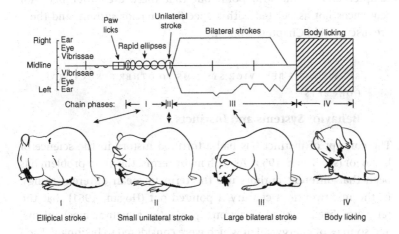

Figure 2.8 A syntactic grooming chain. The pattern is depicted by a choreographic notation system for grooming behavior that displays the trajectories and sequence of facial strokes and the sequential occurrence of other grooming components. Drawings depict action types. Time proceeds from left to right. The horizontal axis represents the center of the nose. The line above the horizontal axis denotes movement of the right forepaw along the face; the line below the axis denotes movement of the left forepaw. Small rectangles denote paw licks. Large rectangle denotes body licking.
From Berridge, 1994.

grooming actions is so predictable, Berridge & Fentress (1986) denote the pattern as a 'syntactic chain' following the lead of Fentress & Stillwell (1973), who called the movement sequences of mouse grooming 'grammar'. That these sequences are not merely chain reflexes, with the occurrence of one behavior pattern being the stimulus for the next, however, follows from the facts that the individual behavior patterns can also be seen to occur by themselves and by the variations seen in the timing and patterning of the sequences.

These results all suggest that in many behavior systems, there must be motor pattern coordinating mechanisms that are independent of the motor mechanisms for the individual motor patterns. In many cases, an alternative phrase might even be 'syntax generating mechanisms'. Developmental evidence (discussed in Chapter 5) and neurophysiological evidence reviewed by Berridge (1994) support this conclusion. Comparison of similarities and differences among motor pattern coordinating mechanisms in closely related species can also give some insight into the evolution of behavior as discussed in Chapter 8. I should also point out that there are other behavior sequences not associated with a specific behavior system, and these are discussed in Chapters 3 and 4.

RELATION OF BEHAVIOR SYSTEMS TO OTHER STRUCTURAL CONCEPTS

Behavior Systems and Instincts

The concept of instinct has had a tortured history in the science of behavior (see Beach, 1955). As with many terms, the main problem has been that different authors use the term to mean different things. In the mid-twentieth century, I pointed out (Hogan, 1961) that the term had been applied to actions, perceptions, neural mechanisms, and sources of energy, all of which were considered to be innate. Each of these kinds of instinct had certain characteristics that varied according to the author. Overall, each author had his own theory of instinct and there was little or no agreement among the theories. Very little has changed since.

William James defined an instinct as an impulse to act in a particular way in the presence of particular perceptions. He also proposed that humans have more instincts than any other animal. He claimed that one reason for the apparent variability in human behavior is that a person has so many instincts that they block each other's path:

"Nature implants contrary impulses to act on many classes of things, and leaves it to slight alterations in the conditions of the individual case to decide which impulse shall carry the day" (James, 1890, vol. 2, p. 392). Sixty years later, Tinbergen (1951, p. 112) defined an instinct as "a hierarchically organized nervous mechanism which is susceptible to certain priming, releasing and directing impulses of internal as well as of external origin, and which responds to these impulses by coordinated movements that contribute to the maintenance of the individual and the species". Both of these definitions have much in common with each other and with my conception of a behavior system, but there are also important differences that will be discussed presently. First, I will mention three of the objections that were made to the instinct concept.

The anti-instinct movement during the first half of the twentieth century, as Lashley (1938) pointed out, was aimed primarily at the postulation of imaginary forces as explanations of behavior – a crusade against a conceptual dynamism. In other words, it was a rejection of instincts as sources of energy. But there were other objections as well. Kuo (1921), in his paper on "Giving up Instincts in Psychology", complained "that there is no general agreement among the students of instincts as to the number and kinds of instincts. Writers on the subject arbitrarily list them in accordance with their own purposes" (p. 648). Kuo also objected to the concept of 'innate' because it sets up a false dichotomy. Ironically, Lashley found genetic transmission to be "the one tie to physiological reality" possessed by instincts. Behavioristic psychologists do not generally object to the concept of innate, but they tend to find the concept of instinct irrelevant because most believe that all of the important aspects of behavior are learned. Of course, as Beach (1955) pointed out, few psychologists have ever seen the behavior that many biologists would call instinctive. These and other objections still exist, and the word 'instinct' is seldom used anymore in scientific theories. I have replaced it with the phrase 'behavior system'.

We can now compare James' and Tinbergen's definitions of instinct with each other and with my definition of a behavior system. All three conceptions have the equivalent of perceptual mechanisms and motor mechanisms that are connected, and all three consider instinct to be the expression of specific physiological mechanisms. James says: "The actions we call instinctive all conform to the general reflex type" (1890, vol. 2, p. 384), and Tinbergen defines instinct as "a hierarchically organized nervous mechanism". Both James and Tinbergen include a motivational element ('impulses') in their

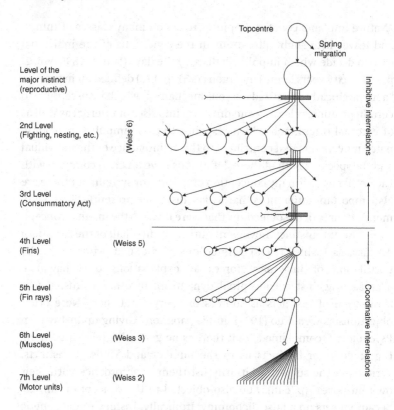

Figure 2.9 Hierarchical organization of behavior. Explanation in text. From Tinbergen, 1951 with permission.

definition. In fact, James defines instinct as "an impulse". My definition does *not* include a motivational element; a behavior system is a structural concept; motivation is considered to be a separate issue and is discussed in Chapter 3. Although neither James nor Tinbergen includes the concept 'innate' in their definition, James does say that "nature implants contrary impulses to act", and the idea of innate pervades Tinbergen's whole discussion of instinctive behavior. My definition of a behavior system specifically excludes the notion of innate, and specifically includes behavior mechanisms and connections among them that are learned or influenced by learning. I discuss the concept 'innate' extensively in Chapter 6. Tinbergen also includes a functional statement in his definition: "movements that contribute to the maintenance of the individual and the species." James assumes that instincts are adaptive, but that is not central to his conception. As I have discussed in Chapter 1, it is better to keep causal and functional questions separate in order to avoid various kinds of confusion. For me, a behavior system can exist even if it is obviously non-functional (e.g. the smoking behavior system in humans).

Finally, Tinbergen introduces the notion of hierarchy in his definition; his diagram of the hierarchical organization of behavior is reproduced in Figure 2.9. I have briefly discussed the concept of hierarchy in Chapter 1, and for a more extensive discussion, see R. Dawkins (1976a). One aspect of hierarchy (embedment) is the grouping of smaller bits into larger units that consist of the smaller bits. In Tinbergen's diagram, the first five levels of Weiss are a series of embedments. Tinbergen's levels 1, 2 and 3 are levels of control: there is a behavior mechanism at level 2 (for example, the hunger coordinating mechanism) that controls a number of independent behavior mechanisms at level 3 (for example, the motor mechanisms for pecking, scratching, etc.). There is good behavioral (and neurophysiological) evidence for the existence of these two levels, but very little evidence for a higher level (and notice that Tinbergen names his level of the major instinct with a functional term).

In my diagram of the hunger behavior system of the young chick (Figure 2.6), there is actually a 3-level hierarchy with object recognition and motor pattern mechanisms at the bottom, the function recognition and motor pattern coordinating mechanisms in the middle, and the hunger coordinating mechanism on top. However, the hunger coordinating mechanism corresponds to Tinbergen's second level, so there is still no higher level corresponding to Tinbergen's level of the major instinct. Baerends' diagram of the incubation system of the herring gull (Figure 2.7) also has a 3-level hierarchy similar to that of the chicks.

Appetitive Behavior

In 1918, the American biologist Wallace Craig published a paper entitled "Appetites and Aversions as Constituents of Instincts". It was virtually unnoticed in North America, but several years later, had an important influence on European ethology. Craig distinguished between appetitive behavior and consummatory act. The former was the behavior of the animal striving toward the goal, which ceased when the latter was attained. The relation between these concepts and reinforcement is discussed in Chapter 7. With respect to behavior systems (or Tinbergen's instincts), the concept of appetitive behavior has always caused some difficulties because appetitive behavior is variable and is defined as goal-directed, whereas behavior systems (instincts) have a definite structure and occur 'blindly' in particular external and internal situations (e.g. the egg-retrieval response). In Tinbergen's hierarchical system (Figure 2.9), appetitive behavior is indicated by the arrows on the right of the diagram heading to spring migration at the top level, and leading nowhere in particular from levels 1 and 2. In effect, we have a system defined functionally (appetitive behavior/consummatory act) and a system defined structurally (behavior mechanisms) that are basically incompatible.

One way to make these two systems compatible is to invoke an exploration system, constructed like all other behavior systems with perceptual, central, and motor mechanisms. There is considerable evidence for the existence of an exploration system in many animals (see Chapter 3), and such a system could be 'co-opted' by other major behavior systems whenever appropriate external situations were not available. However, it seems likely that most behavior systems have 'built-in' mechanisms for appetitive behavior. For example, as we have seen, the hunger system in the young chick has a motor pattern coordinating (foraging) mechanism. This mechanism can be activated by both the food recognition mechanism and the central hunger mechanism. When activated by both (i.e. when the chick is hungry and food is present), the chick will engage in certain of its foraging behaviors. When the foraging mechanism is activated only by the hunger mechanism (i.e. when the chick is hungry and no food is present), it will also engage in foraging behaviors, but will continue until food is discovered. In this case, the foraging mechanism is functioning both as an activating mechanism, but also as a comparator mechanism with respect to some of the motor mechanisms it controls. Foraging in both cases will cease when the hunger mechanism receives the appropriate feedback indicating the chick is satiated. I discuss this issue again in Chapter 4 in the context of expectancies.

It is in the context of appetitive behavior that the importance of von Holst & von Saint Paul's (1960, 1963) concept of 'level-adequate terminology' becomes apparent. (For more information on this study, see below, p. 54.) In one of their experiments, they investigated the behavior of 'locomotor unrest'. General motor unrest is a common occurrence. But usually, unrest actually refers to something quite specific. This can be seen in the results of the experiment depicted in Figure 2.10. The stimulus is electric

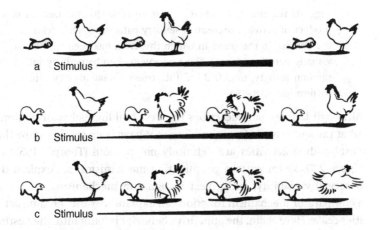

Figure 2.10 Releasing ground-enemy behavior. Explanation in text. From von Holst & von St. Paul, 1963, with permission.

stimulation of a location in the brain. Without a suitable object, the stimulated hen shows only locomotor unrest. Towards a fist, she shows only slight threatening (a). A stuffed, motionless polecat is vigorously threatened and attacked; if the stimulus ends at this moment, the hen remains standing and threatening slightly (b); if it does not end, she checks and flees, screeching (c). The interpretation of this experiment is that the electrode is stimulating the releasing mechanism for ground-enemy behavior.

> The locomotor unrest, which in the example of Fig. [2.10] is the initial element of enemy behaviour, is in another case in the cock directed only toward the head of a conspecific, which is at once pecked (rival fight); in a third case toward water, which is drunk (thirst), or a hiding place, in which the cock calls (leading to the nest), and so forth. If one were always content with the diagnosis 'locomotor drive' and then attempted to study its relationships to other behaviour, the chances of obtaining comprehensible relationships would be slight indeed, since we are in actuality dealing with the most varied goal or appetitive behaviour! The same is true of all movements which sometimes appear 'pure', and at other times as parts of one or another higher drive (von Holst & von Saint Paul, 1963, p. 18).

Another example of appetitive behavior is provided by Lashley's (1938) analysis of a bird building a nest.

> When the hummingbird builds a nest she reacts specifically to lichens and fibrous material but the building to a definite form also

suggests reaction to a deficit ... It is possible that the nest, or other product of activity, presents a sensory pattern which is 'closed' for the animal, in the sense in which this term has been applied to visually perceived forms. The nest might then be built by somewhat random activity, modified until it presents a satisfactory sensory pattern (pp. 448–450).

Although Lashley here describes the nest building behavior as somewhat random, cases of nest building that have been studied show that nest-building activities are definitely not random (Thorpe, 1956; cf. Figure 2.7). Motor pattern coordinating mechanisms can explain the observed variability, and a nest recognition mechanism that serves a comparator mechanism function can explain why the building activities cease. Here again, the appetitive behavior is 'built into' the nesting system. I will have more to say about reactions to a deficit in Chapter 4. I should also emphasize that it is necessary to base conclusions about hypothesized behavior mechanisms on firm experimental evidence in each case.

Behavior Systems and Personality

Personality traits are considered by most psychologists to be relatively permanent characteristics of the individual person, and are thus part of the structure of behavior. Psychologists who study personality have found that personality traits vary on five, and more recently six, independent dimensions: honesty-humility, emotionality, extraversion, agreeableness, conscientiousness, openness to experience (HEXACO: Ashton, 2015). These dimensions are defined by factor analysis of correlations among adjectives describing various behaviors, and one might want to consider these dimensions to represent six human behavior systems. It is, however, obvious that there is little isomorphism between these personality dimensions and behavior systems such as hunger, parenting, sex, aggression, etc. As Ashton says (2015, p. 50): "a person's higher or lower level of Conscientiousness might influence his or her behavior in contexts as varied as eating, sex, aggression, foraging, parenting, and escape." As mentioned above, however, each species and individual animal has its own set of behavior systems, and the six human personality dimensions could represent higher-order (meta-) systems connected to lower-order basic systems. This interpretation raises a number of issues that I discuss in Chapter 4 after I have discussed emotion.

The word 'personality' has also been used in relation to non-human animals. Some personality-trained psychologists have investigated personality traits in dogs, primates, and dolphins in order to compare human and non-human personality structures (see Ashton, 2015, for discussion and references). Recently, however, the word personality (or animal temperament) has also been used by behavioral ecologists who are interested in functional and evolutionary questions (Réale et al., 2007; Carter et al., 2013). These investigators consider traits to be phenotypes that are heritable and that confer differential fitness on individuals and are therefore subject to natural selection. This aspect of personality will be discussed in Chapter 10. Their list of temperaments includes shyness/boldness, explore/avoid, activity, sociability, and aggressiveness. This list also has little overlap with the list of human personality dimensions or with the list of behavior systems, and these differences are discussed in Chapter 4.

Behavior Systems and Modularity

Behavior systems are modules composed of modules (behavior mechanisms). Or are they? The *American Heritage Dictionary* (2015, online) actually lists nine definitions of the word module without giving priority to any of them. And there has been no consensus about its definition among cognitive scientists for more than 30 years. Most authors would agree that a module is a component of a larger system, but agreement generally stops there. Barrett & Kurzban (2006) give a brief history of modularity in cognition as a framework for debating various concepts of modules, but their review is biased in favor of the modularity concepts used by evolutionary psychologists; it omits, for example, all mention of modularity concepts as used by workers in the area of memory research (e.g. Sherry & Schacter, 1987). Bolhuis (2009) gives a critique of the concept of modularity as used in neuroecology. As with behavior units, modules can be categorized structurally, causally, functionally, or historically and at different levels of organization. What is being sought by all authors is a description of the structure of the mind, and that description will vary depending on the question(s) a particular investigator is asking.

In the mid-twentieth century, there was considerable controversy about whether drive systems (behavior systems) had a unitary nature (Miller, 1957, 1959; Hinde, 1959). As understanding of the complexities of behavioral structure developed, this issue became less important. That being said, it is clear that some conceptions of

behavioral structure are better than others in terms of general applic-
ability. Baerends (1976), for example, gave five reasons why he felt the
behavior system concept was worthwhile for the study of complex
behavior. I agree with Baerends that the behavior system concept is
worthwhile, and I will argue throughout this book that the structure of
behavior as I have described it can be applied constructively to understanding a very wide variety of behavioral issues, including cognitive
behavior such as human language as well as the evolution of behavior.

Behavior Systems and Cognition

I define cognition as knowledge (Hogan, 1994a). Knowledge consists of
the representations I have discussed above. Cognitions are behavior
mechanisms, and one can study their structure (units and organization), causes (what activates them), and consequences (what they
accomplish), as well as their development and evolution. I will give
two examples of how this might work.

Cognitive psychologists who study memory are interested in the
relations among the acquisition, retention, and retrieval mechanisms
of memory. These are actually three different kinds of questions about
memory: acquisition is a problem of development or change in structure; retention is a problem of structure – how (in what form) and
where are memories stored; and retrieval is a problem of motivation.
Each question requires its own methods and concepts. And, of course,
one can ask the functional question as well.

One question frequently asked by memory researchers is
whether there are different kinds of memory. Immediately, one can
see that there are at least four different aspects to consider: Are there
different modes of acquisition? Are there different forms and locations
in which memories are stored? Are there different modes of retrieval?
Do memories serve different functions? It is generally agreed that there
are multiple memory systems, but, as with the concept of modularity,
researchers do not agree on what those systems are. One reason for
disagreement is that systems are distinguished on the basis of different
criteria. Sherry & Schacter (1987) proposed, for example, that specialized memory systems can evolve when the functional problems faced
by an organism cannot be solved by a single system. They show that
such 'functional incompatibility' occurs for a number of the distinctions that have been proposed between memory systems, such as
memory for song and memory for spatial location in birds, and
between incremental habit formation and memory for unique episodes

in humans. But using a causal criterion (how are memories acquired) or a structural criterion (where are memories being stored) will lead to different putative systems. This is not the place to pursue these distinctions, but it is important to realize that the answer to the question 'are there different memory systems and what are they?' will depend on the specific question the investigator is asking. I return to the issue of memory in Chapter 7.

My second example is human language. Some years ago, I reviewed a number of studies that compared the development of human language with the development of bird song, as well as several studies of human language development *per se* (Hogan, 2001, pp. 264–267). On the basis of my review, I suggested:

> ... the human language system comprises three basic sets of components at two major levels of organization, and that these components develop largely independently. The sensory-motor components correspond to the perceptual and motor mechanisms depicted in Fig. 1.1 [p. 7], whereas the semantic (meaning) and syntax components correspond to two separate central coordinating mechanisms (p. 267).

At the time, this was an unusual way to characterize language. Recently, however, a very similar structural conception of human language has been proposed by Berwick *et al.* (2013). It can be seen that their scheme (Figure 2.11) has the same basic components as mine. There are minor differences in the organization of the components, especially their emphasis on the syntax component being the essence of language. I do not disagree with this emphasis, and discuss some of these issues in the sections on language development in Chapter 5 and on language evolution in Chapter 8.

In general, I will not be discussing most aspects of human cognition in any detail in this book, but I do want to emphasize that the general framework I am proposing is as applicable to problems of human cognition as it is to other aspects of behavior in both human and non-human animals.

Behavior Systems and Neural Reality

I am concerned in this book with understanding the behavior of individual animals using behavioral concepts and methods. But since behavior (as I have defined it) is the study of the functioning of the brain, it would be gratifying if there was some correspondence between my analysis of behavior and neural mechanisms. I have already alluded to

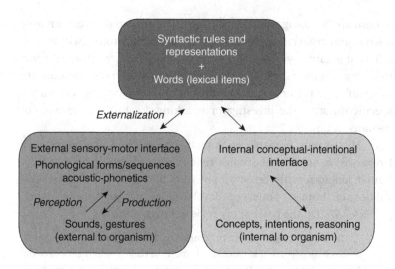

Figure 2.11 The basic design of language.
From Berwick *et al.*, 2013, with permission.

possible neural substrates for motor and perceptual behavior mechanisms, but I will here give two examples of neurophysiological analyses that deal with more complex systems.

My first example is a study published more than 50 years ago by von Holst & von Saint Paul (1960, 1963). These investigators used Plexiglas electrode holders chronically fixed to the skull of adult fowl that were intact and unrestrained. Electrodes could be pushed into the brain of the animal in calibrated steps, and stimulus voltage and frequency varied. Various parts of the brain stem were explored. Many of their experiments and results concern motivational questions, some of which are discussed in Chapter 3. With respect to structure, they found evidence for perceptual, central, and motor mechanisms organized functionally. Their summary figure is reproduced as Figure 2.12.

The upper part of the figure depicts behaviors evoked from various stimulus fields in the brain. Stimulation in either S1 or S2 could produce 'pure' clucking (a motor mechanism). Stimulation in either SI or SII could produce elements of the flight behavior system (flight drive – central coordinating mechanism) such as walking, flying, and clucking. Stimulation in SF (a perceptual mechanism) evokes an 'enemy-response mood' that could result in either attack or flight. Figure 2.13d shows how these systems are connected. Direct stimulation of the motor system leads to behavior that is 'unconcerned' or automatic, whereas stimulation of the perceptual system can lead to

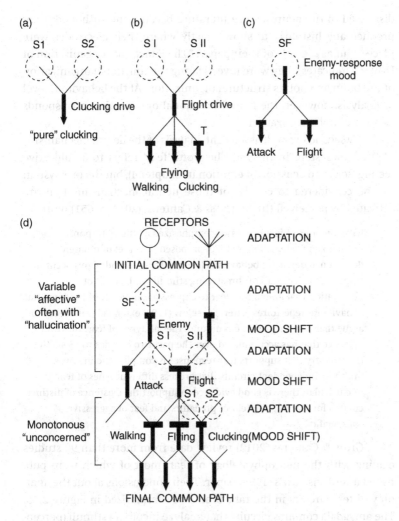

Figure 2.12 Sketch of a fragment from the functional organization of some behavior patterns in the fowl. Explanation in text.
From von Holst & von Saint Paul, 1963, with permission.

the same behaviors, but there is more variability in response and an added 'affective' or emotional component. The concepts of adaptation and mood shift, on the right side of the figure, will be discussed in Chapter 3.

These results show that the nervous system of the fowl is organized in a way that corresponds very closely to the way their behavior is organized under natural conditions. This study has been generally dismissed in the neuroscience literature because the authors did not produce any histology to show exactly where their electrodes were placed. But as the title of their paper, "The Functional Organisation of Drives", indicates, they were investigating the functional organization of the brain and not its structural organization. At the behavioral level of analysis, however, the neural functional organization corresponds to a structural organization.

My second example is a recent analysis of the neural mechanisms of the fear system in the rat. The word 'fear' refers to a subjective feeling (see my discussion of emotion in Chapter 4), but the fear system can be considered to consist of those neural mechanisms that are activated by perceived threat. Gross & Canteras (2012, p. 651) write:

> Early theories of fear processing emphasized a unitary response mechanism for fear. Bolles (1970) proposed that a set of innately determined defensive behaviours (such as freezing and flight) occur in response to all classes of threatening stimuli on the basis of the observation that animals appear to express a limited set of fundamental behavioural repertoires. Later, Fanselow (1994) extended this theory to argue that a unitary brain circuit underlies all types of fear. However, evidence that has accumulated over the past two decades suggests that the circuitry that supports fear responses is complex and involves multiple, independent circuits that process different types of fear. In particular, there is good evidence to support the existence of distinct circuits for fear of pain, fear of predators and fear of aggressive conspecifics.

Gross & Canteras (2012) review data from more than 90 studies dealing with the neurophysiology of fear, most of which were published after Fanselow's (1994) paper. Their conclusions about the anatomy of fear circuits in the rat's brain are summarized in Figure 2.13. The amygdala contains circuits that analyze incoming stimuli (perceptual mechanisms), the hypothalamus contains circuits that coordinate predator-responses and conspecific-responses (central coordinating mechanisms), and the PAG (periaqueductal grey) contains circuits that activate the various defensive responses (motor mechanisms). Although these fear circuits are segregated and independent of each other, they all share a common memory encoding mechanism involving the hippocampus, septum, and cortex (conditioned fear – not shown in this figure). Another conclusion from their review is that each fear circuit also processes or regulates important physiological and behavioral responses that are not related to fear. The predator-responsive nuclei in the medial hypothalamus also have a role in

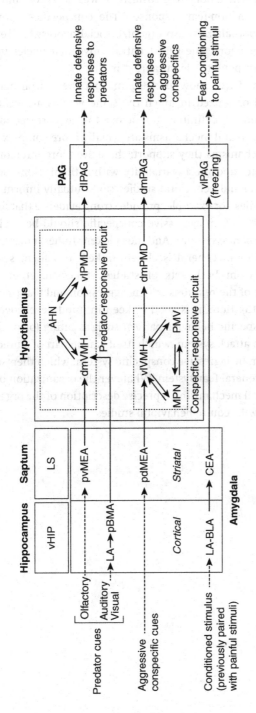

Figure 2.13 Parallel neural circuits mediate fear of pain, aggressive conspecifics, and predators. Explanation in text. From Gross & Canteras, 2012, with permission.

energy balance and metabolic homeostasis, as well as a profound inhibitory influence on non-fear responses. The conspecific-responsive nuclei in the hypothalamus also organize social aggression, sexual, and maternal behaviors. The CEA (central amygdala) nuclei in the pain circuit also respond to pleasurable stimuli.

The studies of fear show, once again, that the central nervous system is indeed organized in much the same way as are behavior systems. But there is definitely not a one-to-one correspondence between the two. Neural mechanisms are usually more complex than the behavior mechanisms they support: there are more interconnections among them, and more variability within them. Nonetheless, neurophysiological and behavioral studies can mutually inform each other. The rat studies, for example, provide strong evidence that fear of a predator and fear of an aggressive conspecific should be analyzed behaviorally as separate systems. And the chicken studies indicate that aggressive behavior in general is closely linked to a flight system. In fact, results from both sets of studies, taken together, give a broader picture of the relations among aggression and fear (see also van Kampen, 2015). Here, it is good to keep in mind 'level-adequate terminology'. A specific behavior pattern such as biting (or pecking) may be part of an attack system, which itself can be part of a predator defense system or an aggressive conspecific system, which themselves can be part of a general fear system. Whatever the organization of the neurophysiological mechanisms, a precise description of the organization of behavior will require behavioral studies.

3

Motivation
Immediate Causes of Behavior

Niko Tinbergen (1907–1988). Photo by Lary Shaffer.

The word 'motivate' means 'to cause to move', and I use the concept of motivation to refer to the study of the immediate causes of behavior: those factors responsible for the initiation, maintenance, and termination of behavior. Causal factors for behavior include stimuli, hormones and other substances, and the intrinsic activity of the nervous system.

A major problem in the study of motivation has been, and remains, that different authors have different meanings for the concept. For example, my use of the term motivation includes external as well as internal causes. Popularly, many people think of motivation as exclusively an internal cause. Causal factors not only motivate behavior, they can also change the structure of behavior; that is, they have developmental effects. The formation of associations and the effects of reinforcement are developmental processes, and developmental processes have played an important role in many theories of motivation, especially in experimental psychology. I restrict the term motivation to the modulating effects causal factors have on the activation of behavior mechanisms. Development refers to the permanent effects causal factors have on the structure of the behavior mechanisms and on the connections among the behavior mechanisms, and is discussed in Chapters 5, 6, and 7.

SOME MOTIVATIONAL ISSUES

In addition to problems concerning the definition of the concept of motivation, there have been four issues that have dominated discussions of motivation. 1) What role should a concept of energy play in motivational theories? 2) What is the relative role of internal versus external causal factors? 3) Do causal factors have specific or general effects? 4) Is the locus of action of causal factors peripheral or central? I will briefly discuss each of these issues.

The Concept of Motivational Energy

One attribute of living matter is its activity, the continuous transformation of energy from one form to another. It was natural, therefore, when people began to seek explanations of their own activity, to invoke some concept of energy. And indeed, the earliest scientific theories of motivation invoked concepts such as instinctual impulses (James, 1890), libido (Freud, 1905, 1915) and psychophysical energy (McDougall, 1923). Within American psychology, these concepts became replaced by the concept of 'drive', but as Lashley (1938) pointed out, drives continued to have all the dynamic properties of the old instinctual urges. A particularly influential theory of motivation was proposed by Lorenz (1937). The core of his theory was an energy variable, *action-specific energy*. For a variety of reasons, all these theories were strongly criticized (Hinde, 1960), and energy concepts quickly disappeared from most accounts of behavior. However, some authors have pointed out that many of the phenomena that used to be explained using energy concepts are still not accounted for by other concepts: they have suggested that an energy concept may still play a useful theoretical role (e.g. Toates & Jensen, 1991; Hogan, 1997). These issues are considered in more detail below.

External versus Internal Causal Factors

As mentioned above, the word 'motivation' often refers only to internal causes of behavior. We speak of an animal's search for food as motivated by hunger, but of chewing and swallowing as reflex actions to stimuli in the mouth. On close inspection, however, hungry animals are clearly guided by environmental cues as they search for food and a thoroughly sated animal will often spit out the same food it would have chewed if it were hungry. In fact, any behavior must be caused by some combination of both internal and external factors.

The motivational model of Lorenz (1950) illustrates the interdependence of internal and external factors. This model is shown in Figure 3.1. According to Lorenz, each behavior pattern (motor mechanism) is associated with a reservoir that can hold a certain amount of 'energy'. Whenever the behavior pattern occurs, energy is used up; but when the behavior pattern does not occur, energy can build up in the reservoir. The higher the level of energy, the more pressure it exerts on the valve. When the valve opens, as a result of external stimulation, energy is released and the behavior occurs. One might imagine that as the pressure in the reservoir increases, it becomes more and more difficult to prevent the energy from escaping through the valve. In fact, behavior does sometimes occur in the absence of any apparent external stimulus. Such behavior has been called a *vacuum activity*. Lorenz described the behavior of a captive starling that performed vacuum insect hunting. This bird would repeatedly watch, catch, kill, and swallow an imaginary insect.

In this model, a particular behavior pattern cannot occur without at least some internal causal factors as well as some external ones. Further, the model makes it clear that internal and external factors can substitute for each other in determining the intensity of a behavior pattern: a strong stimulus can compensate for weak internal factors and vice versa. The fact that both internal and external factors are

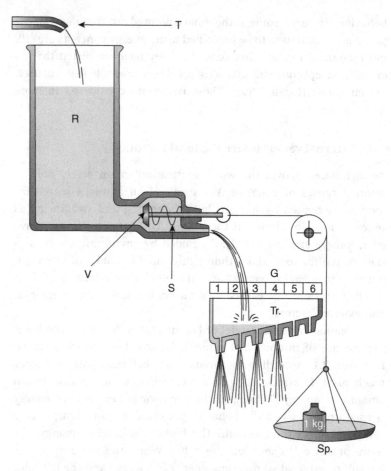

Figure 3.1 Lorenz' model of motivation. T is the tap supplying a constant flow of endogenous energy to the reservoir, R; the valve, V, represents the releasing mechanism, and the spring, S, the inhibitory functions of the higher coordinating mechanisms; the scale pan, Sp, the perceptual part of the releasing mechanism, and the weight applied corresponds to the impinging stimulation. When the valve is open, energy flows into the trough, Tr, which coordinates the pattern of muscle contractions; the intensity of the response can be read on the gauge, G.
From Lorenz, 1950

essential for any behavior to occur does not imply, of course, that one cannot study the effects of internal and external factors separately. The effects of varying various internal factors can be determined if the external situation is kept relatively constant, as can the effects of external factors be determined if the internal state of the animal is held constant.

A classic example of such a study is one by Baerends *et al.* (1955) on the courtship behavior of the male guppy (*Lebistes reticulatus*). Courtship by the male comprises a number of behavior patterns, including posturing in front of the female, a special sigmoid posture, and copulation attempts. These authors were able to derive a scale of internal motivation using the relation of marking patterns on the body of the male to the number of copulation attempts; and external stimulation was considered to be proportional to the size of a female. Figure 3.2 shows the results of an experiment in which females of different sizes were presented to males at different levels of internal motivation. The points plotted on the graph represent the relationship between the measures of internal and external stimulation at which particular patterns of behavior were observed. If it is assumed that the total motivation necessary for a specific behavior pattern to occur is always the same, the patterns 'posturing', 'sigmoid intention', and 'sigmoid' can be seen to represent increasing values of courtship strength. The lines connecting these points of equal motivation have been called *motivational isoclines* by McFarland & Houston (1981).

Specific versus General Effects of Causal Factors

Ever since Hull (1943) postulated a general drive, a recurring question in the study of motivation has been whether causal factors have general or specific effects. Does a hungry dog merely eat its food more quickly and accept less preferred foods more readily, or does it also attack a stranger more fiercely and copulate more vigorously? Common sense suggests that some causal factors are likely to have broad effects, whereas others will have only limited effects. In general, any particular causal factor will most likely have both specific and general effects. Which effects are more important will depend on the question of interest. For example, specific effects of causal factors are implied in Lorenz' model of motivation. The model posits that the fluid in the reservoir is specific to the particular behavior pattern with which it is associated: action-specific energy. On the other hand, the circadian clock will be seen to have an important influence on many behavior systems. I will examine specific and general effects of causal factors in some detail in the section on displacement activities.

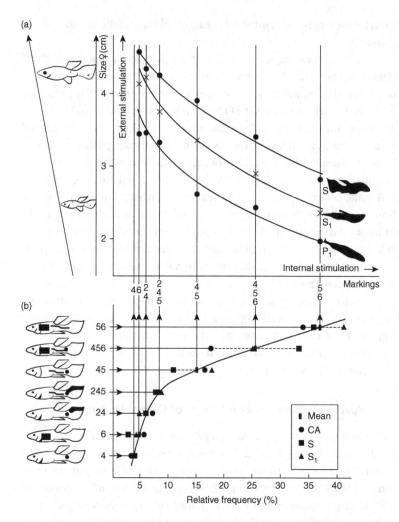

Figure 3.2 Results of an experiment on guppy courtship. (a) Relationship between the intensity of the external stimulation, the intensity of the internal stimulation, and the kind and degree of development of the resulting activity. (b) 'Calibration curve' for determining the place of the different marking patterns on the abscissa of (a). CA = copulation attempt; S = sigmoid posture; Si = sigmoid intention; Pf = posturing in front of the female.
From Baerends et al., 1955

Central versus Peripheral Locus of Action

A fourth pervasive issue in motivation concerns the locus of action of causal factors. Do causal factors operate within the central nervous system (CNS) or at a more peripheral level? Once again, common sense suggests that they must act in both places; nonetheless, this has also been a controversial issue. Historically, the controversy arose as a reaction by the early behaviorist school in psychology to the views of the introspectionists, who thought one could understand behavior by reflecting on one's own experiences. The behaviorists were skeptical of internal causes that could not be investigated directly, and they attempted to explain as much behavior as possible in terms of stimuli and responses that could be measured physically. However, as it has become more and more possible to measure and manipulate events that occur within the CNS, one major objection to the postulation of central factors has been removed. Nonetheless, some researchers continue to emphasize central or peripheral factors, and several examples are presented in later sections.

CAUSAL FACTORS

Motivation is concerned with the factors that control the activity of the behavior mechanisms of the individual. These factors are generally considered to be stimuli (external and internal), hormones and other substances, and the intrinsic activity of the nervous system itself. Each of these factors will be briefly discussed.

Stimuli

Stimuli can control behavior in many ways: they can release, direct, inhibit, and prime behavior. Chapter 2 discussed examples of stimuli that release and direct various behavior patterns. Some stimuli can have exactly the opposite effect: rather than facilitate behavior, they inhibit it. A good example is provided by the nest-building behavior of many species of birds. Birds typically build their nests using specific behavior patterns. The stimuli that release and direct their behavior have been studied in several cases, and conform to the general principles already discussed. However, at a certain point, the birds stop building and no longer react to the twigs, lichens, or feathers with which they construct their nest. There are several possible reasons why they stop, but one reason is that the stimuli provided by the completed nest inhibit further nest building. This can be seen when a bird takes over a complete nest from the previous season and shows very little nest-building behavior. Other birds, in the same internal state, that have not found an old nest show a great deal of nest-building behavior (Thorpe, 1956).

Another example of the inhibitory effects of stimuli is seen in the courtship behavior of the three-spined stickleback (*Gasterosteus aculeatus*), a small fish. Male sticklebacks set up territories in small streams early in the spring, build a nest of bits of plant material, and will generally court any female that may pass through their territory. Courtship includes a zigzag dance by the male, appropriate posturing by the female, leading to and showing of the nest entrance by the male, following and entering the nest by the female, laying eggs, and finally fertilization (see Figure 3.3). The female swims away and the male then courts another female. The male could continue courting egg-laden females for many days, but usually he does not. Experiments in which eggs were removed from or added to the nest have shown that visual stimuli from the eggs inhibit sexual activity: if eggs are removed from the nest, the male will continue courting females, but if eggs are added to the nest, he will cease courting, regardless of the number of eggs he has fertilized (Sevenster-Bol, 1962). This is an especially interesting example because the same visual stimulus that inhibits sexual activity has an activating effect on the parental behavior (fanning the eggs) of the same male.

Stimuli not only control behavior by their presence, but in many cases, continue to affect behavior even after they have physically disappeared. When a stimulus has arousing effects on behavior that outlast its presence, *priming* is said to occur. Aggressive behavior in the male Siamese fighting fish (*Betta splendens*) provides a good example (Hogan & Bols, 1980). This fish shows vigorous aggressive display and fighting toward other males of its species (including its own mirror image). If a fish is allowed to fight with its mirror image for a few seconds and the mirror is then removed, it is very likely to attack a thermometer introduced into the aquarium. If the thermometer had been introduced before the mirror was presented, the fish very likely would have ignored it. Thus, the sight of a conspecific not only releases aggressive behavior, it must also change the internal state of the fish for some time after the conspecific disappears. We can say that the stimulus primes the mechanism that coordinates aggressive behavior or, more simply, that it primes aggression. Similar priming effects have been demonstrated with food and water in rats and hamsters (Van

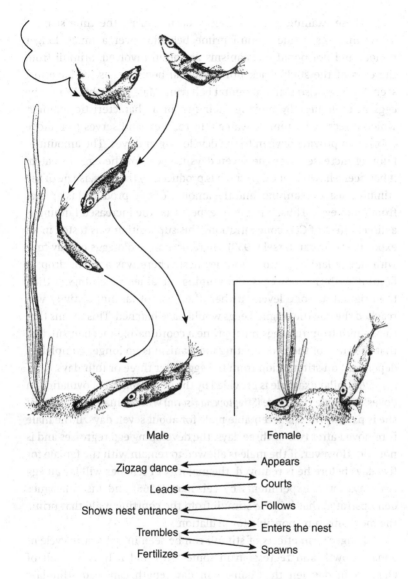

Male	Female
Zigzag dance	Appears
Leads	Courts
Shows nest entrance	Follows
Trembles	Enters the nest
Fertilizes	Spawns

Figure 3.3 Courtship and mating behavior of the three-spined stickleback. The male is on the left and the female, with a swollen belly, is on the right.
From Tinbergen, 1951, with permission.

der Kooy & Hogan, 1978), and with brain stimulation in several species (Gallistel, 1973). An especially elegant mathematical analysis of priming in cichlid fish and crickets is presented by Heiligenberg (1974).

These examples of priming all occur during the time span of a few minutes. Some stimuli prime behavior over a much longer period, and hormonal mechanisms are often involved. Stimuli from the eggs of the stickleback inhibit sexual behavior, as we have just seen, but they also prime parental behavior. Male sticklebacks fan the eggs in their nest by moving their fins in a characteristic manner, which directs a current of water into the nest and serves to remove debris and provide oxygen to the developing embryos. The amount of fanning increases over the seven days it takes for the eggs to hatch. It has been shown that CO_2, which is produced by the eggs, is one of the stimuli releasing fanning, and the amount of CO_2 produced is greater from older eggs. Thus, one might expect that the increased fanning is a direct effect of CO_2 concentration. This supposition was tested in an experiment by van Iersel (1953). He replaced the old eggs on day four with newly laid eggs from another nest. There was a slight drop in fanning with the new eggs, but fanning remained much higher than the original day-one level. Further, the peak of fanning activity was reached the day the original eggs would have hatched. This means that the stimuli from the eggs must prime a coordinating mechanism, and that the state of the coordinating mechanism is no longer completely dependent on stimulation from the eggs after three or four days.

A similar example is provided by the development of ovulation in doves. A female ring dove (*Streptopelia risoria*) will normally lay an egg if she is paired with an acceptable male for about seven days. If the male is removed after two or three days, the developing egg regresses and is not laid. However, if the male is allowed to remain with the female for five days before he is removed, the majority of females will lay an egg two days later. Experiments by Lehrman (1965) and his colleagues demonstrated that it is the stimuli from the courting male that prime the mechanism responsible for ovulation.

Longer-term effects of stimuli can be seen in the yearly cycle of gonad growth and regression in some birds and fish as a result of changes in day length. Changes in day length can also stimulate a host of other physiological changes, including those that prepare migratory birds for their long-distance flight (e.g. Piersma & van Gils, 2011) or various mammals for hibernation in the winter (Nelson, 2016).

Hormones and Other Substances

Hormones are substances released by endocrine glands into the bloodstream; many of them are known to have behavioral effects. Lashley

(1938) suggested that hormones could affect behavior in at least four different ways: during the development of the nervous system, by effects on peripheral structures through alteration of their sensitivity to stimuli, by effects on specific parts of the central nervous system (central behavior mechanisms), and by non-specific central effects. Abundant evidence for all these modes of action has accumulated since Lashley's time, although the mechanisms by which hormones influence behavior have turned out to be more complex and diverse than early investigators realized (Beach, 1948; Nelson 2016).

Both peripheral and central effects of the hormone prolactin are seen, for example, in the parental feeding behavior of the ringdove. Prolactin is responsible for the production of crop 'milk', sloughed-off cells from the lining of the crop that are regurgitated to feed young squabs. Lehrman (1955) hypothesized that sensory stimuli from the enlarged crop might induce the parent dove to approach the squab and regurgitate. His experiments showed that local anesthesia of the crop region, which removes the sensory input, reduced the probability that the parents will feed their young. Later experiments confirmed that prolactin has both peripheral and central effects on the dove's parental behavior (Buntin, 1996).

The maternal behavior of the rat provides an example that illustrates the variety of hormonal effects. The hormones released at parturition change the dam's olfactory sensitivity to pup odors, reduce her fear of the pups, and facilitate learning about pup characteristics; they also activate a part of the brain essential for the full expression of maternal behavior (see Fleming & Blass, 1994).

Substances released from the neuron terminals into the synapse are known as transmitters, many of which are involved in activating specific behavior systems such as feeding and drinking (see Nelson, 2016). Other transmitters such as dopamine are thought to mediate the motivational effects of stimuli for a wide range of behavior systems, especially their reinforcing effects (Wise, 2004; Glimcher, 2011). Examples of these effects are given in Chapter 7. Psychoactive drugs, which are thought to exert their effects by altering neurotransmitter functioning in the brain, are also causal factors for behavior, but will not be considered further in this book.

Intrinsic Neural Factors

In living organisms, the nervous system is continuously active, and this has many consequences for the occurrence of behavior. Adrian *et al.*

(1931) were the first to demonstrate spontaneous firing of an isolated neuron, and von Holst (1935) showed that such nervous activity under-lay the endogenous patterning of neural impulses responsible for swimming movements in fish. One of the earliest attempts to incorpo-rate intrinsic causes into the scientific study of behavior was made by Skinner (1938). His concept of the operant is a unit of behavior that occurs originally due to unspecified, intrinsic causes. It is only as a result of conditioning that the operant comes to be controlled by specific stimuli. The motivational model of Lorenz (1937) was another attempt. This model implies that a continuously active nervous system is kept in check by various kinds of inhibition. The most systematic support for this aspect of the model is provided by the behavior of insects. A particularly striking example concerns the copulatory beha-vior of the male praying mantis (*Mantis religiosa*).

Mantids are solitary insects that sit motionless most of the time, waiting in ambush for passing insects. Movement of an object at the correct distance and up to the mantis' own size releases a rapid strike. Any insect caught will be eaten, even if it is a member of the same species. This cannibalistic behavior might be expected to interfere with successful sex, because the male mantis must necessarily approach the female if copulation is to occur. Sometimes a female apparently fails to detect an approaching male and he is able to mount and copulate without mishap, but very often the male is caught and the female then begins to eat him. Now, an amazing thing happens. While the female is devouring the male's head, the rest of his body manages to move round and mount the female, and successful copulation occurs!

In a series of behavioral and neurophysiological experiments, Roeder (1967) showed that surgical decapitation of a male, even before sexual maturity, releases intense sexual behavior patterns. He was then able to demonstrate that a particular part of the mantis' brain, the subesophageal ganglion, normally sends inhibitory impulses to the neurons responsible for sexual behavior. By surgically isolating these neurons from all neural input, he showed that the neural activity responsible for sexual activity is truly endogenous. More recent work is investigating the role of spontaneous neural activity in operant learning in various invertebrates (Brembs, 2009). These experiments, and many others, point out the importance of intrinsic factors in the motivation of behavior, and support the type of model proposed by Lorenz.

The intrinsic factors just discussed are all related to the motiva-tion of specific behavior patterns. One additional intrinsic factor is the pacemaker or oscillator cells that are responsible for biological clocks (Mistlberger & Rusak, 2005, in press). These clocks do not control any specific behavior pattern, but rather modulate the behavior mechanisms that control many different types of behavior. Most of the experimental work has investigated the oscillators responsible for daily (circadian) rhythms, often at a neurophysiological or genetic level, but there has also been considerable work on the oscillators controlling interval timers (Buhusi & Meck, 2005). Timing mechanisms and biological rhythms are discussed further in Chapters 4 and 7.

CAUSAL FACTORS CONTROLLING SOME SPECIFIC BEHAVIOR SYSTEMS

Hunger

The hunger system comprises perceptual mechanisms for recognizing food, a central mechanism for integrating causal factors for eating (the hunger mechanism), and coordinating and motor mechanisms for locating and ingesting food. A diagram of the hunger system of a young chick (*Gallus gallus spadiceus*) is shown in Figure 2.6. The motor mechanisms used to locate, catch, and ingest food vary greatly among species, as do the causal factors controlling them. I will not review these differences here, but it is an interesting fact about the hunger system of young chicks, and of many other young animals including rats and humans, that ingestion is not controlled by the hunger mechanism. Rather, the motor mechanisms (e.g. pecking or sucking) are relatively independent units, and their activation depends primarily on factors specific to each behavior pattern. For example, the factors that control pecking in young chicks during the first two days after hatching include the specific releasing characteristics of particular objects, the novelty of the objects (stimulus change), and how much pecking the chick has recently engaged in (action-specific energy). Nutritional state (hunger mechanism) does not influence pecking until day three of life, and then only when the chick has had certain kinds of experience (Hogan, 1971, 1984b; see Chapter 5). A similar situation is seen in the control of suckling in rat pups, in which nipple search, nipple attachment, and amount of sucking are not affected by food deprivation until the pups are about two weeks of age (Hall & Williams, 1983). It should be noted that even after nutritional state begins to control ingestion, factors specific to the motor mechanisms continue to have an important influence on their occurrence. This is especially true of the prey-catching motor patterns of many carnivore species (Polsky, 1975; Baerends-van Roon & Baerends, 1979).

The physiological mechanisms controlling the central hunger mechanism have been studied in great detail in a number of species. In the rat, the central mechanism comprises several nuclei in the hypothalamus. Cells in these nuclei are sensitive to a variety of substances in the blood, as well as to specific neurotransmitters. Activation of these nuclei can lead to increased or decreased food intake, as well as to intake of particular types of food (Richter, 1942; Miller, 1957). Nelson (2016) provides a good introduction to the complex workings of the causal factors affecting the feeding system. Signals from the circadian pacemaker also influence activation of these nuclei, which is why rats and other nocturnal animals eat almost exclusively at night, whereas chickens and other diurnal animals eat primarily during the day.

A fundamental problem for young animals of most species is learning what foods they should eat. Because developmental factors are so important in determining an animal's behavior with respect to food (and water), many psychologists have incorporated learning variables into their theories of motivation (see below). Development of food recognition is discussed in Chapter 5.

Parenting

Parental behavior refers to behavior patterns a parent uses to approach and care for its young, although the same behavior is sometimes shown to young that are not its own, or even to young of other species. In some species, parental behavior is shown exclusively by either the male or female parent, whereas in other species, both parents share parental duties. Many species have no parental behavior at all. Parental behavior, of course, is controlled by both internal and external causal factors, and one important internal causal factor is often the hormone prolactin. A few examples illustrate some of these points.

We have already seen that, in sticklebacks, parental behavior is shown exclusively by the male. In other fish species, both parents care for the young. A well-studied example is the behavior of the cichlid fishes *Symphysodon* spp. (discus fishes). In these species, eggs are laid and fertilized on an appropriate substrate, and both parents guard and fan the eggs. After the eggs hatch, the young feed on mucous cells that have developed on the bodies of the parents. Both the fanning behavior and the production of mucous cells are controlled by prolactin, which also inhibits aggressive behavior. The effect of prolactin on mucous cell production seems to be homologous with the production of crop milk in doves, as mentioned above, and with milk production in mammals (Blüm & Fiedler, 1965).

In rats, parental behavior is exclusively maternal behavior. Prior to the birth, the mother-to-be builds a nest. After the birth, the mother licks the pups, adopts a nursing or crouch posture over the pups that allows them to attach to her teats and suckle, and retrieves them if they stray out of the nest. A female that has gone through a normal pregnancy and is presented with a young pup will show this whole set of maternal behaviors whether the pup is her own or not. On the other hand, if the same pup is presented to a virgin female, the female will ignore, avoid, or possibly even attack and eat the pup. The difference between the two females is in their internal state: one is 'maternal' and the other is not. The maternal state is determined in large part by particular levels of the hormones estrogen, progesterone, and prolactin. It is an interesting fact that a virgin female continuously exposed to a litter of young pups will usually become maternal after about eight days. She will then show all the same behaviors – nest building, licking, crouching, and retrieving – as a mother rat (except for lactation). This is actually another example of priming. An important difference between virgin and mother rats is that mother rats find the odor of pups attractive, whereas virgin females do not. Exposure to young pups causes hormonal changes in the virgin female that result in changes to her odor preferences (Fleming & Blass, 1994).

Hormonal state is also known to influence maternal responsiveness in human mothers, but experimental evidence is sparse because it is not permissible, ethically, to manipulate hormonal levels in people for experimental purposes. Nonetheless, it is possible to measure hormone levels in pregnant women and new mothers and ask them about their feelings toward babies. Several studies of this kind have been carried out. The results indicate that although hormones may facilitate a mother's responsiveness to her infant shortly after birth, hormones are neither necessary nor sufficient for maternal responsiveness. A mother's attitudes and prior experience with infants seem to be more important determinants (Fleming et al., 1997).

Aggression

The word 'aggressive' is often used to refer to any behavior that includes attack patterns, such as interactions between predator and prey, or defense of self or young. Here, I will restrict its meaning to attack patterns of the sort directed to members of the same species (cf. Lorenz, 1966). The aggression system is the behavior system controlling these patterns. The aggression system can also be expressed by behavior other than attack patterns: for example, marking behavior and various auditory and visual signals. These non-attack behaviors are discussed below and also in Chapters 8, 9, and 10. Most species possess an aggression system, and much research has been done investigating the causal factors that activate it. An important issue in many of these studies has been the relative role of external and internal causal factors: are animals lured or driven to fight? Experiments with Siamese fighting fish give some insight into this issue.

As we have already seen, a male fighting fish reacts to the visual stimulus of another male or its own reflection with vigorous aggressive behavior. In this case, an external causal factor can be said to lure the fish into fighting. However, a fighting fish will also learn to swim through a passageway at the end of which it finds a goal compartment with a mirror or another male. This fish will not swim through the passageway if the goal compartment is empty; it will chose a goal compartment with a mirror rather than one that is empty; and under some conditions, it will even choose a goal compartment with a mirror rather than one with food (Hogan, 1974). Further, a fish whose aggression has been primed just prior to testing will swim faster to the mirror than a fish that has not been primed (Hogan & Bols, 1980). These results suggest that the fish is being driven to search for a fight in much the same way as a hungry animal is driven to search for food.

However, there is an important difference between the motivation to fight and to eat. When an animal fights until it is exhausted, the tendency to fight will recover to a certain moderate level after a few days, but then it remains fairly constant or may even decline. When an animal eats until it is satiated, the tendency to eat also recovers after a few hours or days; however, if no food is eaten, hunger continues to increase to very high levels. The only way to increase aggression to very high levels is by priming; deprivation, by itself, is not sufficient. In terms of the Lorenzian motivational model, there is very little evidence that the reservoir can be filled endogenously.

Although Lorenz originally made the assumption that endogenous energy for each fixed action pattern is produced continuously, this assumption is not a necessary feature of an energy model. For example, Heiligenberg (1974) used the concept of "behavioural state of readiness" to model attack behavior in a cichlid fish. In Heiligenberg's model, there is no endogenous energy at all; all the energy for attacking is produced by the perception of specific stimuli (in this case, a dummy of a conspecific male). Perception of these stimuli results in a short-term increase in energy (readiness) that lasts for a few seconds, but also a much smaller long-term increase in energy that lasts for days. Continual presentation of appropriate stimuli results in a buildup of energy to high levels. In this example, the view of a conspecific is the stimulus that primes aggression. This model has many of the features of Lorenz' model, but it is external rather than internal factors that affect the level of aggressive motivation. It seems likely that similar considerations apply to the aggression system in most species.

Sex

Sexual behavior is sometimes defined functionally as behavior leading to successful reproduction. The sex behavior system does include behavior patterns that are generally necessary for reproduction, but the organization of the system must be defined independently of reproduction. As with all other behavior systems, the sex system comprises a number of perceptual mechanisms that recognize a potential mate, a central mechanism that integrates information from the perceptual units and from other internal factors, and motor mechanisms that coordinate the performance of sexual activities. As we shall see below, behavior patterns belonging to nonsexual behavior systems are also necessary for successful reproduction in many species.

As with parenting, the sex system is almost always different in the male and the female of the species: different stimuli activate the system, different hormones are important internal causal factors, and different behavior patterns are expressed. A general review of the literature on sexual motivation is impossible here, and I will discuss only one specific example.

Sexual behavior in a male rat consists of a number of ejaculatory series. In each series, the male approaches and mounts a receptive female. Mount bouts typically last a few seconds, during which the male thrusts his penis rapidly and repeatedly, and may briefly insert his penis into the female's vagina. Mount bouts are followed by a period of about one minute in which the male engages in genital grooming. After about eight mount bouts, the male ejaculates. A postejaculatory interval of about five minutes follows, in which the male makes a specific ultrasonic vocalization. A new series then begins. In later series, ejaculation generally occurs after fewer mount bouts. A normal male commonly has six to ten ejaculations before sexual behavior ceases; he is then unlikely to resume sexual behavior for several days. This pattern of copulation has been shown to coincide with the requirements of the female for maximizing the chance of pregnancy (see Sachs & Barfield, 1976 for review).

Analysis of the causal factors responsible for this behavior led early investigators to propose that there are two major central mechanisms in the male sex system: a sexual arousal mechanism (SAM), and an intromission and ejaculatory mechanism (IEM). Beach (1956, p. 20) stated that:

> The main function of the SAM is to increase the male's sexual excitement to such a pitch that the copulatory threshold is attained. Crossing the copulatory threshold results in mounting and intromission ... The initial intromission and those that follow provide a new source of sensory impulses which serve to modify further the internal state of the animal and eventually bring the male to the ejaculatory threshold.

The causal factors that activate the SAM are primarily external stimuli provided by a receptive female, while the factors that activate the IEM are stimuli arising from the performance of sexual behavior.

Later studies have elaborated on this model, and Toates and O'Rourke have proposed a computer model that incorporates these findings and accounts quite well with observed behavior (see Toates, 1983). An analysis of sexual behavior that includes both an SAM and an IEM can be applied to most male mammals, including humans. It should be noted that this analysis suggests that sex and aggression are very similar in that both systems are primarily aroused by external stimuli, and the highest levels of arousal can only be reached while the animal is engaged in the respective behaviors. Further, it is also true for both systems that the external stimuli are only arousing when an animal has an appropriate internal hormonal state (see Nelson, 2016).

Self-maintenance

Most animals possess behavior patterns that can be used for cleaning themselves or for keeping their muscles, skin, or feathers in good condition. These patterns range from simply stretching or rubbing up against some object to complex integrated sequences of behavior used for grooming. One such sequence that has been studied in great detail is the dustbathing behavior of fowl (*Gallus gallus*), and these studies have found that this behavior illustrates a number of general motivational principles.

The dustbathing behavior of fowl consists of a sequence of coordinated movements of the wings, feet, head, and body of the bird that serve to spread dust through the feathers. The sequence of behaviors in a dustbathing bout begins with the bird pecking and raking the substrate with its bill and scratching with its feet. These movements continue as the bird squats down and comes into a sitting position. From time to time, the bird tosses the dusty substrate into its feathers with vertical movements of its wings and also rubs its head in the substrate. It then rolls on its side and rubs the dust thoroughly through its feathers. These sequences of movements may be repeated several times. Finally, the bird stands up, shakes its body vigorously, which releases the dust, and then switches to other behavior (Kruijt, 1964). In adult fowl, bouts of dustbathing last for about half an hour. Dustbathing serves to remove excess lipids from the feathers and to maintain the feathers in good condition (van Liere & Bokma, 1987).

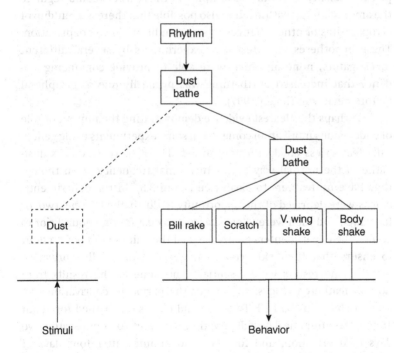

Figure 3.4 Dustbathing behavior system of a young chick. See Figure 2.6 for explanation.

From Vestergaard et al., 1990, with permission.

A diagram of the dustbathing system is shown in Figure 3.4. A complete motivation analysis of dustbathing would investigate the properties of the stimuli that activate the dust recognition mechanism, as well as the factors that control the central mechanism. Some progress has been made in analyzing the stimuli important for dust recognition, as well as the role played by light, heat, and the sight of other birds' dustbathing (Petherick *et al.*, 1995; Duncan *et al.*, 1998), but I will concentrate here on factors intrinsic to the central mechanism and on the role of the circadian clock.

Vestergaard (1982) gave adult fowl limited access to dust and found that the latency to dustbathe was inversely related, and duration of bouts directly related, to the length of dust deprivation. Similar deprivation experiments on young chicks led to similar results (Hogan *et al.*, 1991). One way to explain these results is to postulate the buildup of endogenous motivation during the period of deprivation. Most authors, however, have tried to find an explanation in terms of changes in the external stimulus situation. For example, the density or condition of lipids in the feathers could regulate the amount of dustbathing. It is also possible that there is a buildup of ectoparasites or other changes in skin condition during deprivation. These hypotheses were tested in experiments by several different investigators, none of which were able to provide convincing evidence that increased dustbathing was primarily due to peripheral factors (review in Hogan 1997).

Perhaps the clearest evidence demonstrating the important role of endogenous motivation comes from some experiments using genetically featherless chicks (Vestergaard *et al.*, 1999). These chicks dustbathe in the normal way and with normal frequency, even though they have no feathers to clean (see Figure 3.5). In the experiments, chicks were deprived of the opportunity to dustbathe for one, two, or four days, and then were allowed access to dust for one hour. Prior to access, the experimenters gently cleaned the chicks with potato flour to ensure that their skin condition was similar at all deprivation lengths. The results were essentially the same as the results from normal feathered chicks: the longer the period of deprivation, the more a chick dustbathed. Two-week-old chicks dustbathed for about 15 minutes after one day of deprivation, for about 20 minutes after two days of deprivation, and for about 30 minutes after four days of deprivation.

These results provide one of the best examples of a Lorenzian endogenous process governing behavior in a vertebrate species.

Figure 3.5 Genetically featherless chicks dustbathing in sand.
Courtesy of Klaus Vestergaard.

Nonetheless, the occurrence of dustbathing is also influenced by many other factors. For example, Vestergaard (1982) found that most dustbathing occurred in the middle of the day, and he suggested an influence of a circadian clock; subsequent experiments with young chicks confirmed these results (Hogan & van Boxel, 1993). More recent results demonstrate conclusively that a circadian clock is indeed a causal factor for dustbathing (J. A. Hogan, unpublished results). A behavioral model incorporating a circadian clock is presented in the next section.

Sleep

Sleep is ubiquitous among species, but the form it takes varies widely. Some species have well-defined bouts of sleep at specific times during the day or night, whereas others sleep irregularly. Some species sleep for many hours a day, while others sleep only briefly. Still other species sleep with only one half of the brain at a time (see Webb, 1998 for review). Much of the variation in sleep patterns can be correlated with variation in the ecological requirements of a species (cf. Rial *et al.*, 2010), but I am here concerned with the immediate causes of sleep. I will also only discuss some aspects of human sleep because most research has concentrated on this species.

The physiological causes of sleep are still very poorly understood. Nonetheless, whatever the physiological causes of sleep, the occurrence of sleep is highly predictable if one considers the length of time a person has been awake or asleep, and the time of day. Borbély (1982) originally proposed a two-factor model of the timing of sleep. The first factor, called the homeostatic process, is considered to be a process that increases during wakefulness and decreases during sleep. Thus, the longer one goes without sleep, the more sleepy one becomes. The second factor is the sleep threshold, which varies as a function of time of day and is presumed to be controlled in part by a circadian pacemaker. Under normal light conditions, the threshold is high during the day, but much lower at night. When the sleep-regulating factor approaches the threshold, sleep is triggered; when it decreases to below the threshold, the person awakens. One puzzling aspect about sleep has always been that there are often occasions when one is very tired, but cannot sleep. The two-factor theory can account for many of these occasions, because going to sleep depends not only on how tired one is, but also on a sleep threshold.

The theory predicts that one must be much more tired to fall asleep during the day than at night. It can account for the effects of jet lag on sleep because the circadian rhythm becomes disturbed when the time of light onset changes. It has also been successful in explaining a number of other sleep disorders found in human patients. Of course, as with all behavior systems, the occurrence of any particular behavior depends not only on internal endogenous mechanisms, but also on the environmental context in which the animal finds itself. In the case of human sleep, this includes factors such as conscious decisions to set the alarm clock for a particular time. Much research has been carried out on these and other aspects of sleep and reviews of these studies can be found in Borbély *et al.* (2001), Czeisler & Dijk (2001), and Dijk & Archer (2009).

The functions of sleep are also poorly understood, although recently there is evidence that sleep allows the waste products of neural metabolism to be quickly and efficiently removed from the brain interstitial space (Lulu Xie *et al.*, 2013). Thus, build-up of waste products may be part of the homeostatic process. There is also considerable evidence that sleep facilitates memory consolidation, but the mechanism responsible is currently under debate (Durkin & Aton, 2016).

Fear/Exploration

The relation between fear and exploration has been discussed in a number of papers (see van Kampen, 2015). In this chapter, I consider fear and exploration to be a unitary system that is expressed as approach at low levels, withdrawal at moderate levels, and immobility at high levels. My primary reason for considering this to be a unitary system is that both fear and exploration are aroused by novelty or unfamiliarity – that is, by a discrepancy between an actual situation and the animal's perceptual expectation of that situation (Hebb, 1946). An example is provided by the behavior of a young chick toward a mealworm that it sees for the first time. Some chicks will approach, pick up, and eat the mealworm, others will utter a fear trill and withdraw, whereas still others will stare at the mealworm and fall asleep (Hogan, 1965). It should be noted that in other kinds of situations, there are good reasons for considering fear and exploration as independent behavior systems. Van Kampen (2015) provides an excellent analysis of the relationships among the exploration, fear, and aggression behavior systems, and Gross & Canteras (2012) review the neural circuits underlying the various situations in which fear behavior occurs (see Figure 2.13).

Unlike the other behavior systems I have considered, the fear/exploration system appears to be activated almost exclusively by external stimuli. Nonetheless, intrinsic factors are also important in some cases. For example, monkeys will learn to press a lever that allows them to look out of their cage into an adjoining room (Butler & Harlow, 1954); they will also learn to solve a variety of mechanical puzzles with no extrinsic reward (Harlow, 1950). A wide variety of studies with human subjects have also demonstrated that people seek out novel stimulation and find it rewarding (Berlyne, 1960). Lorenz (1937) even suggested that one reason zoo animals are so often skittish is that their threshold for withdrawal is lowered when they have not recently escaped from a threatening situation. All these cases implicate an important role for intrinsic factors even in a system that is normally activated only by external stimuli. Further, these ideas assume considerable practical importance in the context of animal welfare (Mason, 2010).

MECHANISMS OF BEHAVIORAL CHANGE

What determines when a particular behavior will occur, how long it will continue, and what behavior will follow it? One can imagine that all an animal's behavior systems are competing with each other for expression, perhaps in a kind of free-for-all. For example, if the level of causal factors for eating is very high, the hunger system will inhibit other systems and the animal will eat. As it eats, the causal factors for eating will decline while the causal factors for other behaviors, say drinking, will become higher than those for eating and the animal will change its behavior. If a predator approaches, the escape system will be strongly activated, which will inhibit eating and drinking, and the animal will run away. And so on.

Unfortunately, as attractive as this account appears, it is clearly an oversimplification of reality. Perhaps its most serious shortcoming is that if there were a real free-for-all and only the most dominant behavior system could be expressed, many essential but generally low-priority activities might never occur. If a hungry animal never stopped to look around for danger before the predator was upon it, it would not long survive. Since most animals do survive, this must imply that the rules for behavioral change are more complex than the 'winner take all' model. Lorenz (1966) has compared the interactions among behavior systems to the working of a parliament that, though generally democratic, has evolved special rules and procedures to produce at least tolerable and practicable compromises between different interests. The rules that apply to interactions among behavior systems have only begun to be studied, but a few principles are beginning to emerge (cf. discussion of stabilization of behavior system interaction in Chapter 5, p. 168).

Von Holst & von St. Paul (1963), in their study of behavioral organization in chickens, noted that a particular stimulus situation often leads to a reaction that first becomes more intense and then gradually declines in intensity. They proposed that this was due to two processes: adaptation and mood shift (see Figure 3.6). Their results showed that continued presentation of an electrical stimulus to a specific part of the brain led to cessation of the elicited response (adaptation: similar to 'action-specific energy' running out), while short or repeated presentations of a stimulus led to an increasing intensity of the response (mood shift: similar to priming). These effects could be seen at the level of the response (motor mechanism), the level of central coordinating mechanism, and at the level of the stimulus fields (perceptual mechanisms) – see Figure 2.12. The time course and intensity of the elicited responses depended on the response, as well as on the intensity, frequency, and duration of the stimulation. These two processes can account for many aspects of changes in behavior and are compatible with the 'winner take all' model.

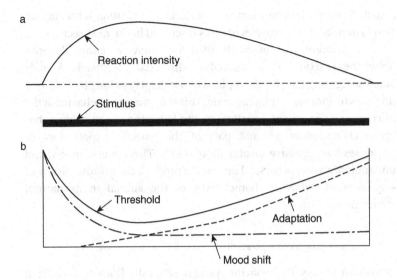

Figure 3.6 Central adaptation and mood shift. *a*. Curve illustrating the
rise and fall of a reaction with constant stimulation. *b*. Initial decline and
subsequent recovery of the threshold, for the same reaction as in *a*.
Further explanation in text.
From von Holst & von St. Paul, 1963, with permission.

Another mechanism for behavioral change arises from the fact
that most behavior systems are organized in such a way that 'pauses'
occur after the animal has engaged in a particular activity for a certain
time. The level of causal factors for the activity may remain very high,
but during the pause, other activities can occur. For example, in many
species, feeding occurs in discrete bouts; between bouts, there is an
opportunity for the animal to groom, look around, drink, and so on.
It appears that the dominant behavior system (in this case, the hunger
system) releases its inhibition on other systems for a certain length of
time. During the period of disinhibition, other behavior systems may
compete for dominance according to their level of causal factors or each
system may, so to speak, be given a turn to express itself. McFarland
(1974) has compared these kinds of interactions among behavior systems to 'time-sharing' that occurs when multiple users share the same
computer system. The mechanism of disinhibition proposed by
McFarland has proved to be controversial, however, and several studies
designed to provide evidence for a time-sharing mechanism have not
succeeded (Hogan, 1989; Hogan-Warburg *et al.*, 1995). Nonetheless, the
empirical facts continue to require a motivational explanation.

A striking example of this sort of behavioral organization is the incubation system of certain species of birds discussed in the next section.

A different type of mechanism for behavior change depends upon the reaction of an animal to discrepant feedback. A male Siamese fighting fish, for example, will not display as long to its mirror image as to another displaying male. This is because the behavior of the mirror image is always identical to the behavior of the subject, but identical responses are not part of the 'species expectation' of responses to aggressive display (Bols, 1977). These mechanisms, and undoubtedly many others, all interact to produce the infinite variety of sequences of behavior characteristic of the animal in its natural environment.

A CASE STUDY: INCUBATION AND BROODING IN HENS

Activation of specific behavior systems generally leads to organized behavior that has a functional outcome. But many functional outcomes depend on the interaction of multiple behavior systems. Several studies on the incubation and brooding behavior of hens provide data that show how interactions among the hen's behavior systems function and are able to lead to the goal of young chicks being able to fend for themselves.

Broody hens of the domestic fowl and of the red junglefowl (*Gallus gallus spadiceus*) lose about 15% of their initial body weight during their three-week period of incubation (Savory, 1979; Sherry *et al.*, 1980). Loss of weight during incubation does not, however, lead to an increased tendency to feed. On the contrary, the incubating female eats only about 30% of the average food intake of non-incubating females. She sits on the eggs almost continuously, but leaves the nest occasionally for short periods of time. Many of these changes are illustrated in the data for the animals labeled 'control' in Figure 3.7. The results of two experiments varying causal factors for hunger are also shown in the figure.

In the first or 'near food' experiment, the food container was placed near enough to the incubating hen so that she was easily able to reach the food without getting off the eggs. These hens did not lose quite as much weight as the control hens that had the food container placed about 50 cm from the nest, though the hens with food nearby did not eat significantly more food than the control hens. The hens with food nearby left the nest twice a day, which is the same as the control hens, but spent less time off the nest than the control hens,

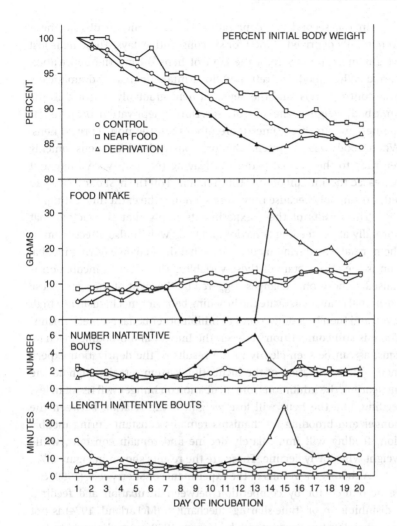

Figure 3.7 Four measures of behavior of junglefowl hens during the 20-day period of incubation. Filled triangles indicate deprivation days. 'Inattentive' refers to behavior off the nest. Further explanation in text. Adapted from data in Sherry et al., 1980.

presumably because they ate some of their food while sitting on the eggs. Overall, however, the differences between the two groups were very small. The differences that did occur are similar to those that would be expected between any two animals, one of which was given easier access to food.

In the second, or 'deprivation', experiment, incubating hens were totally deprived of food for six consecutive days. These hens lost weight more rapidly than the control hens during the deprivation period. When food was returned, food intake increased dramatically over control levels, and the deprived hens gradually returned to the same body weight as the control hens. During deprivation, the hens left the nest two to three times more often than the non-deprived hens. When food was returned, the previously deprived hens quickly returned to the normal pattern of leaving the nest twice a day, but increased the amount of time they remained off the nest once they had left, presumably because they were spending the extra time eating.

The results of these experiments imply that the factors that normally affect feeding behavior and body weight also affect them in the normal way during incubation. What does change during incubation is the steady state of these variables. The onset of incubation is caused by hormonal changes triggered by the sight of the eggs. These hormonal changes activate the brooding behavior mechanism to a high level, and then brooding comes to inhibit other behavior mechanisms. There is still competition between the hunger and brooding mechanisms, as can be seen clearly in the results of the deprivation experiment. Since activation of the incubation system remains high during incubation, the inhibition from incubation on hunger will result in less feeding, and the hens will lose weight. If the relation between the hunger and brooding mechanisms remains constant during incubation, feeding will immediately decline and remain constant, while weight will slowly decline. These are the results shown in Figure 3.7.

Although the results of the experiments of Sherry et al. (1980) can be accounted for by competition between incubation and feeding, a disinhibition or 'time-sharing' mechanism (McFarland, 1974) is not excluded. Further experiments by Hogan (1989) were designed to test for a time-sharing mechanism. In one experiment, broody hens that were deprived of food for 24 hours left their nest about 1.5 hours earlier than non-deprived broody hens. This result also supports a competition model. However, other aspects of the results suggested that time of day might be one of the causal factors for incubation. A subsequent experiment confirmed that the tendency to incubate was indeed modulated by time of day, presumably controlled by a circadian clock; the tendency to incubate is lowest in the middle of the day and highest at night. This result introduced a new and important causal factor for the control of incubation, but all these results can still be accounted for with a competition model (see Figure 3.8). While the hen is off the nest,

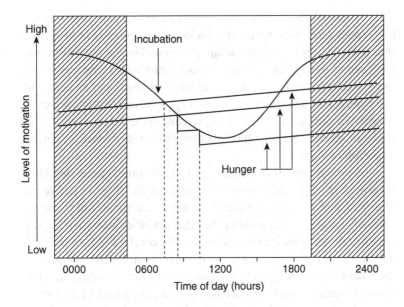

Figure 3.8 The interaction of incubation and hunger. Descriptive model of the presumed variation in the tendencies to incubate and to eat as a function of time of day. Hatched areas indicate hours of darkness. Further explanation in text.

From Hogan, 1989, with permission.

several non-feeding behaviors also occur. Whether pure competition among behavior systems can explain the pattern of occurrence of these other behaviors remains to be investigated. However, further experiments on brooding after the chicks hatch give some insight into this problem.

During the first day after the chicks have hatched, the hen's behavior remains much the same as it was before the eggs hatched. By the second day, however, the hen's behavior changes quite dramatically, and by the third day after hatching, a new pattern of behavior becomes established that lasts for several days. These changes are largely due to the behavior of the chicks, which in turn reflects the chicks' need for thermoregulation (Sherry, 1981). Richard-Yris et al. (1998) have shown that tactile stimuli from the newly hatched chicks cause a reduction in the hormone prolactin and an increase in luteinizing hormone, which is responsible for the change from incubation to other aspects of parental behavior.

Hogan-Warburg noticed that during the first week after hatching, broody junglefowl hens with chicks show a sequence of behaviors that is cyclical in nature: brooding, preening, feeding with drinking, exploration, dustbathing, and then brooding once again. An analysis of the brooding cycle showed that cycle length is about half an hour and the sequence of behaviors within each cycle is remarkably consistent (Hogan-Warburg et al., 1993). The constant cycle length seen in this study suggested that an ultradian (less than a day) clock might be a controlling factor.

A subsequent study (Hogan et al., 1998) investigated this idea. Broody hens with chicks were observed from day three to day seven after hatching. Over the course of the week, the amount of brooding declined while the time spent in the other activities increased; but the length of the cycle still remained remarkably constant. However, when dust was returned after some of the hens had been deprived of dust for three days, dustbathing occurred immediately after brooding and continued for much longer than usual, increasing the length of the brooding cycle. This result is consistent with a competition model and implies that an ultradian clock is unnecessary.

The results of these experiments show how the motivational factors that have been described in this chapter affect the parental behavior of female fowl: external and internal factors, central and peripheral factors, specific and general factors, behavioral and physiological factors. These factors all interact, often in complex ways, to produce the normal flow of behavior that results in a successful functional outcome. Other behaviors of a broody hen, such as 'teaching' the chicks what to eat and where to drink, as well as defense behaviors, have not been considered here, but are also necessary for successful brood care in nature. Many of these other behaviors and the motivational factors controlling them have also been studied, and the results all fit in well with the analysis presented here. A motivational analysis of even a relatively simple system such as parenting behavior in fowl requires an extensive experimental program that will probably never be complete.

INTERACTIONS AMONG BEHAVIOR SYSTEMS

Causal factors for many behavior systems are present at the same time, yet an animal can generally only do one thing at a time. This is a situation of motivational conflict. In this section, I consider the kinds of behavior that occur in conflict situations. There have been two major ways of studying motivational conflict, and I will consider both here. Conflict from an evolutionary fitness perspective is considered in Chapters 9 and 10.

Conflict: Classification by Consequence

One way to classify conflicts is in terms of the direction an organism takes from a goal object: either toward or away. Many psychologists have distinguished three basic kinds of motivational conflict, each designated according to the direction associated with the specific tendencies aroused: approach-approach, avoidance-avoidance, and approach-avoidance.

A classic study on the tendencies to approach or withdraw from a particular goal used rats as subjects and a narrow elevated runway with a goal box at the end as the test apparatus (Brown, 1948). To examine the tendency to approach, a rat first learned to feed in the goal box. Then it was made hungry, placed at the far end of the runway, and allowed to run toward the goal. The rat wore a harness attached to a scale, and the experimenter measured the pulling force that the rat was exerting at 170 and 30 cm from the goal box. The strength of pull increased from about 40 g to about 60 g as the rat approached the goal. The function relating strength of pull to distance from the goal was called the approach gradient. To examine the tendency to avoid, a rat was first given a mild electric footshock in the goal box. Then, while wearing its harness, it was placed on the runway in front of the goal box. The strength of pull declined from about 200 g near the goal box in which it had been shocked to almost 0 g at 170 cm. This was called the avoidance gradient. When hunger was manipulated by varying the hours of food deprivation, and fear by varying the intensity of the shock, the overall strength of pull increased or decreased, but the slope of the gradients remained the same.

These results were explained by noting that fear is primarily aroused by the external cues associated with the goal box in its fixed location, whereas hunger is primarily aroused by internal cues that accompany the animal wherever it goes. Therefore, distance from the goal region will have a larger effect on fear-based avoidance than on hunger-based approach. The reason the approach gradient is not flat is that it is also based to some extent on the incentive value of the food in the goal box, which does decrease with distance. In general, the slope of the approach and avoidance gradients depends on the relative strengths of the internal and external factors activating the motivational systems.

These and similar results were used by Miller (1959) to deduce what will happen in the three types of conflict. When there are two relatively equal positive goals, there will be a point of equilibrium where the tendency to approach one goal is equal to the tendency to approach the other. If the animal is positioned exactly on this point, it will remain there indefinitely. Such a situation is inherently unstable, however, since the slightest movement in one direction or the other will cause the attractiveness of the two goals to become unequal, and lead the animal to approach the most attractive. The other two types of conflict are stable, and the animal should hover around the point of equilibrium. In an avoidance-avoidance conflict, the point of equilibrium is somewhere between the two negative goals; in an approach-avoidance conflict, the point of equilibrium is where the two gradients cross. A series of experiments by Miller and his associates, in which the goal box was associated with both food and shock, and hunger and fear were manipulated, confirmed these predictions. This approach to conflict has been used to bring together concepts from experimental and Freudian psychology (Dollard & Miller, 1950).

Approach/withdrawal concepts have also been used by the comparative psychologist Schneirla (1965). He formulated a theory of "biphasic approach/withdrawal" to analyze the processes underlying vertebrate behavioral development. This theory has not met with wide acceptance, but many of Schneirla's ideas have had an important influence on workers in the fields of development and evolution of behavior (see Aronson et al., 1970). Another area in which approach/withdrawal concepts are being used is the psychology of emotion (Elliot et al., 2013).

Conflict: Classification by Structure or Form

An alternative way to classify conflict situations, used by ethologists, is to look at the specific behavior systems that are activated and analyze the observed behavior. Four major types of outcome have been studied: inhibition and intention movements, ambivalence, redirection, and displacement.

Inhibition and 'Intention' Movements

The most common outcome in a conflict situation is that the behavior system with the highest level of causal factors will be expressed and all the other systems will be suppressed. A male stickleback that is foraging in its territory will stop foraging when a female enters and will begin courting. The male's hunger has not changed, nor has the availability of food. It follows that the activation of the systems responsible for courtship must have inhibited the feeding system. In general, *inhibition* can be said to occur when causal factors are present that are normally sufficient to elicit a certain kind of behavior, but that behavior does not appear as a result of the presence of causal factors for another kind of behavior.

In many cases, inhibition is not complete, and incipient movements belonging to the suppressed behavior system(s) are seen. These provide an indication of the relative strength of the causal factors for other behaviors that are activated in the situation. They have been called *intention movements* because they suggest to an observer, human or conspecific, what behavior might occur next. Intention movements have played an important role in theories of the evolution of motor mechanisms (see Chapter 8).

Ambivalence

When a female stickleback enters the territory of a male, she is both an intruder and a potential sex partner. The appropriate response to an intruding conspecific is to attack it; the appropriate response to a sex partner is to lead it to the nest. The male essentially does both; he performs a 'zigzag dance' (see Figure 3.3). He makes a sideways leap followed by a jump in the direction of the female, and this sequence may be repeated many times. Sometimes the sideways leap continues into leading to the nest, and sometimes the jump toward the female ends in attack and biting. Thus, the zigzag dance can be considered a case of *successive ambivalence*. Ambivalent behavior is behavior that includes motor components belonging to two different behavior systems; in successive ambivalence, these components occur in rapid succession.

A somewhat similar case is provided by the 'upright' posture of the herring gull (Figure 3.9). This display often occurs during boundary disputes when two neighboring gulls meet at their mutual territory boundary. The bird's neck is stretched and its bill points down; the carpal joints (wrists) of the wings are raised out of the supporting feathers; the plumage is sleeked. The position of the bill and wings are characteristic of a bird that is about to attack (fighting in this species includes pecking and wing beating the opponent), and the stretched neck and sleeked plumage are characteristic of a frightened

(a)

(b)

(c)

Figure 3.9 Upright postures of the herring gull. (a) 'aggressive' upright; (b) 'intimidated' upright; (c) 'anxiety' or 'escape' upright.
From Tinbergen, 1959.

bird that is about to flee. Further, actual fighting or fleeing often follows the upright posture. Thus, the upright posture is a behavior pattern that includes motor components belonging to two different behavior systems. Unlike the zigzag dance of the stickleback, however, these components occur simultaneously. The upright posture can be considered a case of *simultaneous ambivalence*. Figure 3.9 also shows that the upright posture can occur in varying forms. In the 'aggressive upright', components of attack predominate, whereas in the 'anxiety upright', components of fleeing predominate.

The simultaneous occurrence of components belonging to different behavior systems greatly increases the number and variety of behavior patterns in a species' repertoire. A technique called *motivation analysis* can be used to explore such ambivalent behavior patterns, which include many of the bizarre displays exhibited by many species. In a motivation analysis, one looks at the form of the behavior, the situation in which it occurs, and other behaviors that occur in association with it (Tinbergen, 1959). An example is provided by Kruijt's (1964, p. 61) analysis of 'waltzing' by the male junglefowl, the wild ancestor of the domestic chicken (Figure 3.10).

It is a lateral display: the waltzing bird walks sideways around or toward the opponent. Back and shoulders are held oblique, the inner side (the side nearest the opponent) lower than the outer side. Both wings are lifted out of the supporting feathers; the upper and lower arms are slightly lowered so that the rump becomes visible. Otherwise, the inner wing and upper and lower arm of the outer wing remain folded. The hand of the outer wing is lowered perpendicularly to the ground and pulled forward, its plane near the body. The primaries touch the ground and the outer

Figure 3.10 'Waltzing' in the male junglefowl.
From Kruijt, 1962a, with permission.

foot makes scratching or stepping movements through the primaries. Head and neck are held at the level of the back and either in the medial plane or slightly turned toward the opponent. The tail spreads and is turned toward the opponent; breast and belly feathers are often ruffled, especially those of the other side.

Kruijt noted that the side of the bird's body near the hen expressed many components of escape behavior, whereas the side further from the hen expressed many components of attack behavior. It was "as if the part of the animal which is nearest to the opponent tries to withdraw, whereas the other half, which is further away, tries to approach" (Kruijt, 1964, p. 65). He also noted that waltzing was always directed toward a conspecific. Somewhat surprisingly, young males directed waltzing equally to males and females, even though adult males almost always direct it only toward females. In about two-thirds of the cases, it was performed immediately before, during, or immediately after fighting, and in some of these cases, behavior associated with escape was also seen. Thus, on the basis of form, situation, and associated behavior, Kruijt could conclude that waltzing is indeed an ambivalent behavior pattern expressing both attack and escape, with attack predominating. Sexual motivation appears to be unnecessary.

Motivation analysis of many complex courtship displays in both birds and mammals has revealed that they are ambivalent activities very frequently involving primarily the attack and escape systems. Such activities are usually essential for successful courtship and reproduction. This means, as mentioned above, that the sex system by itself is often insufficient for achieving these ends, and illustrates clearly why causal and functional analyses need to be kept separate.

Redirection

When two herring gulls meet at their mutual boundary, causal factors for both attack and escape behavior are present. As we have just seen, the birds usually adopt the ambivalent upright posture in this situation. A common occurrence during this mutual display is that one of the birds viciously pecks a nearby clump of grass and then vigorously pulls at it. In form, 'grass pulling' resembles the feather pulling seen during a heated fight between two gulls. This behavior can be considered a case of *redirection* because the motor components all belong to one of the behavior systems for which causal factors are present (i.e. aggression), but it is directed toward an inappropriate object. The causal factors for the other behavior (in this case, escape or fear) must be responsible for the shift in object. Redirection of aggressive behavior seems to be especially common in many species, including humans (Lorenz, 1966).

Displacement

Ambivalent behavior and redirected behavior are appropriate responses to causal factors that are obviously present in the situation in which the animal finds itself. Sometimes, however, an animal shows behavior that is not expected, in that appropriate causal factors are not apparent. A male stickleback meets its neighbor at the territory boundary and shows intention movements of attack and escape; then it suddenly swims to the bottom and takes a mouthful of sand (which is a component of nest-building behavior). A young chick encounters a wriggling mealworm and shows intention movements of approach to eat the mealworm and of retreating from the novel object; then while watching the mealworm, the chick falls asleep. A pigeon, actively engaged in courtship, suddenly stops and preens itself. A student studying hard for an exam puts down her book, walks to the kitchen, and makes herself a sandwich. These behaviors are all examples of *displacement* activities that are controlled by a behavior system different from the behavior systems one might expect to be activated in a particular situation.

In the case of the stickleback, it is reasonable to show components of attack and escape behavior at the boundary of its territory because the neighboring fish is an intruder when it crosses into our subject's territory, and our subject loses the security of home when it ventures into its neighbor's territory. But why should it engage in nest-building behavior? The stickleback has probably already built its nest elsewhere and, in any case, would not normally build it at the edge of its territory. What are the causal factors for nest building in this situation? Similar considerations apply to the other examples as well. In all cases, causal factors for the displacement activity appear to be missing. It is this apparent inexplicableness of displacement activities that has caused so much attention to be focused on them. Why does this unexpected behavior occur?

There have been two main theories put forward to account for displacement activities: the overflow theory and the disinhibition theory. The original theory was proposed independently by Kortlandt (1940) and by Tinbergen (1940), and is usually called the overflow theory. They proposed that when causal factors for a particular behavior system (e.g. aggression) were strong, but appropriate behavior was prevented from occurring, the energy from the activated system would 'spark' or flow over to a behavior system that was not blocked (e.g. nest building) and a displacement activity would be seen. The appropriate behavior might be prevented from occurring because of interference from an antagonistic behavior system (e.g. fear or escape) or the absence of a suitable object or thwarting of any sort.

This theory was formulated in the framework of Lorenz' model of motivation, which accounts for the graphic metaphor of energy sparking over or overflowing. In more prosaic terms, this is actually a theory in which causal factors have general as well as specific effects. Many examples of displacement activities are described as being incomplete or hurried – the stickleback does not calmly proceed to build a nest during a boundary conflict – and such observations give support to a theory that posits general effects of causal factors. It can be noted that Freud's (1940/1949) theory of displacement and sublimation of sexual energy (libido) is basically the same as the overflow theory: sexual energy is expressed in nonsexual activities such as creating works of art.

The alternative theory is called the disinhibition theory. In essence, it states that a strongly activated behavior system normally inhibits weakly activated systems. If, however, two behavior systems are strongly activated (e.g. sex and aggression), the inhibition they exert on each other will result in a release of inhibition on other behavior systems (e.g. parental) and a displacement activity will occur. The general idea was proposed by several scientists, but the most detailed exploration of the theory was made by Sevenster (1961). He studied displacement fanning in the male stickleback, which often occurs during courtship before there are any eggs in the nest. The sex and aggression behavior systems are known to be strongly activated during courtship. By careful measurements, it was possible to show that fanning occurred at a particular level of sex and aggression when their mutual inhibition was the strongest. Of special importance for the disinhibition theory, the amount of displacement fanning that occurred depended on the strength of causal factors for the parental behavior system. When extra CO_2 was introduced into the water, there was an increase in fanning.

The primary difference between the two theories is that according to the disinhibition theory, the displacement activity is motivated by its own normal causal factors and the conflict between systems merely serves a permissive role; whereas according to the overflow theory, the displacement activity is motivated by causal factors for one or both of the conflicting systems. In the disinhibition theory, causal factors always have specific effects; in the overflow theory, they have general effects. Which theory is correct? As is so often the case, neither theory, by itself, is able to account for all the phenomena associated with displacement activities. The disinhibition theory is in many ways more satisfying because it only requires that causal factors have their normal and expected effects on behavior. Nonetheless, more general effects of causal factors must be invoked to account for the frantic or excited aspects of displacement activities seen in many situations.

It is frequently true that the causation of a behavior pattern is even more complicated. For example, ground pecking occurs as a displacement activity during aggressive encounters between two male junglefowl. Arguments for considering this activity as a displaced feeding movement include the fact that it is often directed to food pieces on the ground and the fact that it occurs more frequently when the animals are hungry. This same activity can also be considered redirected aggression, and experimental evidence also supports this interpretation. Thus, one behavior pattern can be both a displacement activity and a redirected activity at the same time (Feekes, 1972). Each contribution to the causation of a behavior pattern can be analyzed separately, but the list of causal factors affecting the behavior pattern can be very long. Indeed, multiple causation of behavior is the rule rather than the exception. The causation of behavior is a very complex question, and it is unreasonable to expect a simple answer.

Finally, I want to point out that the two ways of analyzing conflict behavior discussed here are alternatives. Any conflict situation could be analyzed either way, and in my mealworm experiments (Hogan, 1965, 1966), for example, both ways were used. Which way is best depends on the question being asked.

THEORIES OF MOTIVATION

As Bolles (1967) discusses, prior to the end of the nineteenth century, theories of motivation were unnecessary because Western thought was dominated by the idea that man, unlike other animals, was a creature whose actions were controlled by rational thought; actions of other animals were basically reflexes. The work of Darwin, however, showed that humans and animals had evolved from common ancestors, and Freud showed that much human behavior was caused by unconscious drives. It thus became necessary to formulate broader theories of the causes of behavior. As mentioned above, most of the early theories of motivation employed some concept of instinctual energy. In North

America, there was strong reaction to and criticism of the idea of instinctual energies, especially by the behaviorists (Watson, 1919; Kuo, 1921). Part of the criticism was related to inferred entities that could not be observed directly, as could stimuli and responses. But many other issues were involved as well. Unfortunately, the criticisms did not separate structural from causal issues, so that, for example, the notion of energy was dismissed (a causal issue) because instincts could be multiplied at will (a structural issue). Further, the behaviorists were strong proponents of environmental, as opposed to hereditary, influences on behavior (a developmental issue). Because the instinct concept had strong hereditary connotations, all other aspects of instinct were considered suspect. The controversy was resolved by banning the word 'instinct' (see Beach, 1955) and replacing 'instinctual energy' with 'drive'. However, as Lashley (1938) pointed out, drives quickly assumed all the characteristics that the old instinctual energies had; only the name had changed.

One advantage of a drive concept for the study of motivation was that it encouraged measurement. Measuring the strength of different drives became the occupation of several laboratories, especially that of Warden and his coworkers (Warden et al., 1931). Using an apparatus called the Columbia obstruction box, they measured the number of times a rat would cross an electric grid in a 20-minute observation period when a particular object could be found in the goal box. Animals were tested under different levels of deprivation for a wide range of incentives, including food, water, a member of the opposite sex, young, and an empty goal box. The purpose of these studies was to develop some yardstick for comparing the strength of different drives. This attempt eventually turned out to be unsuccessful because it was shown that the observed behavior depended on much more than the degree of deprivation. Factors such as prior training, length of test session, and type of interaction with the incentive object could influence grid crossing as much as, or even more than, deprivation. In this case, an operational definition of drive was unsatisfactory because there was no underlying theory of drive (see Bolles, 1967).

The most influential of the drive theories was that of Hull (1943). Hull clearly separated energy (drive) from structure (habit) and postulated that the strength of observed behavior was proportional to the product of drive strength and habit strength. Unlike the energy variable in most instinct theories, which differed according to the specific instinct with which it was associated, Hull's drive variable was general. Drive arising from all sources summated to produce total drive

strength, and drive strength energized all habits equally. Which particular behaviors occurred depended on the stimulus factors (external
and internal) specific to those behaviors (Hull, 1933). Hull's theory
stimulated a great deal of research, but the experimental results con-
tradicted the theory as often as they supported it (see Bolles, 1967).
Evidence supporting the notion of a generalized drive was especially
difficult to find, and this was an important reason for the gradual
abandonment of the drive concept. In its place, associative (structural)
theories of motivation attempted to explain various motivational phenomena without an energy concept (Hull, 1952; Bindra, 1959; Bolles,
1967). In fact, some psychologists went so far as to state that motivation
itself was an irrelevant concept (Herrnstein, 1977).

Parallel developments occurred somewhat later in Europe.
The theories of the ethologists Lorenz (1939) and Tinbergen (1951) used
energy concepts similar to those of Freud and McDougall. These theories,
and especially the energy concept, were also strongly criticized, primarily on the grounds that the relation between behavioral energy and
physical energy was unclear (Hinde, 1960). Rather than suggesting that
the phenomena be explained in terms of associative processes, however,
Hinde proposed that analysis of motivational phenomena should be
carried out at a physiological level. One result of these criticisms was
that, for about 25 years, motivation became merely a variable that might
be necessary for certain associations to form, and its study was relegated
to physiologists. Only toward the end of the twentieth century did
motivation reemerge as a behavioral problem in its own right (Mook,
1987; Colgan. 1989; Toates & Jensen, 1991; Hogan, 1997).

The vicissitudes of the energy concept notwithstanding, it remains
necessary to have a way to describe and analyze the motivation of
behavior at the behavioral level. I will begin by noting that in many
ways, the occurrence of both dustbathing in fowl and sleep in humans
can be described by a motivational model similar in principle to the
original Lorenz (1937, 1950) psycho-hydraulic model (see Figure 3.1).
In this model, the level of fluid in the reservoir represents the 'energy'
available, or the 'behavioral state of readiness' for performing a specific
behavior pattern such as dustbathing or sleep. The reservoir itself,
together with the trough, represents the structure of the motor mechanism, while the spring valve, which prevents the continuous outflow of
fluid, can be considered to represent the threshold. Lorenz originally
proposed that the energy available increased as a function of time due to
accumulation of endogenous neural activity, and decreased when the
behavior was actually performed. This formulation can account for

some of the results for behaviors such as dustbathing, sleep, and feeding, but is not so successful with other behavior systems such as sex and aggression. However, if one allows other factors such as the priming effects of stimuli (Lorenz, 1980) and various substances in the blood to contribute to the level of activation, and factors such as performance of other behavior (e.g. displacement activities) and passage of time (a leaky reservoir – Roeder, 1967) to lower the activation, then many more results can be accounted for (Hogan, 1997).

The generality of the model can also be increased by expanding the factors contributing to the threshold. In the original model, the fluid is held in check by the releasing mechanism (the spring valve), which is activated by appropriate external stimuli. One can imagine that the strength of the spring is also influenced by factors such as specific stimuli, the time of day (circadian clock), and inhibition from other behavior systems. For example, specific stimuli such as increased light intensity make behaviors such as dustbathing or sleep more or less likely, respectively. As well, the circadian clock has been shown to have a very strong influence on sleep and feeding in most species, as well as on dustbathing and incubation in fowl and parenting in female ringdoves.

In this model, structure (threshold) and activation (energy) are the important concepts. They are distinguished because of their usefulness in thinking about causal factors. For example, even with the highest values for all the structural factors (best time of day, supernormal releasing stimulus, no inhibition from other systems, etc.) no behavior will occur if factors contributing to the activation variable are absent. On the other hand, even with low values of all the structural factors, behavior will occur if the activation factors are sufficiently strong (e.g. as a vacuum activity). It is necessary, however, to realize that structure and activation are relative concepts. Mathematically, these two variables are actually interchangeable (Nelson, 1964). Nonetheless, following Tinbergen (1951, p. 123), these two variables represent different functions at the behavioral level, which is sufficient justification for maintaining the distinction. At the neurophysiological level, this conception may no longer be useful because it is physically impossible to separate structure from energy (Lashley, 1938; Hebb, 1949). This is a clear example of behavioral concepts not being isomorphic with the neurophysiological mechanisms underlying them, which underlines the need to study behavior at different levels of analysis. The motivation of behavior is a major problem and even simple general principles can lead to highly complex outcomes.

4

Motivational Consequences of Behavior
Emotion, Homeostasis, Expectancies, Orientation, Rhythms

William James (1842–1910). Photo by Paul Thompson/Getty Images.

Activating perceptual and central behavior mechanisms can lead to a variety of consequences. Some of these are purely internal to the nervous system. In humans, we talk about perceptions, thoughts, and feelings: we see a tree; we remember an appointment; we feel sad. Introspectively, we are aware of these consequences, and there is no reason not to suppose that other species experience analogous consequences. These are consequences that are not directly observable. Nonetheless, as we have seen in Chapter 2, there are simple methods for studying many of these internal events, and there are more modern neurophysiological and imaging technologies as well. Some studies using these newer technologies will be discussed later. Activation of perceptual and central behavior mechanisms can also lead to activation of motor mechanisms, and these consequences can be observed directly. In this chapter, I discuss motivational consequences of activating behavior mechanisms – that is, consequences that do not cause changes in the structure of behavior. Consequences that do cause changes in the structure of behavior are discussed in the chapters on ontogeny, Chapters 5, 6, and 7.

HUMAN EMOTION

The concept of emotion is problematic because there is no consensus about its definition. Books have been written on the subject (e.g. Ekman & Davidson, 1994; Barrett & Russell, 2015) and two major journals, *Emotion* and *Emotion Review*, publish studies on emotion. The term usually refers to certain subjective experiences called feelings, but observable features often accompany them. In Chapter 1, I defined a feeling as the activation of a specific central behavior mechanism. But characterizing those behavior mechanisms that subserve emotion is as intractable as the concept of emotion itself. I think insight into the problem can be gained by considering some of the views of William James (1890, vol. 2).

> In speaking of the instincts it has been impossible to keep them separate from the emotional excitements which go with them. Objects of rage, love, fear, etc., not only prompt a man to outward deeds, but provoke characteristic alterations in his attitude and visage, and affect his breathing, circulation, and other organic functions in specific ways. When the outward deeds are inhibited, these latter emotional expressions still remain, and we read the anger in the face, though the blow may not be struck, and the fear betrays itself in voice and color, though one may suppress all other sign. *Instinctive reactions and emotional expressions thus shade imperceptibly into each other. Every object that excites an instinct excites an emotion as well.* Emotions, however, fall short of instincts, in that the emotional

reaction usually terminates in the subject's own body, whilst the instinctive reaction is apt to go farther and enter into practical relations with the exciting object.

Emotional reactions are often excited by objects with which we have no practical dealings. A ludicrous object, for example, or a beautiful object are not necessarily objects to which we *do* anything; we simply laugh, or stand in admiration, as the case may be. The class of emotional, is thus rather larger than that of instinctive, impulses, commonly so called. Its stimuli are more numerous, and its expressions are more internal and delicate, and often less practical. The physiological plan and essence of the two classes of impulse, however, is the same (p. 442).

He goes on to state his famous theory of emotion.

Our natural way of thinking about these coarser emotions [grief, fear, rage, love] is that the mental perception of some fact excites the mental affection called the emotion, and that this latter state of mind gives rise to the bodily expression. My theory, on the contrary, is that *the bodily changes follow directly the perception of the exciting fact, and that our feeling of the same changes as they occur* is the emotion (p. 449).

In other words, when we meet a bear and run away, we are not running away because we are frightened; we are frightened because we are running away – fright is our perception of all the bodily changes that occur when we run away. This theory has been subject to much criticism, to which I will return shortly. But first I will quote some more of James' views.

If such a theory is true, then each emotion is the resultant of a sum of elements, and each element is caused by a physiological process of a sort already well known. The elements are all organic changes, and each of them is the reflex effect of the exciting object. Definite questions now immediately arise – questions very different from those which were the only possible ones without this view. Those were questions of classification: "Which are the proper genera of emotion, and which the species under each?" or of description: "By what expression is each emotion characterized?" The questions now are *causal*: "Just what changes does this object and what changes does that object excite?" and "How come they to excite these particular changes and not others?" ... Now the moment the genesis of an emotion is accounted for, as the arousal by an object of a lot of reflex acts which are forthwith felt, *we immediately see why there is no limit to the number of possible different emotions which may exist, and why the emotions of different individuals may vary indefinitely*, both as to their constitution and as to objects which call them forth ... such a question as "What is the 'real' or 'typical' expression of anger, or fear?" is seen to have no objective meaning at all. Instead of it we now have the question as to how any given 'expression' of anger or fear may have come to exist; and that is a real question of physiological mechanics on the one hand, and of history on the other . . . (pp. 453–454).

I have quoted from James so extensively because the ideas he expressed have continued to dominate studies of emotion from his time to the present. Historically, the idea that has been most discussed is his idea that emotion is the perception of the feedback one gets from bodily changes in response to some arousing situation. Everyone agrees that most emotional situations cause a multitude of visceral changes such as increases in heart rate, vasoconstriction, sweating, etc., all of which are caused by sympathetic nervous system action. However, Cannon (1927) challenged the notion that perception of these changes *is* the emotion for several reasons. He pointed out that the viscera are relatively insensitive structures and that visceral changes are too slow to account for emotional feelings that occur demonstrably quicker. He also cited results of Marañon, who injected epinephrine, a sympathetic nervous system stimulant, into human subjects and asked them to describe their feelings. The results showed a clear distinction "between the perception of the peripheral phenomena of vegetative emotion (i.e. the bodily changes) and the psychical emotion proper, which does not exist and which permits the subjects to report on the vegetative syndrome with serenity, without true feeling" (tr. by Cannon, 1927, p. 113).

Cannon also pointed out that the same visceral changes occur in very different emotional states as well as in non-emotional states. What distinguishes the different emotional states from each other? What is responsible for the 'feeling' of each emotional state? James was aware of this problem, as we have seen, because he showed that there is no typical expression of each emotion. But James does list four 'coarser' emotions and discusses a large number of 'subtler' emotions, so presumably, he thought that we can introspectively distinguish among them, perhaps on the basis of the non-organic emotional expressions such as facial features or postures.

Schachter & Singer (1962) proposed that the quality of an emotion is arrived at by a process of cognitive appraisal. In one of their experiments, they injected epinephrine into human subjects. Some of the subjects were told what effects they might expect and others were not told anything about the effects. All subjects were then observed in the presence of a confederate of the experimenters who acted in either an elated or angry manner. The subjects were later asked about their emotional reactions. The informed subjects reported very little emotion (as also the subjects of Marañon); but the uninformed subjects did report emotional feelings, and the kind of emotion they felt tended to mimic that of the confederate. In other words, in precisely the same state of physiological arousal, emotional labels depended on the cognitive aspects of the situation.

Since 1962, there have been hundreds of studies on emotion. Gendron & Barrett (2009) review the history of scientific ideas about emotion and posit three major approaches to its study: basic emotion, appraisal, and psychological construction. The basic emotion approach has been a major focus of studies in *affective science*, as studies of emotion have come to be called. The goal of workers in this area is to discover and characterize the 'basic emotions', which are considered to be inherent in our biological endowment. However, in spite of years of research, there is still no consensus about the identity of the basic emotions. William James would not be surprised!

Tracy & Randles (2011) recently reviewed four models of basic emotions proposed by four prominent researchers in the area (Ekman, Izard, Levenson, Panksepp). Their lists of basic emotions are somewhat similar (fear is included in all four lists, and sadness, anger, and disgust are included in three of the four lists), but there are still many differences and many problems of definition of terms remain. Panksepp's (2005) list is the most divergent from the other three, which probably reflects the fact that he bases his list on his analysis of "the neurodynamics of brain systems that generate instinctual emotional behaviors" in various mammalian species, whereas the others base their lists on experimental studies of human subjects.

The appraisal approach assumes "that emotions are not merely triggered by objects in a reflexive or habitual way, but arise from a meaningful interpretation of an object by an individual" (Gendron & Barrett, p. 317). It considers the identification of emotional quality (its meaning) to require appraisal of the object and situation; it is the meaning that then leads to internal state changes: we are afraid of the bear when we see it (we appraise the situation) and then we become aroused and flee. Although Schachter and Singer use the word 'appraisal' in their theory, Gendron and Barrett consider their theory closer in content to the psychological construction approach.

The psychological construction approach posits that emotions are constructed out of more basic psychological ingredients that are not themselves specific to emotion. Two such basic components were proposed by Russell (2003): *core affect* and *affective quality*. "Core affect is that neurophysiological state consciously accessible as the simplest raw (nonreflective) feelings evident in moods and emotions" (p. 148). It is a single integral blend of two dimensions: pleasure-displeasure (which can range from elation to agony) and activation-deactivation (which can range from frenetic excitement to sleep). The feeling is an assessment of one's current condition. Affective quality is a property of the stimulus: its capacity to change core affect. Perception of affective quality together with core affect allows a person to construct the emotion.

Barrett (2013) believes that psychological construction constitutes a paradigm for the scientific study of emotion that is different from the 'faculty' psychology paradigm of the basic emotion and appraisal approaches. She points to three principles of psychological construction that define this difference: the principles of variation, of core systems, and of emergentism and holism. I will not discuss these principles here, but I will say that, in many respects, the details of these principles bear a striking resemblance to many of the ideas originally expressed by William James (and acknowledged by Barrett and Russell). Whether these ideas will actually change the way researchers on emotional issues behave remains to be seen. Barrett also calls psychological construction the Darwinian approach to the science of emotion, primarily because pre-Darwin, species were considered fixed, whereas post-Darwin, the variety within species could be exploited by natural selection and lead to new kinds. Although there are some similarities between the two approaches, I doubt that most biologists would be impressed with the analogy. It is my opinion that there is a much closer analogy between psychological construction and Developmental Systems Theory that I discuss in Chapter 6.

In another recent development, LeDoux (2012) has proposed rethinking the emotional brain in terms of survival circuits. The survival circuits proposed by LeDoux correspond almost exactly to behavior systems as I have defined them, although he places more constraints on which brain circuits would be considered survival circuits. His survival circuits are considered to be 'innate' and to have functional [survival] significance for the organism. So, for example, my smoking behavior system, which is neither 'innate' nor functional, would not be included in his list of survival circuits. I discuss the concept of innate in Chapter 6. With respect to function, I pointed out in Chapter 1 that invoking functional ideas when dealing with causal (or in this case, structural) issues can lead to serious confusion. A final quote from William James puts this problem in perspective.

To sum up, we see the reason for a few emotional reactions; for others a possible species of reason may be guessed; but others remain for which no plausible reason can even be conceived. These may be reactions which are purely mechanical results of the way in which our nervous centres are framed, reactions which, although permanent in us now, may be called accidental as far as their origin goes ... It would be foolish to suppose that none of the reactions called emotional could have arisen in this *quasi-accidental* way (p. 484).

LeDoux also notes that his list of survival circuits does not align well with human basic emotions. I discuss this issue below.

In considering these various approaches to the study of emotion, I would propose that it is the activated behavior system that determines the quality of the emotion. The study of emotion then becomes the study of what behavior systems exist in any organism, what motivational factors activate them, and how they are expressed. *Emotions are the subjective aspect of strongly activated behavior systems.* A corollary of this conceptualization is that the felt emotion becomes an epiphenomenon: like the whistle of the steam engine, it has no causal significance – which is, of course, consonant with James' viewpoint. Much of the research on emotion in the past 50 years can be understood in these terms.

NON-HUMAN EMOTION

I have defined emotion as the subjective aspect (feeling) of strongly activated behavior systems. Since we have no access to the subjective experience of any animal (except ourselves), any discussion of non-human emotion must rely on investigation of the expression of such strongly activated behavior systems. One of the first systematic studies of the expression of the emotions in man and (other) animals was that of Darwin (1872). Darwin was primarily interested in similarities between animal expression of presumed emotional states such as anger, terror, and joy and human expression of these and other emotions. Darwin assumed that animals such as dogs, cats, horses, and monkeys had such emotional states, and tried to show that the expression of these emotions in humans could be traced to their expression in various animals as support for his theory of evolution.

Since we know that the nervous systems of all animals have similar components, it should be possible to infer the emotional state of an animal from observations of its behavior. We would be inferring the state of activation of an animal's various behavior

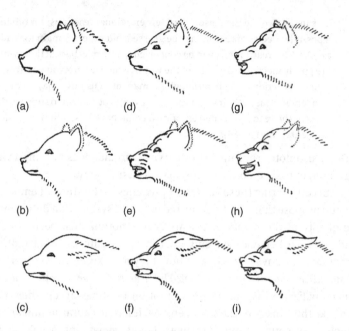

Figure 4.1 Facial expressions of fear and aggression in dogs. Explanation in text.
From Lorenz, 1966.

systems, irrespective of whatever subjective experience the animal might be having. In effect, we would be performing a motivation analysis (p. 93). In Chapter 3, we have already seen examples of this with respect to the upright posture of the herring gull, the zigzag dance of the stickleback, and waltzing in junglefowl. A similar example is Lorenz' (1966) analysis of the facial expressions of fear and aggression in dogs (Figure 4.1). In this figure, increasing aggression goes from left to right and increasing fear goes from top to bottom. In (a) the dog is calm and unemotional; in (b) and (c) it is becoming more afraid; in (d) and (g) it is becoming more aggressive. The other images depict ambivalent expressions. It can be seen that as fear is increasing, the ears and the corners of the mouth are drawn backward and downward; as aggression is increasing, the upper lip is raised and the mouth opened. These examples show that it is possible to ascertain which behavior systems in an animal are activated. But how strongly does the system have to be activated in order to be considered an emotion? And how do we measure strength?

Strength has been measured both behaviorally and physiologically, and recently, 'cognitively' as well. An early investigator of 'emotionality' in animals (the rat in this case) was Hall (1934). He showed that defecation and urination in a standard situation were valid measures of individual differences in emotionality. Hall considered emotionality a trait, a characteristic of an individual. He felt that attempts to differentiate specific emotions were extremely speculative, a view still held by many investigators of human emotion, as we have seen above. Viewing emotionality as an individual trait is consonant with studies of animal 'personality' that I discuss in Chapter 10. Hall, as also most prior and subsequent investigators of animal emotion, was really interested in using animal studies as a model for understanding human emotion. And soon thereafter, many other measures of bodily changes in animals, both behavioral and physiological, began to be used in investigations of various aspects of emotion (see Paul *et al.*, 2005)

Most studies of animal emotion are directed to understanding human emotion, but the rise of interest in animal welfare has led many investigators to study animal emotion *per se*. In the context of welfare, it is crucial to discover what makes an animal 'feel good' (or, at least, not suffer). However, feelings are subjective and we can never know what an animal feels (see Panksepp, 2010 and M. Dawkins, 2015 for recent discussions of animal consciousness). M. Dawkins (2008) suggests that a scientific study of animal suffering and welfare can be based on answers to two questions: Will the situation improve animal health? And, will it give animals something they want? The answer to the second question can be determined by discovering what the animal finds positively and negatively reinforcing (what they want and do not want) in a learning situation. Even here, a difference between 'wanting' and 'liking' (Berridge, 1996; Berridge & Robinson, 2003) makes interpretation of the results not straightforward (you may like something, but not want it at the moment). Nonetheless, Dawkins' approach seems the most reasonable for the time being. (I discuss the properties of reinforcers in Chapter 7). Theoretically, Mendl *et al.* (2010) have proposed a framework that integrates the *discrete emotion approach* (i.e. the basic emotion approach above) with the *dimensional approach* (i.e. the psychological construction approach above) for the study of animal emotion and mood. The cognitive aspects of the dimensional approach allow one to experimentally dissociate merely 'liking' something from actually 'wanting' something, which solves some problems. Nonetheless, in all cases, the feelings of the animal remain a conjecture.

Addendum: Mendl & Paul (2016) have just reported emotion ('happiness') in bumblebees! See for yourself.

A COMPARISON OF BEHAVIORAL CATEGORIES

I pointed out in Chapter 2 that workers studying human and non-human personality use categories that are different from the behavior system categories that I am using, and all of these are different from the list of basic emotions. Table 4.1 lists the behavioral categories proposed by workers in the fields of human personality, animal personality, human emotion, and ethology. The most obvious impression from comparing the lists is how different they are. Are workers in these fields really studying different things? The answer is yes, but . . . ! To begin, they are studying different aspects of behavior. Behavior systems, personality traits, and ecological temperaments are structural entities that presumably reflect the organization of the central nervous system. Basic emotions are consequences of activating behavior mechanisms. If emotions are the subjective aspect of strongly activated behavior systems, as I have suggested, an immediate problem is that there is no direct correspondence between activated behavior systems and the list of the 'basic' emotions. Activation of the aggression system may lead to a feeling of anger, but activation of the fear system, as discussed in Chapter 2, does not necessarily lead to the emotion of fear. And, as we will see in Chapter 7, activation of all the behavior systems listed in the table can

Table 4.1 *Behavioral Categories Used by Workers in Various Fields*

Human personality	Ecological temperaments	Basic emotions	Behavior systems
honesty-humility	shyness/boldness	fear	hunger
emotionality	explore/avoid	anger	thirst
extraversion	activity	happiness	aggression
agreeableness	sociability	sadness	sex
conscientiousness	aggressiveness	disgust	filial
openness to		surprise	parental
experience			exploration
			fear
			self-maintenance
			nest-building
			sleep

be reinforcing, which suggests that they could all be associated with happiness, while inhibition of the expression of a behavior system could lead to sadness. It is quite likely that the emotions of happiness and sadness correspond to the pleasant-unpleasant dimension of core affect as postulated by Russell and Barrett and discussed above. Insofar as this is true, happiness and sadness become very generalized categories of emotion. Finally, the categories of animal temperament and especially human personality also do not correspond to the list of behavior systems. They clearly cut across the list. It is possible that the temperament list will become more congruent with the behavior system list as workers in the area collect and analyze more data. But the human personality list presents a different way of looking at behavior. As Ashton (2015) points out, there are literally hundreds of personality traits, and the categories in the list represent groupings of these traits that are independent of each other. If one considers the individual personality traits to correspond to specific representations (as defined in Chapter 2), the broader categories in the list could correspond to 'chunks' of representations that function in parallel to, but in many ways like, behavior systems. This is a subject that needs more integrative thinking.

FEEDBACK MECHANISMS AND HOMEOSTASIS

When the presentation of a specific stimulus to an animal elicits a specific response, we can speak of a reflex. The response that occurs in a reflex can terminate the behavioral episode, but it can also provide the stimulus for further behavior. As was mentioned in Chapter 2, early accounts of insect locomotion considered such locomotion to be a series of chain-reflexes: the movement of one leg provided the stimulus for the movement of the next leg, and so on. Chain reflexes turned out not to be the correct explanation of insect locomotion, but there are many cases in which chain reflexes do describe the behavioral situation. However, the response that occurs in a reflex can also change the situation in such a way that the original eliciting stimulus situation is altered: when the weight of the egg being retrieved increases on one side of the bill, the bill pushes the egg to the other side of the bill, reducing the weight on the first side, and vice-versa. This is a simple feedback mechanism that serves to keep the egg balanced in the middle of the bill. Reafference, also discussed in Chapter 2, is another example of a simple feedback mechanism. Mechanisms that tend to preserve such a steady state are frequently considered under the concept of 'homeostasis' (see Hogan, 1980; Berridge, 2004).

The Concept of Homeostasis

The word homeostasis is derived from two Greek words that mean 'similar to standing still'. The word was coined by Cannon (1929) to refer to the constant conditions or steady states that are maintained in the body by coordinated physiological processes. As defined by Cannon, homeostasis is a functional concept. The definition specifies a goal or outcome, the steady state, but it does not specify the mechanisms or causes by which the steady state is attained except in a very general way – coordinated physiological processes. This definition is not peculiar to Cannon. An important antecedent of Cannon's concept of homeostasis is Bernard's concept of 'milieu interieur'. Bernard (1878, tr. by Cannon, 1932, p. 38) wrote: "All the vital mechanisms, however varied they may be, have only one object, that of preserving constant the conditions of life in the internal environment." Later, a group of experimental biologists agreed on a definition that is practically a restatement of Cannon's definition and that concluded: "Its essential feature is the interplay of factors which tend to maintain a given state at a given time" (Hughes, 1964, p. vii). In all these cases, the essence of the concept is a function or goal, whereas the mechanisms by which the goal is achieved are varied.

It is worth noting that Cannon's concept of homeostasis is the motivational analog of the developmental concept of canalization introduced by Waddington (1942 – see Chapter 6). Both are functional concepts that emphasize the tendency for a system to attain a constant result (i.e. a steady state or normal form, respectively) through varying means. In both cases, when a factor exists that tends to shift the system in one direction, other factors are present that automatically oppose the first factor and shift the system back. The actual mechanisms responsible for both homeostasis and canalization are of many different kinds.

It is the mechanisms that bring about the steady state, of course, in which most of the investigators who study homeostasis are interested. This is especially true of physiologists and students of behavior interested in motivation. A surprising fact is that in recent years, almost all these investigators assume that the mechanisms that bring about the steady state conform to a negative feedback model. That is, they assume that departures from the steady state create conditions that effect a return to that state in the same way that a thermostat controls a heater in order to maintain a constant temperature in a room. The assumption of negative feedback mechanisms in homeostasis is so strong that many authors use the term homeostatic as a synonym for negative feedback (e.g. McFarland, 1970). Logically, negative feedback mechanisms are only one type of mechanism that is responsible for homeostasis. The other possibilities are positive feedback mechanisms and mechanisms in which feedback plays no role.

In his discussion of the general features of bodily stabilization, Cannon (1932, Ch. 17) proposed that "two general types of homeostatic regulation can be distinguished dependent on whether the steady state involves *materials* or *processes*". The homeostasis of materials is accomplished through storage and overflow; temporary storage is a consequence of inundation, and more permanent storage is a consequence of segregation. The mechanisms responsible for storage by inundation and overflow are automatic, and their "ability to maintain homeostasis is marvellous", yet they generally involve no feedback at all. Rather, they appear to be relatively simple biochemical reactions dependent on the relative concentrations of the substances concerned. Of course, many of the mechanisms responsible for storage by segregation (e.g. storage of glycogen in the liver) and for the homeostatic regulation of processes (that can be speeded up or slowed down) could be classified as negative feedback mechanisms. But the point is that homeostasis is the result of all these factors interacting together, and negative feedback may or may not play an important role in any specific case. Certainly if one were asked to characterize Cannon's conception of homeostatic mechanisms, it would be fairer to do so in terms of buffered solutions in chemistry than in terms of feedback mechanisms in engineering.

Homeostasis and Behavior

Behavior is by nature episodic. Over time periods of minutes or hours, it is difficult to imagine any steady state applicable to behavior. Animals eat, drink, copulate, walk, sleep, and do many other things, but neither doing nor not doing some or all of these activities can be identified as a steady state of the sort described by Cannon. Nonetheless, there are two ways in which homeostasis and behavior have been traditionally linked (Hogan, 1980).

First, one can consider a behavior, such as eating, as one of the mechanisms used to bring about homeostasis of a system such as blood sugar level (Richter, 1942). It is often possible to show, in such a case, that the behavior belongs to a clearly defined feedback loop that works in parallel with a number of purely physiological mechanisms (cf. McFarland, 1970). Used in this way, the linkage between homeostasis and behavior seems perfectly justified. Two points should be noted, however. First, a behavior can properly be considered to be a part of a homeostatic system (i.e. a system that maintains a relatively steady state with respect to a particular variable) even if it is not part of a feedback loop. One can imagine, for example, an animal that drinks every time it eats, but has no independent mechanism for drinking as a response to dehydration. And one can imagine environments in which it will thrive. Drinking certainly plays an essential part in maintaining this animal's water balance, even though the level of hydration has no direct control of drinking. In fact, rats do drink while eating dry food, and this drinking is not controlled by the state of body hydration (Kissileff, 1969). In some learning situations, rats even drink many times their normal intake while eating (see Chapter 7). The second point is that a particular behavior can be used as a tool or mechanism for several different homeostatic systems: eating is an essential mechanism for bringing about homeostasis of blood sugar level, blood calcium, body temperature, etc.

A second way in which homeostasis and behavior have been linked is with respect to the mechanisms that control specific behaviors or behavior systems, such as eating, drinking, sex, aggression, etc. For example, behaviors that are controlled by negative feedback mechanisms have been called homeostatic. It seems undesirable to me to use the term homeostatic as a synonym for negative feedback for the reasons given above, but this is basically a problem of semantics and need not concern us further as long as our meanings are understood. Real problems do exist, however, when behaviors are classified according to the mechanisms that control them. Perhaps the most serious problem arises from the fact that there is no simple relationship between neurophysiological mechanisms and behavior – there is certainly not a one-to-one correspondence between them, as we have seen in Chapter 2 with the examples of object recognition (releasing) mechanisms and of fear. Thus, a particular behavior such as eating may be controlled by many factors originating in different homeostatic systems (blood sugar, blood calcium, body temperature, etc.), as well as by stimulus factors (taste, smell, texture, etc.) or by the previous experience of the animal. If eating is to be called a homeostatic behavior because it is controlled by negative feedback mechanisms, how many of the mechanisms that are known to control eating must conform to the negative feedback model?

A somewhat different relation between homeostasis and behavior has been suggested by Young (1961), Grossman (1967), and others, also on the basis of controlling mechanisms (see also Berridge, 2004). Grossman distinguished between homeostatic and non-homeostatic drive mechanisms: "... homeostatic drives (such as hunger, thirst, sleep, and elimination) which are elicited and reduced directly by changes in the *internal* environment and non-homeostatic drives (such as sexual or emotional arousal, and perhaps, activity itself) which appear to be elicited and reduced by changes in the *external* environment of the organism" (p. 605). Here, the term homeostatic seems to be being used as a synonym for internally elicited. There is often a correlation between feedback mechanisms and internal elicitation, but this is certainly not always the case. The thermoregulatory system, for example, is usually brought into operation by changes in temperature in the external environment, but many of the physiological and behavioral mechanisms that maintain thermal homeostasis fit a feedback model. Thus, here is a further confusion in the definition of terms.

An important point about Grossman's discussion of motivation is that the terms homeostatic and non-homeostatic are applied to drives and not to behaviors. Although behavior is by its nature episodic, drive states underlying behavior are presumably continuous variables, and it might thus be possible to apply the concept of homeostasis as a steady state to drive states. Grossman's theory is basically a drive reduction theory, akin to theories of Freud (1905), Hull (1943), and Neal Miller (1959). For Grossman, drive reduction is the goal of homeostatic and non-homeostatic drives alike: The non-homeostatic drive states elicited by environmental stimuli "persist until the behavioral responses have terminated the stimulation or removed the organism from the environment" (p. 610). Thus, one would presume that the steady state to be maintained would be the 'nil' drive state. There are many objections to this type of theory, some of which follow from examples described later. Nonetheless, Cannon's notion of homeostasis may be more applicable to drive states than to behavior itself. The question is whether it is possible to define and agree upon a 'steady state'. The data reviewed in Chapter 3 show that all behavior systems that have been investigated are influenced by factors of all types: internal and external factors, as well as positive (priming) and negative feedback (see also Hogan & Roper, 1978). Thus, it is not clear how either

a behavior or the drive state underlying behavior could be characterized in terms of a steady state or equilibrium point: behavior occurs
and drive states vary in strength, but there is no optimal level of either.
A possible solution to this problem is considered in the next section on
expectancies.

EXPECTANCIES

In Chapter 2, I discussed comparator mechanisms, appetitive behavior,
and reactions to a deficit. As implied there, these three concepts can
actually be brought together under the concept *expectancy*. In many
cases, central coordinating mechanisms (CCMs) can, in addition to
activating behavior mechanisms, also inhibit them. One can suppose
that a CCM summates input from all its motivational variables in the
same way as Baerends proposed for the summation unit of the releasing mechanism, and as suggested by the motivational model I proposed
in Chapter 3. When the total motivation reaches a threshold level, the
CCM sends excitatory output to other central and/or motor mechanisms (appetitive behavior or reaction to a deficit). The CCM at that point
can be considered a type of expectancy or comparator mechanism:
excitatory output continues until the CCM receives the appropriate
feedback necessary to inhibit further behavior (cf. the comparator
unit near the top of Baerends' model of the incubation system of the
herring gull – Figure 2.7). A young man may not have eaten for some
time, or he may have eaten lunch quite recently but sees his friends
eating, and in both cases he eats. A full stomach or the departure of his
friends (expected feedback) will inhibit eating. We might want to talk
about the output of the CCM as an eating drive or an ingestion drive,
but we are not talking about a hunger drive as normally conceived.
We might even want to consider the system homeostatic: the steady or
optimal state is the absence of an expectation or of a discrepancy
between intention and feedback. However, Cannon's concept refers
to physiological processes, while expectancy refers to behavioral processes, and explicitly considers both external as well as internal motivational inputs. Further, the expectancy concept can also easily
incorporate the effects of experience.

Expectancy is a structural concept and a question arises as to
whether the CCM as an activating mechanism should be considered to
be independent of the CCM as a comparator mechanism. In von Holst's
conception of the reafferance principle (Figure 2.5), the 'higher center'
could correspond to the CCM as an activating mechanism, but the

efference copy is conceived of as a separate mechanism. It seems likely that in most complex behavior systems, such as feeding, the CCM consists of several interconnected behavior mechanisms, some of which activate motor mechanisms, and others of which (comparator mechanisms) inhibit motor mechanisms. Although a full stomach may inhibit eating, the ingested food does not remove many of the factors that initiated eating for a considerable time after ingestion. This implies the existence of an inhibitory behavior mechanism. In the mammalian feeding system, many of the neural structures and their connections responsible for eating behavior have been analyzed (Nelson, 2016), and the results are concordant with the behavioral analysis just given.

Van Kampen (2015, p. 12) has recently analyzed the concept of expectancy with respect to the cause and function of exploration, fear, and aggression. He points out that animals continuously create representations of their environment

> from which they derive expectancies regarding current and future events. These expected events are compared continuously with information gathered through exploration to guide behaviour and update the existing representation. When a moderate discrepancy between perceived and expected events is detected, exploration is employed to update the internal representation so as to alter the expectancy and make it match the perceived event. When the discrepancy is relatively large, exploration is inhibited, and animals will try to alter the perceived event utilizing aggression or fear. The largest discrepancies are associated with a tendency to flee.

Van Kampen's analysis is an expansion of the ideas of Hebb (1946) regarding the nature of fear, and is based on an extensive review of the many relevant behavioral studies published since that time. His conclusions also fit well with the neurophysiological analysis of fear by Gross & Canteras (2012) discussed in Chapter 2.

It should be noticed that expectancies are an alternative way of thinking about motivation in contrast to the drive theories discussed above. Causal factors for new behavior arise because the animal finds itself in a situation that is unexpected. Food deprivation leading to hunger could be one such situation, but many other unexpected situations have no connection to any form of deprivation. Although removing the discrepancy could be viewed as a form of drive reduction, it is distinct from traditional views. Reducing discrepancies can be considered a form of homeostasis. Expectancies (prediction errors) are also an important variable in modern learning theories (Chapter 7).

Orientation is generally considered to be a change in the position in space of an organism as a response to an external stimulus. It is therefore one category of the consequences of activating behavior mechanisms. Orientation reactions range from simple reflexes to highly integrated sequences of behavior. I will consider three types of orientation reactions here: taxes, navigation, and migration.

Taxes

In Chapter 2, I described the balancing movements of the bill of the greylag goose during the egg retrieval response. These movements are called *taxes* and are defined by the fact that their form is continuously determined by stimuli from the external environment. Most behavior patterns have a taxis component, and there is also a large older literature on taxes in simple invertebrates. One of the most influential figures in this field was the German-American physiologist, Jacques Loeb (1918), who was active at the end of the nineteenth and the beginning of the twentieth centuries. He led a crusade against the anthropomorphic and teleological explanations of invertebrate behavior then current in the writings of men such as Romanes (1883), and attempted to understand all behavior in physical and chemical terms. Although many of his ideas were oversimplified or incorrect, his work stimulated a great deal of experimental work on the behavior of simple animals.

Much of the behavior of most simple organisms can be described as movement toward or away from particular stimuli. Loeb used the word 'tropism' to denote this movement. A tropism is an involuntary orientation by an organism or one of its parts that involves turning or curving by movement or by differential growth and is a positive or negative response to a source of stimulation (note that this definition can include plants). Later workers pointed out that not all reactions are directed in relation to a stimulus. They use the word 'kinesis' to describe such reactions (kinesis is a movement that lacks directional orientation and depends upon the intensity of stimulation) and reserve the word 'taxis' for directed movements.

Kühn (1919) proposed a classification of orientation reactions of animals that was revised by Fraenkel & Gunn (1940/1961, pp. 133, 317) and that is still in use today. Each of the categories refers to the mechanism used to achieve orientation: a single intensity receptor (klino-taxis); paired intensity receptors (tropo-taxis); a number of elements pointing in different directions (telo-taxis); etc. Fraenkel and Gunn's book provides an exhaustive presentation of the early experimental work, including discussion of mechanistic versus teleological (functional) interpretations of the experimental results.

Navigation

Our ability to easily find our way about in space raises many additional questions about the mechanisms of orientation. In a recent article entitled "Do We Have an Internal Model of the Outside World?" Land (2014) answers the question positively. He argues that

> the ability to interact with objects we cannot see implies an internal memory model of the surroundings, available to the motor system. And, because we maintain this ability when we move around, the model must be updated, so that the locations of object memories change continuously to provide accurate directional information. The model thus contains an internal representation of both the surroundings and the motions of the head and body. In other words, a stable representation of space.

In everyday life, we subjectively take a stable representation of space for granted. But how the nervous system might accomplish such a representation poses huge problems. This is not the place to discuss these problems and their solutions in humans, but many other species from insects to birds and other mammals behave in ways that support the idea that they too have a stable representation of space. I will consider three lines of evidence: path integration, cognitive maps, and migration.

Path Integration

> It is often assumed that navigation implies the use, by animals, of landmarks indicating the location of the goal. However, many animals (including humans), are able to return to the starting point of a journey, or to other goal sites, by relying on self-motion cues only. This process is known as path integration, and it allows an agent to calculate a route without making use of landmarks (Etienne & Jeffery, 2004, p. 180).

The term 'path integration' was first used by Mittelstaedt & Mittelstaedt (1980) with respect to pup retrieval in the gerbil (*Meriones unguiculatus*), but the principle was already suggested by Darwin (1873) with respect to humans and derived from 'dead reckoning' as used by mariners in navigating across open seas with no visible landmarks. The path integration mechanism is often assumed to be a neural accumulator that continuously monitors directional cues and distances traveled and that integrates them so that the animal always 'knows' the direction from its current position to its home or other starting position. Such a mechanism implies the existence of some sort of neural representation of space (Gallistel, 1990a).

There have been many experiments studying navigation of insects, especially ants and honeybees, and many of these experiments have demonstrated the use of path integration. Von Frisch (1955), for example, showed that honeybees (*Apis mellifera*) communicate the distance and direction of a food source to other bees by means of the 'wagging dance' (see Chapter 9). In one experiment, he demonstrated that the subject bee communicated the straight-line direction and distance from the hive to the food source even though the subject bee had only visited the food source by detouring around a large rock and never by flying directly herself; this information could only have been acquired through path integration. Other experiments by Wehner and his colleagues (e.g. Wehner & Menzel, 1990) on the desert ant (*Cataglyphis fortis*) also demonstrate the use of path integration in navigation. Collett & Collett (2000) reviewed the results from desert ants and honeybees and discuss three models of how an accumulator might work. They also provide new experimental results that show that insects can use both path integration and landmark navigation, and which is employed depends on conditions in the specific case.

As mentioned above, path integration is also used by mammals and birds. In a series of very careful experiments, Ariane Etienne and her colleagues (1996) demonstrated that foraging golden hamsters (*Mesocricetus auratus*) can return to their home using only information derived from path integration. Her results also showed that the return home need not even be direct. Nonetheless, under normal environmental conditions, path integration information is combined with landmark information in navigation. Etienne & Jeffery (2004) discuss how these behavioral results complement neurophysiological results on neural activity in the hippocampus and other areas of the brain in suggesting how an internal representation of space ('cognitive map') may be organized. See also Gallistel (1990b) and Gallistel & Matzel (2013).

Cognitive Maps

Tolman (1948) introduced the term 'cognitive map' into psychology to interpret the results of his experiments on maze learning in rats. He suggested "that in the course of learning, something like a field map of the environment gets established in the rat's brain" (p. 192). Such a map would be an internal representation of the geometric relations among noticeable points in the animal's environment. Evidence for such a map came from a number of sources, but especially from observations that, when physical barriers were removed, a rat would run directly to the source of food. Thirty years later, O'Keefe & Nadel (1978) presented neurophysiological evidence for such a view in their book, *The Hippocampus as a Cognitive Map*.

Since then, there has been an explosion of experimental work, both behavioral and neurophysiological, exploring the characteristics of such maps. It soon became apparent, however, that different investigators had differing views of what was meant by the concept of a cognitive map and whether insects, for example, did or did not have such a map (Wehner & Menzel, 1990). Shettleworth (2010, pp. 296–310) provides a thoughtful and comprehensive review of the behavioral evidence for and against the idea of a cognitive map. She concludes:

> in the case of cognitive mapping, there is little if any unambiguous evidence that any creature gets around using a representation that corresponds to an overall metric survey map of its environment.
> The exceptional cases in which animals satisfy one or another classic criterion for mapping-like behavior by taking novel short cuts in the absence of direct cues from the goal ... or finding their way home when displaced ... are better explained by reference to what cues the animals are actually using, how they are using them, and how they come to do so than to the ill-defined notion of a cognitive map (p. 310).

Neurophysiological research on the hippocampus has also continued unabated, and here too, there have been and still are many controversial issues. Whether the hippocampus is the seat of a cognitive map or of a memory store has been resolved by realizing that it is both (see Redish, 2001, for a review). The reality of the hypothesized existence of a cognitive map has been supported by the recent discovery of spatial grid cells in the entorhinal cortex (a neighboring structure of, and with direct connections to, the hippocampus) that encode both distance and direction information independent of the animal's immediate location (Hafting *et al.*, 2005; Moser & Moser, 2016). Together with 'place' information from the hippocampus, this system could be used by the animal to find its way about in space. (John O'Keefe, May-Britt Moser, and Edvard Moser were awarded the Nobel Prize for physiology in 2014 for this work.) However, as Redish (2001) pointed out, there are different kinds of maps: a local map of one's surroundings; a map of the city in which one resides; a map of the world. How are these maps related? And how is space represented in animals such as birds, fish, insects, and cephalopods that do not have the same brain structures as mammals? Humans have an internal model of the outside world as Land (2014) concluded, but whether 'cognitive map' is the best metaphor for how the outside world is represented is moot. It is perhaps better to follow Shettleworth's advice and look for the mechanisms each species is actually using to find its way about in space.

Spatial Memory

My final example of how animals navigate comes from experiments on food storing in birds.

> Black-capped chickadees (*Parus atricapillus*) store food in concealed locations scattered throughout their home range. They store seeds and invertebrate prey, and they put each food item in a separate cache site. Hollow stems, crevices in bark, moss, leaf buds, dry leaves, and natural cavities are used as storage sites. Chickadees ... may store as many as several hundred food items per day and place neighboring caches at least several meters apart. They recover their stored food a few days after caching it ... Stored food is recovered by remembering the spatial locations of caches (Sherry & Vaccarino, 1989, p. 308).

Experimental evidence for these statements is reviewed by Sherry (1985). Further evidence for the role of spatial memory in the retrieval of stored food is provided by Sherry & Vaccarino (1989). They removed (aspirated) the hippocampus of black-capped chickadees after the birds had had an opportunity to cache sunflower seeds and found that recovery of stored seeds was reduced to the chance rate, even though the rate of searching behavior was not affected. The results of a second experiment showed that both memory for places and working memory were disrupted by hippocampal damage, and that both these memory capacities were essential for cache recovery.

Place memory – that is, remembering where something is located, is a basic requirement for view-based navigation. Such a mechanism is presumably being used by the chickadees when recovering food, and there is abundant evidence that species from insects to humans also use view-based navigation. But what is actually being remembered? In early studies of digger wasps, Tinbergen (1932) and Baerends (1941) saw that the wasps made an orientation flight above the nest before departing on a foraging trip; experiments using displacement of landmark arrays near the nest showed that the wasps relied on visual local landmarks to relocate their nests. In food-storing birds, similar experiments showed that the "birds rely on visual information from nearby landmarks to locate concealed caches. The appearance of the cache sites themselves seems to be relatively unimportant in cache retrieval" (Sherry & Duff, 1996, p. 165).

How the landmark memories are used is another question. In a study of landmark learning in bees, Cartwright & Collett (1982, p. 560) proposed:

> bees trained to forage at a place specified by landmarks do not construct a Cartesian map of the arrangement of landmarks and food source. Instead they store something like a two-dimensional snapshot of their surroundings taken from the food source. To return there, bees move so as to reduce discrepancies between the snapshot and their current retinal image.

Some variation of image matching is likely to be a general mechanism used in spatial navigation. Gallistel (1990b), Collett & Collett (2002), and Collett et al. (2013) discuss many of these issues in detail.

There is one important issue that seems to have been overlooked in studies of animal navigation: the question of motivation. Why do bees, ants, and hamsters go searching for food when, in most cases, there is plenty of food in the hive or in the burrow? Why do wasps provide prey to undeveloped eggs? There are obvious functional reasons for these behaviors, but the causal factors instigating the behavior are seldom considered. The chickadee studies show that motivation is a separate issue from memory. This would be an excellent question to study.

Migration

The word migration (as applied to animals) can evoke four different but overlapping concepts: (1) a type of locomotory activity that is notably persistent, undistracted, and straightened out; (2) a relocation of the animal that is on a much greater scale, and involves movement of much longer duration, than those arising in its normal daily activities; (3) a seasonal to-and-fro movement of populations between regions where conditions are alternately favorable or unfavorable (including one region in which breeding occurs); and (4) movements leading to redistribution within a spatially extended population (Dingle & Drake, 2007, p. 114).

Dingle and Drake point out that definition 1 refers to processes, whereas the other three definitions refer to outcomes, and that failure to make this distinction has been responsible for considerable confusion and controversy among biologists. In the context of orientation, I will be concerned with definition 1. Further, migration occurs throughout the animal kingdom, from insects to fish and mammals. Dingle and Drake provide many examples, but I will here consider only some aspects of bird migration.

The phenomenon of bird migration has been known since ancient times, although explanations of it have radically changed since then (Wiltschko & Wiltschko, 2003). Bird migration can vary in distance from a few hundred to several thousand kilometers. Many species interrupt their migratory flights in places where they can rest and refuel, but at least one species, the bar-tailed godwit (*Limosa lapponica*) has been tracked making an 11,000 km-long nonstop flight from Alaska to New Zealand and Eastern Australia (Battley *et al.* 2012). However, regardless of distance, there are three questions all migrants face with respect to orientation: In which direction do I go? How far do I go? How do I know I have arrived in the right place? Another question, 'When do I go?', is a question of motivation, and is not considered in detail here. Most of these questions have been investigated using birds kept in aviaries, although some studies have observed free-flying birds.

'In which direction do I go?' is the aspect of migration that has been most studied. During the migratory season, most aviary birds show an increase in activity (*Zugunruhe* or migratory restlessness). This *Zugunruhe* has been shown to be due to physiological changes controlled by the photoperiod and a circannual rhythm (Gwinner, 1986, 1996). Early experiments by Kramer (1952; see review by Hinde, 1970) on starlings (*Sturnus vulgaris*) showed that the birds were able to orient their flights using a sun-compass (see p. 175). But since most birds, including starlings, migrate at night, later experiments have concentrated on other means of orientation. Emlen (1967) tested captive indigo buntings (*Passerina cyanea*) for celestial orientation under the natural night sky and inside a planetarium and showed that the birds displayed a consistent tendency to orient in the direction appropriate for the migration season in question (i.e. fall or spring migration). Outside periods of *Zugunruhe*, the birds oriented in random directions. Other experiments by the Wiltschkos and their colleagues (review in 2003) provide convincing evidence that most birds can also use earth's magnetic fields for orientation.

The sun-compass, the starry sky, and magnetic fields are all tools that birds can use for orientation, but they do not actually answer the question of which direction is correct. Experiments by Emlen (1969, p. 716) provided evidence that "changes in the internal physiological state of the bird rather than differences in the external stimulus situation are responsible for the seasonal reversal of preferred migration direction in this species". Further, many experiments with birds of various species, migrating for the first time, demonstrate that they set off in the correct direction (review in Wiltschko & Wiltschko, 2003). These results all suggest that migratory direction is determined prefunctionally. Other experiments show that experienced birds also use memory to determine direction of flight. Another intriguing possibility is suggested by the results of Verkuil *et al.* (2012). They observed that large numbers of ringed migratory ruffs (*Philomachus pugnax*) that arrived on their arctic breeding grounds via the western flyway through the Netherlands departed for their wintering grounds in Africa via the eastern flyway through Poland. The fuelling possibilities in the Netherlands are deteriorating due to intensive farming and the number of visiting birds is declining rapidly, while the number of birds using the eastern flyway is increasing. Somehow, information is being transmitted between birds in the arctic such that some western arriving birds now leave to the east.

The question 'How far I should fly?' has not been much studied, but there are some data that suggest possible answers. Berthold (1973) found, in various species of migratory warblers that he and his colleagues were studying, that the amount of *Zugunruhe* shown by aviary birds during the migratory period correlated highly with the distance that each species flew on its migratory trips. Gwinner (1986, 1996) proposed that a circannual rhythm (see below) produced timing programs that could play a major role in determining migratory distance. Another possible mechanism is related to the magnitude of the energy stores built up prior to migration (see Piersma & van Gils, 2013). The bird would fly until its store was used up. In experienced birds, memory of previous flights would be sufficient to determine distance.

The third question, 'How do I know I have arrived?', has also not been extensively addressed experimentally. One study of habitat preference in the dark-eyed junco (*Junco hyemalis*) suggests a possible general mechanism. Roberts & Weigl (1984) recorded the time juncos spent in front of pictures of their summer and winter habitats. "Juncos caught in and kept on winter or summer photoperiods preferred winter and summer habitat [pictures] respectively. A gradual increase in photoperiod up to natural summer conditions produced a switch from winter to summer habitat preference" (p. 709). A decrease in photoperiod produced mixed results. Insofar as habitat preference is based on visual features of the environment, birds would know that they have arrived when the actual habitat matches their mental picture of the appropriate habitat. Experiments demonstrating a seasonally changing habitat preference in migrating birds would provide evidence for this hypothesis.

Bird migration has provided many challenging questions for biologists. One conclusion from the work that has been done is that the mechanisms used by migrating birds are many and complex. Bird brains have often been ridiculed as being tiny (which they are in comparison to many mammalian brains), but accumulating evidence shows that small brains can accomplish highly intricate and complex tasks that many larger brains are unable to achieve (cf. Chittka & Niven, 2009; Olkowicz *et al.*, 2016).

BIOLOGICAL RHYTHMS AND OTHER TIMING MECHANISMS

Biological Rhythms

In Chapter 2, we saw that oscillators are one of the basic central mechanisms of behavioral structure. Oscillators perform many functions in the body, including controlling aspects of behavior. Apart from being pattern generators for many specific motor mechanisms, they also influence the timing of the occurrence of much behavior. Rhythms of about 24 hours – circadian rhythms (Aschoff, 1960, 1989) – have been the most studied and are now the centerpiece of a separate scientific field: chronobiology. But there are also rhythms of less than 24 hours (ultradian rhythms) and more than 24 hours (infradian rhythms) including rhythms synchronized with the tides (circatidal), the moon (circalunar), and the solar year (circannual).

Circadian rhythms have been demonstrated in single cells, plants, and most invertebrate and vertebrate species that have been studied, though there are species in which a circadian system is missing or at least not expressed (Bloch et al., 2013). Clearly, the mechanisms that underlie these rhythms must be different. In mammals, a 'master clock' is located in the suprachiasmatic nuclei (SCN) of the hypothalamus, but there exist many other clocks (oscillators) located in other parts of the brain and in most organs of the body that control various physiological and behavioral functions (see Figure 4.2). These clocks are normally synchronized with the master clock, but they can also function independently.

With respect to behavior in mammals, the circadian system is usually considered to be an important controlling factor for sleep (see Chapter 3) and activity. In fact, in some studies, the pattern of activity in a running wheel defines the circadian system. Nonetheless, many other behaviors are partially controlled by the circadian system, including feeding in many species (for reviews, see Stephan, 2001; Mistlberger, 2011) and human performance in attention and memory tasks (Dijk & von Schantz, 2005). The feeding studies are especially interesting because they demonstrate that the time of day the animal feeds can be learned. Developmental aspects of circadian rhythms are considered in Chapter 5.

In birds, the central circadian pacemaking system is less consolidated, comprising the retina and the pineal gland, as well as a hypothalamic oscillator (Gwinner & Brandstätter, 2001; Underwood et al., 2001). In spite of these differences, the pacemaker systems of mammals and birds generally function in a similar way. They are one of the controlling factors of many physiological reactions and much behavior.

In chickens, a number of studies have looked at the role of the circadian system as a controlling factor for activity, sleep, feeding, tonic immobility, incubation, and dustbathing. As discussed in Chapter 3, Vestergaard (1982) noted that his adult hens engaged in dustbathing almost exclusively in the middle of the day. Later experiments with three-week-old chicks confirmed his observations (Hogan & van Boxel, 1993). Further experiments conducted in my lab (unpublished) advanced or delayed the light cycle by three hours, and by three days later, all the chicks had readjusted their dustbathing to the middle of the new day.

Other experiments (Hogan et al., unpublished) have looked in more detail at circadian control of sleep, dustbathing, and feeding in

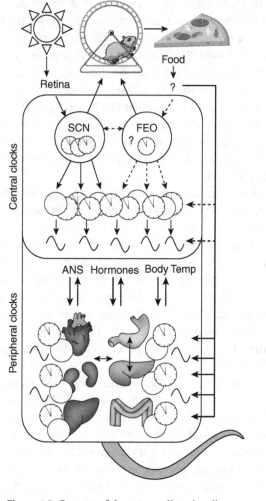

Figure 4.2 Cartoon of the mammalian circadian system, illustrating
known (solid arrows and circles) and hypothesized (dashed arrows and
circles) components and pathways. Circles without clock hands represent
processes directly driven by other clocks or external stimuli. Sine waves
represent overt rhythms. Light-dark cycles entrain a circadian pacemaker
in the suprachiasmatic nucleus (SCN). SCN output via polysynaptic
pathways drive daily rhythms of activity and feeding, autonomic efferents,
hormones, and body temperature, all of which contribute to phase control
of circadian clocks and driven processes in peripheral organs. If food is
temporally restricted, locomotor activity comes under control of a timing
mechanism (most likely neural) with circadian properties that generates

young chicks. The standard method for studying circadian rhythms is to monitor the occurrence of some physiological or behavioral variable under constant conditions to see if these variables show a 'free-running rhythm' of about 24 hours. Although most studies using mammals study free-running rhythms under conditions of constant darkness, chickens do not dustbathe in the dark, so many of our experiments used conditions of constant light. In our experiments, chicks were kept under normal light conditions (12 hours light: 12 hours dark) from hatching until two weeks of age. They were then exposed either to constant dark, constant light, or constant dim light for seven or eight more days, and their behavior was recorded. The results clearly demonstrated that feeding, dustbathing, and sleep in young chicks are partially controlled by a circadian clock. The free-running period (τ) of the clock was about 24.5 hours in constant dark conditions and about 25.5 hours in constant light conditions.

Results of one experiment that recorded both sleep and dustbathing in constant light are shown in Figure 4.3. The free-running rhythm is clear to see, but of special interest, circadian control of sleep declines during the week, but it continues with dustbathing. A similar decline of control of feeding was also seen in dark, light, and dim light conditions. I should point out that the decline of circadian control seen in Figure 4.3, and also seen in other experiments on feeding and sleep, are representative of individuals; they are not due to any results of averaging. The role of experience in the development of circadian control of behavior is considered in Chapter 5.

Other Timing Mechanisms

Our subjective experience tells us that we can easily distinguish the time taken to brush our teeth from the time taken to shower. A host of

---

Caption for Figure 4.2 (cont.)

food anticipatory activity. There may be a central food-entrainable oscillator (FEO) coordinating behavioral and physiological rhythms to mealtime, or there may be a distributed system of central and peripheral FEOs, entrained in parallel by feeding related stimuli.
From Mistleberger, 2011, with permission, re-drawn using artwork by Ylivdesign and kathykonkle.

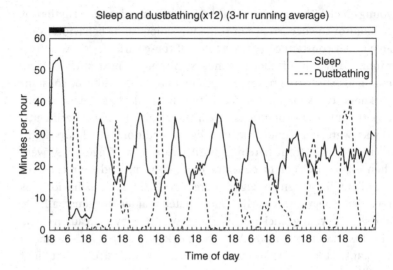

Figure 4.3 Circadian control of sleep and dustbathing. Data from nine pairs of three-week-old junglefowl chicks observed under constant light. Hogan, unpublished.

psychophysical experiments confirm that, in general, human sub-jects are very good at estimating relatively short time intervals. And most other animals can do the same. There is a large literature on interval timing in human and non-human animals (see, for example, Buhusi & Meck, 2005). With respect to learning and memory, Shettleworth (2010, pp. 323–339) reviews this literature and discusses the theories that have been proposed to explain the results. Oscillators figure prominently in most of the theories, but as Shettleworth concludes, "circadian and interval timing can be seen as two functionally and causally distinct information-processing sys-tems or modules" (p. 338).

Further, most animals also have another interval timing system (sometimes called an hourglass timer) that does not require the use of oscillators. The hourglass system differs from oscillator systems in that it is affected by the occurrence of behavior (it measures time since the behavior occurred), whereas the oscillator systems are not. These systems can be thought of in the terms discussed in Chapter 3: threshold and energy as causal factors for behavior. In many cases, increases in the energy or activation variable with time can be due to a build-up of particular substances in the blood or various neural processes. Changes in the threshold variable can be due to the circadian system. Together, such changes can determine the interval at which a particular behavior occurs. We have already seen an example of how the interaction of a circadian process (the threshold) and a homeostatic process that increases with time can determine the cycle of sleep and wakefulness in humans (p. 80). Experiments by Derk-Jan Dijk and his colleagues (Dijk *et al.*, 1992) show how this model can be used to understand the relatively stable level of alertness and performance characteristic of most people during the waking day.

These experiments made use of a technique called the forced desynchrony protocol. It is possible to uncouple the circadian cycle (measured by a subject's temperature) from the sleep-wake cycle if the subject is kept under conditions of low light intensity and told when to sleep and when to wake. Using this technique, subjects in these experiments were placed on a 28-hour sleep-wake cycle, in which they slept for one-third of the time and were kept awake for the other two-thirds. Their subjective alertness and their performance on simple math problems were measured. Because the subjects' circadian cycle (about 24 hours) and sleep cycle (28 hours) were uncoupled, it was possible to test their performance at different times of the circadian cycle when they had been awake for varying lengths of time.

The results showed that subjective alertness was highest just after waking and declined monotonically as a function of the time they were awake. With respect to the effects of the circadian cycle on alertness, subjects were least alert at the part of the cycle that coincides with normal waking time, but became more alert during the course of the day until there was a sharp drop in alertness at the part of the cycle corresponding to normal sleep time. When these two results were combined (as suggested by the two-process model of sleep), subjective alertness was high in the morning and remained high until the end of the normal waking day, which is the actual result from subjects tested under normal conditions. The results further suggested that the two processes each made an equal contribution to alertness.

A final example of the important role played by the interaction of the circadian clock with an interval timing mechanism is provided by the parental behavior of ring doves. Both parents share in the building of the nest and in the incubation, brooding, and feeding of their young. However, doves partition their time on the nest, with the male incubating and brooding in the middle of the day, and the female caring for the young the rest of the time. A series of experiments by Silver and her colleagues (reviewed in Silver, 1990) has shown that the female's tendency to sit on the nest varies in a circadian manner, with the lowest level during the middle of the day. On the other hand, the male's tendency to sit on the nest builds up with the time he spends off the nest, and declines as a function of time sitting. The interaction of these two timing mechanisms results in the pattern of nest attendance typical of the species.

5

Ontogeny of Structure
Development of Behavior Systems

Daniel Lehrman (1919–1972). Photo by Rae Silver.

In this chapter, I will present and discuss examples showing how behavior systems develop during the period between birth (hatching) and maturity. Many of these examples are based on my own work on chickens, but I also show how the behavior systems of chickens can be considered to be typical of behavior systems in other species, including the language system in humans. I discuss the general principles of development that emerge from the data in Chapter 6.

The thesis of this chapter is that perceptual, central, and motor mechanisms are the building blocks out of which complex behavior is formed. A developmental analysis requires looking for the factors causing the development of the building blocks themselves, as well as for the way connections among these building blocks become established. In some cases, the building blocks and/or their connections appear for the first time *prefunctionally* (Schiller, 1949/ 1957); that is, functional experience is not necessary for their development. A building block (e.g. the pecking motor mechanism) is functional when its associated response (i.e. pecking) occurs in its adaptive context (i.e. grasping small objects). If the pecking response occurs in its normal form before the chick has ever grasped an object, the development of the pecking motor mechanism can be said to occur prefunctionally: experience grasping an object is not necessary for the development of a normal pecking response.

It should be noted that saying a behavior mechanism develops prefunctionally implies only that particular kinds of experience play no role in its development; there is no implication about the role of other kinds of experience. For example, the development of the pecking motor mechanism in the chick may be influenced by events associated with beak movements that occur in the egg before hatching or with head and beak movements that occur during hatching (cf. Kuo, 1967). The pecking motor mechanism would nonetheless still be regarded as appearing prefunctionally. This concept is discussed in greater detail in Chapter 6. However, even in cases in which behavior develops prefunctionally, developmental questions arise. I begin with an example of such a system. I then consider several examples of how individual behavior mechanisms develop, and finally, some examples of the development of more complex systems.

THE 'GUSTOFACIAL REFLEX': A PREFUNCTIONALLY
DEVELOPED SYSTEM

Steiner (1979) showed that newborn human infants have at least three gustofacial reflexes. A sweet stimulus to the tongue elicits a 'smile' reaction, a sour stimulus elicits a 'pucker' reaction, and a bitter substance elicits a 'disgust' reaction. The identification of these reactions by even inexperienced observers is highly reliable. In terms of the framework of this book, we can posit that the newborn infant has three perceptual mechanisms for particular tastes (a sweet, a sour, and a bitter mechanism). These mechanisms and the specific connections between them and their motor mechanisms are formed prefunctionally – that is, before the consequences of ingesting sweet, sour, or bitter substances have been experienced and before any social (or other) reactions to these facial expressions can have been perceived. Nonetheless, there are still many questions of developmental interest that can be asked about these results.

With respect to the motor mechanisms, there is a large literature on the form and development of human facial expressions. Ekman and Friesen (see Ekman, 1982) devised a facial action coding system that analyzes all human adult facial expressions as combinations of about 50 basic action units, and Oster (1978) reported that almost all of these discrete action units can be identified in the facial movements of newborn infants. In this system, the smile, pucker, and disgust patterns discussed by Steiner consist of particular combinations of the basic action units. One can ask how these motor patterns are organized, how they change as the infant grows older, and what experience is necessary for the changes to occur. Thelen (1985) used this framework of hierarchical organization of coordinative structures for understanding the development of motor mechanisms in general, and I return to some of her ideas in a later section.

The perceptual mechanisms that recognize sweet, sour, and bitter are probably the basic perceptual units, and developmental interest would focus on connections between them and other behavior mechanisms rather than on the development of the perceptual mechanisms themselves. Some of these connections develop before birth, and may depend on specific experiences of the fetus. These would include possible effects of tasting and swallowing amniotic fluid or feedback from movements of facial or other muscles. I am not concerned here with such prenatal experiences, but it is important to realize that there is a complex developmental history before the emergence of even a prefunctionally developed system.

Other connections develop after birth. For example, many adults will smile at the taste of coffee (a bitter substance). In such a case, presumably neither the perceptual nor the motor mechanism has changed over time. What has changed is the connection between them. Further, the change is not simply one in which the bitter mechanism becomes attached to the smile mechanism, because other bitter substances still elicit a disgust expression. Identification of the changes that occur and the experience that is necessary requires experimental analysis (Rozin, 1984), but this formulation of the problem makes that analysis easier to tackle.

A related question has to do with connections between the motor mechanisms and higher-level coordinative structures. People smile not only in response to sweet tastes, but also in response to a wide range of stimuli associated with the hunger, sexual, parental, and other systems. How does the smile mechanism become attached to these various systems? This question also requires experimental analysis, and several examples of this type are considered below.

DEVELOPMENT OF PERCEPTUAL MECHANISMS

Three of the most studied examples of behavior development, song learning and imprinting in birds, and speech development in human infants, are all cases that involve perceptual mechanisms that develop independently of connections with central and motor mechanisms. Several aspects of the bird studies will be discussed here and aspects of speech development will be discussed later in the section on human language, as well as in Chapter 7. The development of food recognition mechanisms serves as a final example.

Song Recognition Mechanisms

Many years ago, Thorpe (1958, 1961) showed that the male chaffinch, *Fringilla coelebs*, had to learn to sing its species-specific song, and that this learning occurred in two stages. First, the young bird had to hear the normal song (or, within limits, a similar song); later it learned to adjust its vocal output to match the song it had heard when it was young. Similar results have also been found for the white-crowned sparrow, *Zonotrichia leucophrys*, (Konishi, 1965; Marler, 1970a), though not necessarily for other species of songbirds (Marler, 1976; Logan, 1983). The first, or memorization, stage of learning involves the development of a perceptual mechanism, and that is discussed here; the second, or selection, stage involves the development of a motor mechanism, and that is discussed later. There have been many reviews of the bird song literature (e.g. Catchpole & Slater, 2008; Bolhuis & Everaert, 2013), and only highly selected aspects are mentioned in this chapter.

Konishi (1965) and Marler (1976, 1984) proposed that the results of studies of song learning imply the existence of an auditory template, which was conceived of as a sensory mechanism that embodies species-specific information. The normal development of the template requires auditory experience of the proper sort at the proper time. In my terms, the template becomes a song-recognition (perceptual) mechanism that is partially formed at hatching.

One question that has been asked about templates concerns constraints on the kinds of experience that can affect development. Thorpe (1961) found that chaffinches would learn to sing normal or rearranged chaffinch songs heard when young, but exposure to songs of other species resulted in songs no different from those sung by birds raised in auditory isolation. Other species, however, such as the nightingale, *Luscinia megarhynchos*, can easily learn songs of different species (Hultsch, 1993). Still other species can learn and sing the songs of other species, but they have a bias for their own species song. When fledgling male song sparrows, *Melospiza melodia*, and swamp sparrows, *Melospiza georgiana*, were exposed to taped songs that consisted of equal numbers of songs of both species, they preferentially learned the songs of their own species. Males of both species are able to sing the songs of the other species; thus it appears that they are predisposed to perceive songs of their own species (Marler, 1991). Another constraint is the age at which the young birds are exposed to song. Young chaffinches are able to memorize their species' song any time before sexual maturity (Thorpe, 1961), but white-crowned sparrows are unable to learn after the age of about three months (Marler, 1970a). Other species are able to learn new songs throughout their lives (Catchpole & Slater, 2008).

A second question concerns the processes that are involved in development. In the memorization stage, it is often assumed that mere exposure to an adequate stimulus is sufficient for perceptual learning to occur. In a restricted sense, this is probably true, but what makes a stimulus adequate often depends critically on the conditions under which the bird is exposed. For example, in many cases, memorization is more likely to occur when exposure occurs during social interaction with another bird (Nelson, 1997). Interestingly, social interaction with another human is also necessary for perceptual learning of speech sounds in infants (Kuhl, 2015).

A third question that has been asked is whether there is one template, or many. Originally, it was thought that the young bird memorized a single song and that later variation in produced song came about because of mismatches during the selection stage. More recently (Marler & Peters, 1982; Nelson, 1997), it has become clear that the bird memorizes a variety of species-specific songs when young, but only one (or one subset) of these is selected later for production. How this choice is made is not known, but it appears that the template used in the development of song production is stored in a different part of the brain from the other song memories (Jarvis & Nottebohm, 1997; Gobes *et al.*, 2013).

Although most research on song learning has focused on males, there is also considerable evidence that females develop song recognition mechanisms too. For example, Riebel (2003) and colleagues found that early exposure of female zebra finches, *Taeniopygia guttata*, to the song of a zebra finch male led to a preference for that song over the song of an unfamiliar male when the females were tested in adulthood. Such female song preferences are comparable to song preferences in males. And, as with male zebra finches, there is a sensitive period between 35 and 65 days of age for this exposure to be effective.

A final point about song recognition mechanisms is that the same perceptual mechanism, once it has developed, serves several different functions: in the male, it serves as a standard against which the bird's own song develops, and it also releases aggressive behavior when the male becomes territorial in the spring; in the female, it releases sexual behavior. Finally, as we have seen, the range of stimuli that affect development turns out to depend crucially on such factors as the species, the age at which the bird is exposed, the previous experience of the bird, and the conditions under which the bird is exposed (Nelson, 1997; Catchpole & Slater, 2008). There are no easy causal generalizations.

Parent and Partner Recognition Mechanisms

Many species of birds do not recognize conspecifics on the basis of their song. These species have analogous perceptual mechanisms that analyze visual or other sensory input. The development of such perceptual mechanisms has usually been studied in the context of imprinting. The phenomena of imprinting have been known since the early experimental work of Spalding (1873/1954) on the following response of newly-hatched domestic chicks. Spalding's work was cited extensively by William James (1890) in his chapter on instinct. James, in fact, proposed two principles to account for Spalding's results: the inhibition of instincts by habits, and the transitoriness of instincts. However, the concept of imprinting only attracted wide scientific interest after the publication of Lorenz' paper on *Der Kumpan in der Umwelt des Vogels* (1935; tr. as: Companions as Factors in the Bird's Environment, 1970). This concept, as originally elaborated by Lorenz, was primarily concerned with the process by which early experience affects development. Lorenz proposed that "through imprinting, the bird acquires a schema of the conspecific animal" (1970, p. 133). There is a very large literature on imprinting (for reviews see Bolhuis, 1991; ten

Cate, 1994; van Kampen, 1996), and only very selected aspects are discussed here.

The 'schema', as discussed by Lorenz, is basically identical to the template proposed by Konishi and Marler. Lorenz (1935) noted that the young of some species such as the curlew, *Numenius arquata*, require no visual experience in order to recognize members of their own species, whereas the young of other species such as the greylag goose, *Anser anser*, direct all their species-typical social behaviors to the first moving object they see. Imprinting was relevant only to the acquired aspect of the schema. In my terms, we would say that most, and perhaps all, species have a preassigned perceptual mechanism that serves a species recognition function. In a species such as the curlew, this perceptual mechanism develops prefunctionally. In a species such as the greylag goose, this perceptual mechanism requires various kinds of experience for its development.

Lorenz (1935/1970) posited that a greylag gosling became imprinted on the first moving object that it viewed, regardless of its other features. In this respect, he believed that the schema (perceptual mechanism) for parent recognition was largely a blank slate, with essentially no constraints on what stimuli were 'imprintable'. However, he acknowledged that greylag geese are "an extreme, to the extent that so *few* characters of the adult companion are innately determined in the freshly-hatched bird" (1970, p. 126). He went on to give examples of species such as the golden pheasant, *Chrysolophus pictus*, in which characters such as the species call are 'innately' determined. Later studies on other species, especially on young chicks, have shown that, although the range of stimuli that can become imprinted is very large, some stimuli are more imprintable than others (for review, see Bolhuis, 1996).

As well as constraints on what stimuli can be imprinted, the age at which the animal is exposed to the stimuli can be crucially important. Lorenz stated that for nidifugous species, in which the young leave the nest soon after hatching, the period during which imprinting can occur is very short – a critical period, perhaps only a few hours in duration – and that once imprinted, the imprinted object can no longer be changed. He did note, however, that for some species, there was a second period in which a 'change of determination' was still possible. Many studies have shown that the phase of greatest sensitivity in many species is not as restricted as Lorenz originally supposed, and Bateson & Hinde (1987) have argued that the term sensitive period is much better to describe such phases of development; this term is now widely, but by no means universally, used.

Although Lorenz based his notions of imprinting and critical periods on accepted concepts from embryology, many behavioral scientists, especially in North America and the United Kingdom, were skeptical of his ideas. An important reason for this skepticism was that very few Anglophone scientists could actually read German, and early accounts of Lorenz' work in English were much more categorical than the original in German. In any case, after the Second World War, many laboratories carried out numerous studies testing various aspects of Lorenz' ideas. One especially controversial idea was that *"the process of imprinting of the object of instinctive behaviour patterns oriented towards conspecifics possesses a series of features which are basically different from learning"* (1970, p. 127). Lorenz had Pavlov's conditioned reflexes in mind when he used the term learning, and it is true that imprinting in the greylag goose has many features that are different from a typical conditioned reflex. Nonetheless, van Kampen (1996), after reviewing a large number of studies by him and his colleagues, concluded that the development of the perceptual mechanism for recognizing the imprinting stimulus follows the same rules of learning as are found in studies of perceptual learning in general (see below).

Other studies have shown that there must be at least two perceptual mechanisms involved in the development of the filial system in chicks: one recognizes the imprinting object and the other responds to general features of conspecifics. These mechanisms are neurally and behaviorally dissociable; the latter mechanism has been called a 'predisposition' (Horn, 1985; Bolhuis 1996). The predisposition has a number of interesting features (see Hogan & Bolhuis, 2009 for a review), but it appears primarily to serve as a mechanism to focus the animal's attention on particular stimuli, which are then incorporated into the parent recognition mechanism (cf. van Kampen, 1996).

There are also interesting similarities between the development of face recognition in human infants, and the development of filial preferences in chicks (Blass, 1999; Johnson & Bolhuis, 2000). Newborn infants have been shown to track a moving face-like stimulus more than a stimulus that lacks these features, or in which these features have been jumbled up. Similarly, in both human infants and young precocial birds, the features of individual objects need to be learned. Once learned, both infants and birds react to unfamiliar objects with species-specific behavior patterns that tend to bring them back to the familiar object.

One final aspect of developing parental recognition mechanisms deserves mention. In an extensive series of experiments published between 1975 and 1987, Gottlieb (for review, see 1997) investigated the mechanisms underlying the preferences that young ducklings of a number of species show for the maternal call of their own species over that of other species. He found that differential behavior toward the species-specific call could already be observed at an early embryonic stage, before the animal itself started to vocalize. However, a post-hatching preference for the conspecific maternal call was only found when the animals received exposure to embryonic contact-contentment calls, played back at the right speed and with a natural variation, within a certain period in development. Thus, the development of the perceptual mechanism responsible for recognizing the maternal call in ducklings is dependent on specific kinds of experience before hatching.

In many species, a 'species recognition' perceptual mechanism is actually a misnomer, because in these species, there are different perceptual mechanisms for filial, social, and sexual behavior. Lorenz (1935) discussed the fact that parental, infant (filial), sexual, social, and sibling behavior in jackdaws, *Corvus monedula*, could be directed to different objects that he called companions (*Kumpan*). He also pointed out that "imprinting of different conspecific-oriented functional systems to the relevant object occurs at different points of time in individual ontogeny" (1970, p. 131). These comments have been borne out in subsequent research on sexual imprinting.

Sexual imprinting is the process by which early experience affects sexual preference later in life. Although sexual imprinting was one of the earliest subjects studied by ethologists, there have been many fewer studies of it than of filial imprinting, primarily because experiments require very much more time to do, and proper control conditions are difficult to arrange. Early studies were relatively limited and anecdotal (for review, see Immelmann, 1972), but more recent studies have been conducted in a systematic fashion. In sexual imprinting experiments, birds are usually reared up to a certain age with parents of their own species, or with foster parents of a different species. When the animals are sexually mature, their sexual preferences are tested. Measures of preference include time near the test stimuli, amount of courtship shown, and copulation choices.

Some of the first systematic studies were carried out with mallard ducks, *Anas platyrhynchos*, investigating the question of whether there was a conspecific bias, that is, whether the developing sexual perceptual mechanism preferentially accepted species-specific information. The conclusion of several studies was that there was no evidence for a conspecific bias (Kruijt, 1985): female mallards chose males on the basis of their early experience. Similar experiments were carried out by ten Cate and his collaborators on male zebra finches, *Taeniopygia guttata*, that were cross fostered in situations that included Bengalese finches, *Lonchura striata*. Sexual preferences of the male zebra finches are more complex than the preferences of the mallard females, but the conclusion reached from a series of studies was the same: it was not necessary to invoke species-specific predispositions to explain the behavior (review in ten Cate, 1994).

Ten Cate (1987) also examined the sexual preferences of zebra finch males raised by zebra finch parents for the first 30 days, then placed individually in groups of Bengalese finches for the next 30 days, and thereafter raised in social isolation. Under these conditions, many of the birds will court females of both species when tested after sexual maturity at 100 days; such birds are considered to be double imprinted. Ten Cate was interested to discover whether the double-imprinted birds had developed two separate internal representations (perceptual mechanisms) of the rearing species, or had developed one representation that combined the imprinted information of the two rearing species. He gave the birds preference tests in which zebra finch, Bengalese finch, and hybrids of the two species were used. The results showed the female hybrids to be more attractive than females of either species. This outcome suggests that zebra finch males combine the imprinted information of the two rearing species into one internal representation. A review of the relevant literature (Hollis *et al.*, 1991) led to the conclusion that the rules by which combinations of stimulus features are represented in memory are the same for the formation of the stimulus-recognition mechanisms studied in imprinting studies as for associations formed in perceptual learning.

The final aspect of sexual imprinting that I will consider is the stability of sexual preferences. It was long thought that once a significant sexual preference was established, it would remain stable (Lorenz, 1935; Immelmann, 1979). Under some conditions, this is clearly not the case. In two independent studies (Immelmann *et al.*, 1991; Kruijt & Meeuwissen, 1991, 1993), male zebra finch young were reared with Bengalese finch foster parents for 40 days and subsequently isolated from them until day 100. Half of the males were then given a preference test and all showed a strong preference for females of the foster species. All the birds then received breeding experience with a female of their own species for several months. Thereafter, the birds were tested in two series of preference tests, one immediately after the end of the breeding experience, the other some months later. The results of both studies were remarkably similar. Whereas in the group that received a pretest, most of the birds retained their original preference for the foster species, the majority of the males in the group without pretest preferred females of their own species. Results such as these led Bischof (1994) to suggest that there was a second sensitive period at sexual maturity that led to either consolidation or modification of the perceptual mechanism for recognition of the appropriate sexual partner depending on the animal's first sexual experience. Bischof thus describes sexual imprinting in the zebra finch as a two-stage process.

Food Recognition Mechanisms

The work of Steiner (1979) suggests that newborn infants have well-developed perceptual mechanisms for recognizing sweet, sour, and bitter. Most substances that humans (and other animals) treat as food, however, are recognized on the basis of more complex properties and require specific experience for recognition to develop (for review, see Hogan 1973, 1977). I will here discuss two examples of how food recognition mechanisms develop in chicks and cats. I should emphasize that I am considering 'recognition mechanisms' in a strictly (behaviorally) causal sense. That is, stimuli that activate a food recognition mechanism, for example, are those stimuli that the animal treats as food. We infer that an animal is treating a stimulus as food from the occurrence of behavior that belongs to the hunger system. Such stimuli may or may not be nutritious and could even be poisonous.

Newly hatched chicks peck at a wide variety of objects, although even at the first opportunity, certain colors and shapes are preferred (Hess, 1956; Fantz, 1957). These preferences need not be a reflection of an undeveloped food recognition mechanism, however, for at least two major reasons. First, pecking is a component of aggressive, sexual, and grooming behavior as well as of feeding behavior, and the stimuli that release and direct pecking in these various contexts are quite different. Second, chickens continue to peck a wide variety of objects throughout their lives, even after the objects toward which they direct their feeding, grooming, aggressive, and sexual behavior have become quite specific. Thus, one could view these early preferences as being due to a perceptual mechanism directly connected to the pecking mechanism in the same way that the various taste mechanisms are connected to specific motor mechanisms in humans. This 'independent' pecking might be regarded as serving an exploratory function, and it also has many of the characteristics of play.

The putative food recognition mechanism in newly hatched chicks must be largely unspecified because of the very wide range of stimuli that are characteristic of items that chicks will come to accept as food. Certain taste and tactile stimuli are more acceptable than others (for review, see Hogan, 1973), but these stimuli can be effective only after the chick has the stimulus in its mouth. In some cases, taste and tactile feedback seem to be sufficient to cause an item to become recognized as food. For example, as early as one to two days of age, a chick that has eaten one mealworm (*Tenebrio* larvae) will treat all subsequent mealworms as food. Presumably, the taste of the mealworm is sufficient for subsequent visual recognition to occur, because a second mealworm will be accepted immediately after the first, and thus long before any effects of digestion could be expected to play a role (Hogan, 1966). Taste is also sufficient for a chick to develop visual recognition of a stimulus to be rejected: a one-day-old chick will learn to reject a distasteful cinnabar caterpillar in just one trial (Morgan, 1896; see also Hale & Green, 1979). Nutritive factors do not gain control of pecking until day three (see below). The fact that mealworms can come to be recognized as food (i.e. are avidly ingested) and other insects can come to be rejected as food before day three is evidence that the food recognition mechanism is independent of the central mechanism of the developing hunger system.

The food recognition mechanism also develops under the influence of the long-term (one to two hours) effects of ingestion. Experiments by Hogan-Warburg & Hogan (1981) provide evidence that chicks gradually learn to recognize food particles as a result of the reinforcing effects of food ingestion. In these experiments, visual stimuli from the food gained significant control over the chicks' behavior after one substantial food meal, though oral stimuli gained control of ingestion more slowly.

The development of food recognition in young kittens is similar in many ways to that in chicks (Baerends-van Roon & Baerends, 1979). Kittens begin ingesting their first solid food at about four weeks of age. Some items are immediately recognized as food, whereas others require various kinds of experience before being accepted (or rejected) as food. Fish odor appears to be attractive to all cats, even those with no experience of fish. Fish is ingested as early as a kitten is able to eat solid food, but the main problem for the kitten is learning to catch a fish. This topic is discussed in the next section. Mouse odor, on the other hand, does not appear to have an inherent attractiveness for cats (Berry, 1908). Mice become recognized as food only after a kitten has eaten a mouse. This can happen if a mother cat presents a dead (and opened) mouse. It can also happen if a kitten attacks and bites a live mouse by itself. It is not yet possible to say whether the taste of the mouse is sufficient experience for its subsequent recognition as food (as in the chicks) or whether nutritional effects of digestion are necessary. The Baerends' did observe that a shrew may be caught and ingested by a naïve kitten, but it is vomited within 15–20 min. Thereafter, kittens may catch and 'play' with shrews, but they never ingest them. This finding suggests that the effects of ingestion may be the critical experience for food recognition to develop. Such observations also indicate considerable independence of catching and eating behavior, a topic discussed later.

In a functional sense, the nutritional effects of ingestion should be the ultimate factor determining which objects are recognized as food. But sometimes other factors override the effects of nutrition and lead to the development of a food recognition mechanism that is maladaptive. Two observations made on chicks' food preferences are relevant here (Hogan, 1971). First, many chicks that were fed mealworms on the first few days after hatching died at about six or seven days. These chicks could generally be characterized as mealworm fanatics because of their excited, positive behavior toward mealworms. These mealworm fanatics never learned to eat the regular chicken food that literally surrounded them, and they apparently died of starvation. Second, many chicks that were raised on a mixture of chicken food and aquarium gravel also died at about six days of age, also apparently of starvation. In this case, the gravel seemed to be an exceptionally good releasing stimulus for pecking and swallowing. Both these examples show that factors other than the nutritional effects of ingestion can play an important role in the development of food recognition.

DEVELOPMENT OF MOTOR MECHANISMS

Many motor mechanisms develop prefunctionally. For instance, young chicks show normal locomotion and pecking movements almost immediately after hatching. Within the next few days, ground scratching and various grooming movements appear. Kruijt (1964) showed that the proper functioning of these and other movements in the posthatching situation is not a necessary causal factor for their development. Of course, prehatching conditions obviously influence the development of these movements, though many of the processes responsible for behavioral organization remain largely unknown (Oppenheim, 1974). Studies of the responses of young rat pups to electrical stimulation of the medial forebrain bundle provide an additional example (for review, see Moran, 1986). Three-day-old rat pups show a number of organized behavior patterns such as licking, pawing, gaping, and lordosis in response to such stimulation. These patterns are not seen in their normal functional context until later in development. Thus, these motor mechanisms must also be organized prior to their functioning.

It should be emphasized here that, although motor patterns are visible to an observer, motor mechanisms are not. An example should make this point clear. Kuo (1967) noted that chicks that developed with their yolk sac in an abnormal position were often crippled when they hatched. He interpreted these results to mean that the development of normal walking movements required functional experience in the egg: the legs had to push actively against the yolk sac for normal development to occur. Such experience is indeed necessary for the development of normal joints (Drachman & Sokoloff, 1966), and without properly functioning joints, a chick cannot move normally. Nonetheless, the movements of a crippled chick cannot provide evidence for whether or not the motor mechanism for walking has developed normally. Such evidence, moreover, does not contradict the conclusion of Hamburger (1973) that the neural patterning underlying the walking movements of a chick develops without functional experience (Lehrman, 1970). (For a related example concerning human locomotion, see Thelen & Fisher, 1982.)

Perhaps the best-studied cases of how a motor mechanism develops on the basis of functional experience are the development of bird song and human speech. I will discuss the development of bird song here, but will discuss the development of human speech in the human language section (p. 169). I then consider the development of some displays in birds; experience is effective in a surprising way in this example. Finally, I discuss aspects of the development of behavior sequences in dustbathing of chickens, grooming in rats, and prey catching in cats. These examples all give insight into the development of more complex behavior systems.

Song Learning

As seen above, in many species, the young bird forms an auditory image of the song it will learn to sing. Learning to sing the song does not happen until later, when the bird's internal state (e.g. the level of testosterone) is appropriate. At this point, it appears that the bird learns to adjust its motor output to match the image it has previously formed. This adjustment must involve the bird's hearing itself, because deafened birds never learn to produce any song that approaches normal song (Konishi, 1965). Experiments by Stevenson (1967) showed that hearing its species-specific song could serve as a reinforcer for an operant perching response in male chaffinches. On the basis of these results, Hinde (1970) suggested that song learning might involve matching the sounds produced by the young bird with the stored image: sounds that matched the image would be reinforced, whereas other sounds would be extinguished. In this way, a normal song could develop in much the same way as an experimenter originally trains a rat to press a lever (Skinner, 1953). Subsequent experiments with both male (ten Cate, 1991) and female (Riebel, 2003) zebra finches have confirmed the reinforcing properties of hearing species-specific song.

In most species, three stages in the production of song can be distinguished: subsong, plastic song, and crystalized song (Thorpe, 1961). During the subsong phase, the bird essentially babbles, and slowly adjusts its production to match phrases and songs it heard during the memorization stage; it may also invent new combinations of phrases during this phase. These changes presumably come about in the manner suggested by Hinde. In the plastic song phase, the bird may be singing a number of songs that resemble songs it previously heard. Which of these songs becomes chosen as the crystalized song depends, in many species, on the songs it hears from other birds at this time. In some species, the bird selects a similar song (which probably accounts for the occurrence of local dialects) and in other species, the bird selects a dissimilar song. In either case, a selection is made from songs already developed (Marler & Peters, 1982; Nelson, 1997). The selection process presumably also involves some kind of reinforcement, often provided by the behavior of conspecifics. A particularly interesting example is the song of the brown-headed cowbird, *Molothrus ater*. Males of this species increase their performance of those songs that are associated with a wing stroke display given by the females (West & King, 1988). Once the song has crystalized, it had long been thought that auditory feedback from singing was no longer necessary to maintain the song, at least in most species. It now appears that auditory feedback after crystalization continues to affect the song in most species, and is necessary for the song in many species (Brainard & Doupe, 2000).

Displays

A *display* is a motor pattern that is adapted to serve as a signal to a conspecific (Tinbergen, 1952). The mechanism controlling the display is thus the motor counterpart of the species-recognition perceptual mechanisms discussed above. Displays are often complex, yet they typically develop prefunctionally. For example, waltzing is a courtship display in chickens that essentially involves the male circling a female in a characteristic manner (see Figure 3.10). Kruijt (1964) showed that the form of this display can be derived from components of behavior that belong to the aggression and escape systems, and that these systems are activated when waltzing first appears. Nonetheless, waltzing appears even in animals that are reared in social isolation, so social experience cannot be a necessary causal factor in its development.

One example of a display in which social experience has been implicated as a causal factor in its development is the 'oblique posture with long call' of the black-headed gull, *Larus ridibundus*. Groothuis (1992) raised gulls to the age of one year either in social isolation, in small groups of two to four individuals, or in large groups of 12. Black-headed gulls are colonial breeders, and large groups are the normal social environment for the developing young. All the birds raised in large groups, 50% of the birds raised in small groups, and 35% of the birds raised in social isolation developed the normal display in the first year. Of particular interest is that about 40% of the birds raised in small groups developed an aberrant display in which the head was held in an abnormal posture. Further experience in large groups for more than one year had no effect on the form of this aberrant display. This result contrasts with the finding that all of the isolated and other birds that originally showed only fragmentary forms of the display subsequently developed a normal display when placed together in a large group. A separate experiment (Groothuis & Meeuwissen, 1992) showed that isolated birds that were injected with testosterone at ten weeks of age all developed a normal display within a few days of injection.

One process underlying the development of this display may be the same as that suggested by Hinde for the development of bird song.

There may be some sort of template sensitive to proprioceptive feed-back from the display (a comparator mechanism) that selects out the correct forms from all the transitional forms that normally occur. Such a cognitive structure that recognizes proprioceptive feedback has been proposed to explain results of experiments on imitation by human infants (Field et al., 1982). Nonetheless, the results from the testosterone experiments, in which essentially no transitional forms were seen, do not support such a process in the gulls. Further, the fact that many isolated birds developed a normal display means that social experience cannot be a necessary causal factor for normal development. However, the aberrant displays that developed in some birds raised in small groups suggest that social interactions can be of importance in special circumstances. Groothuis (1992, 1994) discusses several hypotheses to explain these results, one of which is that abnormal experience encountered in the small groups could have distorted normal development. I return to this idea in Chapter 6.

Dustbathing and Grooming

Dustbathing in the adult fowl has been described in Chapter 3 and a diagram of the dustbathing system is shown in Figure 3.5. Dustbathing does not appear fully formed in the young bird. Rather, individual elements of the system appear independently, and only gradually do these elements become fixed in the normal adult form. Pecking is seen on the day of hatching, but the other motor components appear gradually over the first seven or eight days posthatch (Kruijt, 1964). Vestergaard et al. (1990) asked whether the rearing environment influenced the organization of the motor components. They observed small groups of chicks that were raised either in a normal environment containing sand and grass sod or in a poor environment in which the floor was covered with wire mesh. A comparison of the dustbathing motor patterns of two-month-old birds raised in the two environments showed surprisingly few differences. The form and frequency of the individual behavior patterns, as well as the temporal organization of the elements during extended bouts of dustbathing, developed almost identically in both groups. There were some differences in the microstructure of the bouts that could be related to the presence or absence of specific feedback, but the motor mechanisms and their coordination developed essentially normally in chicks raised in a dustless environment (see also van Liere, 1992). Clearly, the experience of sand in the feathers removing lipids or improving feather quality is not necessary for the integration of the motor components of dustbathing into a normal coordinated sequence.

Somewhat later, Larsen *et al.* (2000) studied in detail the development of dustbathing behavior sequences in chicks from hatching to three weeks of age. They found that the individual behavior elements, as soon as they appeared, were incorporated into the normal adult sequence structure; this occurred even though the form of the elements themselves is not yet fixed. These results support the conclusion that separate mechanisms are responsible for the form of the individual behavior elements and for the organization of these elements into recognizable sequences. A similar conclusion was reached by Berridge (1994, see Figure 2.9) on the basis of results on the development of grooming sequences in young rodents (Colonese *et al.*, 1996). As mentioned before, these investigators called certain sequences of grooming movements 'syntactic chains' to emphasize the rules controlling natural action sequences (see also Fentress & Gadbois, 2001). I return to these ideas in the section on human language.

Prey Catching

My final example concerns the development of sequences of motor patterns is the prey-catching behavior of cats (Baerends-van Roon & Baerends, 1979). Locomotion, pouncing, angling (with one paw), and biting are the basic motor patterns out of which effective prey-catching develops, and all these behaviors can be seen, prefunctionally, by the time the kitten is about four weeks old. The way these behaviors become integrated depends primarily on the type of prey being caught. If a mouse is the prey, locomotion and biting are sufficient to catch and kill, whereas with larger prey, pouncing is necessary in addition. If a fish is the prey, angling and biting are the necessary motor patterns. The evidence suggests that the 'correct' behavior sequences are selected on the basis of the effects of the behavior. In other words, an operant shaping process can account for all the results, with the proviso that the basic elements – locomotion, pouncing, angling, and biting – are not themselves shaped. This conclusion is supported by the fact that the course of development can vary considerably among individuals even though the final result is quite stereotyped.

Two further points should be mentioned. The first is that the outcome, or reinforcer, that shapes prey-catching behavior is probably not related to eating (nutrition). This conclusion follows from the fact that prey catching in general is often independent of nutritional state (e.g. Polsky, 1975). It seems more likely that prey catching is an end in itself. This conclusion is also attested to by the experience of many cat owners who have fed their pets from weaning on the best cat food, but who, nonetheless, are often presented with dead birds that have been caught but not eaten.

A second difference between prey-catching sequences and many of the other examples discussed above is that prey-catching sequences in cats do not 'crystalize'; that is, functional experience continues to be effective in shaping new sequences. For example, kittens that have developed proficient fish-catching behavior can subsequently learn to catch mice, although there is some interference from the previous learning in that such kittens take longer to learn to kill the mouse with a bite than do naïve kittens. Learning to catch a fish after the kitten has already developed mouse-catching behavior turns out to be considerably more difficult. The primary problem here is that older kittens have a stronger tendency to avoid getting wet than younger kittens. If the fear of water can be overcome, the fish-catching sequence can be easily acquired (Baerends-van Roon & Baerends, 1979). This last example indicates an important problem in the study of development: Certain cases of learning may seem to be irreversible when, actually, indirect factors (such as fear of water in this case) obscure the fact that functional experience can still have direct effects on development.

DEVELOPMENT OF CONNECTIONS BETWEEN CENTRAL
AND MOTOR MECHANISMS

We can say that a central and a motor mechanism are connected when the occurrence of a behavior varies directly with the presence of factors known to affect the central mechanism. Consider the behavior of pecking and the central hunger mechanism of a chicken. If the amount of pecking varies directly with the amount of food deprivation, then we have evidence that hunger and pecking are connected. On the other hand, if variations in food deprivation have no effect on the amount of pecking, we have evidence that hunger and pecking are not connected – that is, are independent mechanisms. For many behavior systems in many species, connections between the central and motor mechanisms develop prefunctionally, though these connections are often subject to change as a result of various kinds of experience. But there are also many examples in young animals in which these mechanisms require functional experience to become connected. The developmental problem is how to get from the state of independence to the state of connectedness. I will begin by considering the normal development of feeding behavior in several species, and will then give some further examples from the operant conditioning literature.

Hunger

As mentioned in Chapter 3, a surprising fact about the feeding behavior of many neonatal animals is that their early feeding movements are relatively independent of motivational factors associated with food deprivation. Hinde (1970, p. 551ff) reviewed a variety of evidence from studies on kittens, puppies, lambs, and human infants that show that the amount of suckling by a young animal is very little influenced by the amount of food it obtains. Since then, a series of studies on the development of feeding in chicks and in neonatal rats has been published, and these are reviewed briefly here.

A chick begins pecking within a few hours of hatching, but its nutritional state does not influence pecking until about three days of age (Hogan, 1971). When chicks were one or two days old, five hours of food deprivation did not influence the subsequent rate of pecking at food, whereas by the time the chicks were four or five days old, five hours of deprivation led to a large increase in pecking at food. A very similar change in control of feeding has been found in rat pups (Blass et al., 1979; Cramer & Blass, 1983; Hall & Williams, 1983). Before the age of about two weeks, the occurrence of behaviors such as nipple search and nipple attachment, as well as the amount of sucking itself, was not influenced by food (i.e. maternal) deprivation of as long as 22 hours. After two weeks, however, deprived pups attached to the nipple more quickly and sucked longer than nondeprived pups. Similarly, when tested in a spatial discrimination task in a Y maze, nutritive suckling provided a greater incentive than nonnutritive suckling only after the pups were older than two weeks (Kenny et al., 1979).

The developmental question with respect to these results is: How do the motivational factors associated with food deprivation come to control feeding behavior? For chicks, early experiments (for review, see Hogan, 1977) led to the hypothesis that it is the experience of pecking followed by swallowing that causes the connection between the central hunger mechanism and the pecking mechanism to be formed. In other words, a chick must learn that pecking is the action that leads to ingestion; once this association has been formed, nutritional factors can directly affect pecking. Subsequent experiments have shown that the association of pecking with ingestion is indeed the necessary and sufficient condition for pecking to become integrated into the hunger system (Hogan, 1984). Experiments on the development of pecking in ringdoves, *Streptopelia risoria*, also indicate that experience is necessary for hunger to gain control of pecking; though in this case, the necessary experience apparently involves interaction with the parents, as well as with food (Graf et al., 1985; Balsam et al., 1992; Balsam & Silver, 1994).

Similar experiments with rat pups have not been done, though the problem with mammals in general is more complex because the suckling response drops out altogether at weaning and is replaced by different behaviors (Hall & Williams, 1983). Hall and his colleagues have shown that, under special conditions, rat pups ingest food away from the mother very soon after birth, but these experiments have not asked the same questions being asked here (see also Johanson & Terry, 1988). In the study by Kenny et al. (1979), the infant rats received their nourishment through intragastric feeding between days 12 (before their ingestion was influenced by hunger) and day 17. When tested at day 17, motivational control of their ingestion was the same as in normally reared pups, which implies that experience eating solid food is not necessary for motivational control to develop. However, there are some results from guinea pigs that are also relevant (Reisbick, 1973). Guinea pigs normally begin ingesting solid food within one day of birth, and Reisbick found that experience of ingesting and swallowing was necessary before the guinea pigs showed evidence of discriminating between nutritious and nonnutritious objects. These results are very similar to the results from the chicks and have been discussed in more detail elsewhere (Hogan, 1977).

Operant Conditioning

A second source of evidence for the development of connections between central and motor mechanisms is the operant conditioning literature. The process of reinforcement, in general, can be regarded as influencing the development of connections between central and motor mechanisms. For example, the response of lever pressing is an easily recognizable motor pattern in a rat. Reinforcing lever pressing with food leads to a connection of the motor mechanism for lever pressing with the hunger system, and reinforcing with water leads to a connection with the thirst system (cf. Moore, 1973).

Schiller (1949/1957) reported the results of studies of problem solving by chimps. He noted that many of the behavior patterns used by his chimps to procure food that was placed out of reach were apparently the same manipulative patterns that had first appeared spontaneously and prefunctionally. The patterns included 'weaving', 'poking and sounding', and 'joining sticks'. Schiller suggested that these patterns could be considered operant responses that were used to solve the problem, and that they were reinforced when the chimp was successful. In the terminology used here, we could say that the originally independent motor mechanisms responsible for the various observed behavior patterns became connected to the hunger system as a result of operant reinforcement. The test for 'connection' here, as elsewhere, is to see if the occurrence of a behavior varies directly with the presence of factors known to affect the central mechanism. Do hungry chimps engage in these behaviors more than sated chimps? Schiller's results suggest that they do.

DEVELOPMENT OF CONNECTIONS BETWEEN PERCEPTUAL AND CENTRAL MECHANISMS

We can say that a perceptual mechanism and a central mechanism are connected when a stimulus that activates the perceptual mechanism can lead to the occurrence of the set of behaviors known to belong to the central mechanism. For instance, an egg recognition mechanism is connected to the incubation system in many birds because the presentation of an egg (or other appropriate stimulus) can lead to approach, retrieval, and settling on the nest (cf. Figure 2.7). Additional evidence to show that a perceptual mechanism is in fact connected to a central mechanism, and not directly to a motor mechanism, is to show that the presentation of an adequate stimulus has the same effect on the central mechanism as a direct manipulation of the relevant internal factors, for example, by deprivation or the injection of hormones. Such evidence is most easily provided by demonstrating priming effects (Van der Kooy & Hogan, 1978; Hogan & Roper, 1978, p. 231). For example, the presentation of food may make an animal hungrier (the 'appetizer' effect), or the presentation of a sexual stimulus may increase its sexual appetite, which leads to more eating or more persistent courtship, respectively. As with the development of connections between central and motor mechanisms, the developmental question is what kinds of experience are necessary to bring about connections. I will discuss evidence from studies of dustbathing and hunger, and from classical conditioning studies. At the end of this section, I will consider further evidence from studies of filial imprinting.

Dustbathing

In chickens, functional experience plays an essential role in the development of the perceptual mechanism for recognizing dust and of the connection between it and the central mechanism (dashed lines in Figure 2.6). Sanotra et al. (1995) review the evidence for the role of experience in the development of the dust recognition mechanism itself. Factors responsible for the connection between the dust recognition mechanism and the central mechanism are reviewed here.

Young chicks can be seen engaging in dustbathing movements on almost any surface that is available, ranging from hard ground and stones to sand and dust. In fact, Kruijt (1964) found that making the external situation as favorable as possible for dustbathing was insufficient for releasing the behavior. This result implies that early dustbathing may be controlled exclusively by internal factors (see below). It also implies that the connection between the dust recognition perceptual mechanism and the central mechanism is not formed until well after the motor and central mechanisms are functional.

Vestergaard & Hogan (1992) found that early dustbathing is most likely to occur in whatever substrate is pecked at most. They point out that pecking is a movement that functions as exploratory, feeding, dustbathing, and later aggressive behavior. They suggest that perceptual mechanisms specific to each system develop gradually out of exploratory pecking on the basis of functional experience. It remains to be determined whether removal of lipids, the sensory feedback from the substrate in the feathers, or facilitation of the dustbathing behavior itself is the crucial factor.

Other evidence shows that early experience can lead to stable preferences for particular stimuli (Petherick et al., 1995; Vestergaard & Baranyiova, 1996). As an extreme example, Vestergaard & Hogan (1992) raised birds on wire mesh, but gave them regular experience on a substrate covered with coal dust, white sand, or a skin of junglefowl feathers. In choice tests given at one month of age, some of the birds that had had experience with junglefowl feathers were found to have developed a stable preference for dustbathing on the feathers. This example is important because it shows how a system can develop abnormally. It also suggests that the pecking associated with dustbathing may be a cause for 'feather pecking', a common pathological condition in which some hens pull out the feathers of their cage mates, which is seen in many commercial groups of fowl (Rodenburg et al., 2004; Chow & Hogan, 2005).

Hunger

The results with the chicks discussed above show that a mealworm recognition mechanism can become connected to the motor mechanisms for pecking at least one day before nutrition (i.e. the central 'hunger' mechanism) gains control of pecking. The evidence indicates that the ingestion of mealworms remains semi-independent of hunger, probably throughout life: satiated chicks avidly ingest many mealworms, and the ingestion of a substantial number of mealworms, at least in the first week after hatching, has no effect on the amount of other food subsequently ingested (Hogan, 1971). This semi-independence of mealworm ingestion and hunger is probably the same phenomenon as the semi-independence of suckling and hunger in rats and prey catching and hunger in cats and most other predators.

Evidence that a food recognition mechanism becomes connected to the central hunger mechanism comes from the fact that food particles develop incentive value between three and five days posthatching (Hogan, 1971); development of incentive value probably reflects the same process involved in the development of food recognition discussed above. More direct evidence of perceptual mechanisms' becoming connected to central mechanisms is provided by several examples from the learning literature.

Classical Conditioning

There are now numerous examples of complex, species-typical behaviors that become released by previously neutral stimuli that develop their effectiveness by means of a classical conditioning procedure. For instance, Adler & Hogan (1963) paired the presentation of a weak electric shock with a mirror to a male Siamese fighting fish and showed that full aggressive display could be conditioned to the shock. In a similar way, Farris (1967) conditioned the courtship behavior of Japanese quail, Coturnix japonica, to a red light. Moore (1973) showed that a small, lighted key followed consistently by food elicited a food peck in a pigeon; when followed consistently by water, it elicited a drinking peck. Blass et al. (1984) were able to condition the ingestive behaviors of head orientation and sucking in human infants (which are unconditioned responses to the oral delivery of a sucrose solution) to gentle forehead stroking. These and many other cases exemplify the development of connections between a perceptual mechanism and a set of behaviors as a result of a classical conditioning procedure. They do not, however, distinguish between a connection between a perceptual mechanism and a central mechanism or directly between a perceptual mechanism and a complex motor mechanism.

There are also cases, however, where a connection between a perceptual mechanism and a central mechanism is directly implicated. Wasserman (1973) looked at the behavior of young chicks tested in a cool environment. The chicks were trained by being exposed to a lighted key for several seconds and then to presentation of heat from a heat lamp. After several pairings of the light and the heat, the chicks began to approach the key when it lighted up and showed pecking and 'snuggling' movements to it. These behaviors were never shown to the heat lamp itself (which was suspended above the chicks, out of reach). Pecking and snuggling movements are behaviors shown by young chicks when soliciting brooding from a mother hen (Hogan, 1974). Wasserman's results imply that the recognition mechanism for the lighted key becomes connected to a thermoregulatory system in the young chick (Sherry, 1981), and that the presentation of this stimulus to a cold chick elicits brooding solicitation movements (cf. discussion of causal reasoning in rats in Chapter 7, p. 238).

A second example comes from the work of Hollis (1984) on aggressive behavior in the blue gourami fish (*Tricogaster tricopterus*). She paired the presentation of a red light with a rival behind glass during the training phase of her experiment. The fish were tested by being presented the red light and then being allowed to engage in a real fight. Fish that had been conditioned responded sooner and more frequently with the aggressive behaviors of biting and tail beating than control fish that had not been conditioned. These results support the idea that the conditioned stimulus, the red light, primed the central aggression mechanism of the fish, which allowed it to respond appropriately to the rival fish.

Similar examples have multiplied. The systems include hunger, aggression, sex, and fear, in species ranging from insects through fish and birds to mammals, including humans (for reviews, see Timberlake, 1994; Domjan & Holloway, 1998; Fanselow & De Oca, 1998). What many of these examples show is that previously neutral stimuli can, as a result of classical conditioning procedures, develop control of entire behavior systems. The studies by Fanselow and his colleagues, using the species-specific defense reactions of rats as their behavior, have even shown that the conditioned stimulus has its effects through the same neural structures as the unconditioned stimulus.

Imprinting

The studies of imprinting, discussed above, might suggest that an 'empty' perceptual mechanism is connected prefunctionally to the central coordinating mechanism in both the filial system and the sex system of many birds. Experience then 'fills in' the schema, subject only to the constraints noted. However, the reality is much more complex. Van Kampen (1996) has proposed a framework for the study of filial imprinting that is based on the behavior system concept developed in this book. His scheme is shown in Figure 5.1. The dashed line leading to S, the stimulus recognition mechanism that is being imprinted, indicates that this representation of the stimulus is not prefunctional, but needs to be developed on the basis of the chick's experience. Likewise, the dashed line from S to C, the central

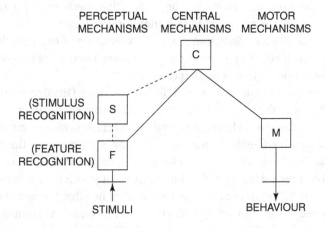

Figure 5.1 Basic scheme of the filial system of a chick. Perceptual mechanisms include feature recognition in F (such as color, shape, size, and movement) and stimulus recognition in S (representations of stimuli on which a chick is imprinted). The motor mechanisms in M are responsible for the execution of motor patterns such as shrill calling, searching, approach, and twitter calling. The central mechanisms in C can be observed as the internal state of the chick, for instance its motivation to seek contact with the imprinting stimulus.
From van Kampen, 1996, with permission.

mechanism, indicates that this connection also needs to be developed. In a very sophisticated analysis of the imprinting and related literature, van Kampen was able to show that S develops according to the principles of perceptual learning, with exposure to a stimulus being the only requirement, but that the connection between S and C develops according to the principles of association learning (cf. Chapter 7). With respect to the development of S, the chick appears to be directed to appropriate stimuli by the 'predisposition' (probably some aspect of F) mentioned above, which van Kampen interpreted as a preexisting stimulus-recognition mechanism. The formation of the connection between S and C follows all the rules of classical conditioning. He also showed that the central mechanism consists of two parts, a search mechanism and a contentment mechanism. S is connected to the contentment part: when the chick is motivated to search (because it finds itself separated from its mother or imprinted stimulus), various search behaviors are activated (M), which continue until a stimulus that matches S (its mother or the imprinted stimulus) is found. The factors that initiate search are discussed in Chapter 6.

DEVELOPMENT OF CONNECTIONS AMONG PERCEPTUAL, CENTRAL, AND MOTOR MECHANISMS

The previous sections have presented evidence about the effects of various kinds of experience on the development of connections between pairs of building blocks.

The principles of development that emerge from those results are sufficient to allow us to understand much of the development of more complex systems. As we shall see, however, some new principles seem also to be involved in these more complex cases. A review of some examples of the development of hunger, aggressive, sexual, and play systems will illustrate how these principles operate, and the final section will consider how they could be applied to the study of human language acquisition.

Hunger

The hunger system of a chicken has been described and depicted in Chapter 2 (Figure 2.6). Earlier in this chapter, I have shown how the food recognition mechanism develops and discussed what experience is necessary for the central mechanism to develop its modulating effects on feeding. With respect to motor mechanisms, the previous discussion has focused almost entirely on pecking. There are, however, several other motor mechanisms that are normally associated with the hunger system, such as those controlling ground scratching and locomotion. As with dustbathing, both the individual motor mechanisms of the system and the coordination of these mechanisms into effective foraging behavior appear prefunctionally. Unlike with dustbathing, however, the integration of the motor mechanisms in the hunger system disintegrates in the absence of effective functional experience (Hogan, 1971). Hogan (1988) reviews the evidence that suggests that new connections are formed between the central hunger mechanism and individual motor mechanisms on the basis of specific experience of the individual chick (dashed lines between H and P and between H and S in Figure 2.6), and that these new connections effectively block the expression of the original prefunctional connections. A similar explanation may apply to the disappearance of pecking for food in chicks that have been raised in the dark and force-fed food, and for the disappearance of the ability to suck in human infants that have been fed from a cup.

The general picture that emerges from all the data is summarized in Figure 2.6. A young chick has a number of feature recognition perceptual mechanisms, an undeveloped food recognition mechanism, an independent central hunger mechanism, an integrated complex of motor mechanisms, and some connections between the perceptual and motor mechanisms; these mechanisms are available prefunctionally. The food recognition mechanism develops (perhaps simultaneously with a number of object recognition mechanisms) under the influence of experience with certain tastes or positive nutritious aftereffects of ingestion. The food recognition mechanism probably has connections to the motor mechanisms prefunctionally. A connection between the central hunger mechanism and the complex of motor mechanisms develops as a result of the experience of pecking followed by swallowing, and between the central hunger mechanism and the food recognition mechanism as a result of experience of the nutritive aftereffects of ingesting particular particles (the incentive value of food crumbs). More specific connections develop between the central hunger mechanism and particular motor mechanisms on the basis of nutritive feedback as well. These specific connections are in evidence especially when the chick is hungry, but the original prefunctional connections among perceptual mechanisms and motor mechanisms remain operative and can be seen especially when the chick is not hungry.

It should be noted that the picture for the development of prey catching in kittens is not essentially different from the picture just presented for the chicks. Although the individual behavior patterns used in prey catching (pouncing, angling, biting) are originally independent in the sense that the precise ordering of components is not determined, nonetheless, these behavior patterns do not occur at random. The specific patterns of these components that develop with respect to particular stimuli can be considered subsystems of the sort that chicks develop with respect to mealworms or to grainlike objects. These subsystems in kittens – a mouse-catching system or a fish-catching system – also have a relationship to the central hunger mechanism that is very similar to the relationship between hunger and pecking or ground scratching in chicks.

A final point is that the development of a hunger system can be greatly influenced by factors that are basically irrelevant to feeding or nutrition. The factor that has been mentioned here is fear, with respect to both the development of recognition of mealworms as food in chicks and the development of fish catching in kittens. A chick that is too afraid of a mealworm will never pick one up (Hogan, 1965), and a kitten that is too afraid of water will never learn to catch a fish. Such indirect motivational factors play an even more important role in the development of social behavior, as will be seen below.

Aggression

The aggression system of an adult chicken consists of perceptual mechanisms that serve an 'opponent' recognition function, various motor mechanisms that are used in fighting (including those that control threat display, leaping, wing flapping, kicking, and pecking), and a central mechanism that is sensitive to internal motivational factors (such as testosterone) and that coordinates the activation of the motor mechanisms. Kruijt (1964) showed that fighting develops out of hopping, which is a locomotor pattern that is not initially released by or directed toward other chicks. While hopping, chicks sometimes bump into each other by accident, and in the course of several days, hopping gradually becomes directed toward other chicks. Frontal threatening starts to occur, and by the age of three weeks, pecking and kicking are added to aggressive interactions. Normal, well-coordinated fights are not seen until two to three months.

The various behavior patterns comprising adult fights can be seen to occur, independently, in the one- to two-week-old chick, well before their integration into fighting behavior. This means that functional social experience could be a necessary factor guiding development. This is, however, not the case. In other experiments, Kruijt (1964) raised chicks in social isolation for the first week of life and then placed them together in pairs. Many of these chicks showed aggressive behavior toward each other within seconds. Further, the fights that developed were characteristic of the fights of one-month-old, socially raised chicks. Such results suggest that the organization of the motor components of the aggression system, as well as the connections between the central and motor mechanisms, develop prefunctionally, and that the occurrence of aggressive behavior requires only the proper motivational state. Similar results and conclusions apply to the development of aggressive displays in gulls (Groothuis, 1994). In this way, the aggression system is more like the dustbathing system than it is like the hunger system.

The opponent recognition perceptual mechanism must be partially formed prefunctionally because a chick as young as two and a half days old will respond with frontal threat and aggressive pecks to the stimulus of a 6-cm green wooden triangle moved directly in front of it (Evans, 1968). Likewise, as already mentioned, socially isolated chicks showed fully coordinated aggressive behavior when confronted with another chick at the age of one week. But isolated chicks of the same age can also direct aggressive behavior to a light bulb hanging in the cage, and older isolated males often come to direct their aggressive behavior to their own tails (Kruijt, 1964). Presumably, the complete development of the perceptual mechanism depends on the proper experience at the proper time (just like the templates for song learning in many species), but the experiments necessary to explore this idea have not yet been done.

The development of normal aggressive behavior in kittens does require specific functional experience. Baerends-van Roon & Baerends (1979) described early attack behavior, which included most of the same behavior patterns previously discussed with respect to prey catching, including pouncing and biting. These patterns are apparently the same when originally directed to either a prey or another kitten, but they become modified in different ways as a result of feedback from the opponent. In particular, the force of the pounce, the extension of the claws, and the strength of the bite all become reduced after a nestmate responds in kind. The occurrence of 'play' behavior, especially in the period from four to eight weeks, seems to provide the kitten with essential experience for the development of normal social behavior. Two kittens that were raised in social isolation (after weaning at seven weeks) showed either unrestrained attack or total avoidance when confronted with a normally reared cat at the age of several months. These two cats also showed abnormal maternal behavior when they later had their own litters. The Baerendses suggested that normal development requires a proper balance of attack and escape motivation.

Sex

The sex system of a normal adult rooster consists of perceptual mechanisms that serve a 'partner' recognition function; motor mechanisms for locomotion, copulation (which includes mounting, sitting, treading, pecking, and tail lowering), and various displays, such as waltzing, wing flapping, tidbitting, and cornering; and a central mechanism that is sensitive to internal motivational factors such as testosterone and that coordinates the activation of the motor mechanisms. In small groups of junglefowl, Kruijt (1964) saw mounting and copulatory trampling (treading) on a model in a sitting position as early as three or four days, but such behavior was not common until weeks later. Full copulation with living partners did not occur before the males were four months old.

Many of the components of the copulatory sequence, including mounting, sitting, and pecking, are seen independently in young chicks, and there is ample opportunity for social experience to influence the occurrence and integration of these components. As with aggression, however, several lines of evidence suggest that the motor mechanisms are already organized soon after hatching, and that their expression merely requires a sufficiently high level of motivation. For example, Andrew (1966) was able to elicit well-integrated mounting, treading, and pelvic lowering in socially isolated domestic chicks as young as two days old by using the stimulus of a human hand moved in a particular manner. Andrew also found that injection of testosterone greatly increased the number of chicks that responded sexually in his test during the first two weeks (see also Groothuis, 1994, for similar evidence on the expression of sexual displays in young gulls, and Williams, 1991, for evidence on the expression of sexual behavior in rat pups). Further, junglefowl males that had been raised in social isolation for six to nine months copulated successfully with females within so few encounters that it was clear that the motor mechanisms had been integrated before testing (Kruijt, 1962).

The occurrence of the courtship displays presents a somewhat different picture. For example, waltzing is first seen at two to three months of age, when it always appears in the context of fighting. As already mentioned, the form of the display seems to develop independently of social experience. The factors controlling the occurrence of waltzing, however, appear to be largely determined by social experience. Waltzing to a female often has the effect that the female crouches, and a crouching female is the signal for mounting and copulation. Experiments reported by Kruijt (1964) showed that the frequency of waltzing increased when mating was contingent on its occurrence and decreased when mating was not allowed. This finding suggests that, in normal development, the switch that is seen from the occurrence of waltzing in a fighting context to a sexual context may require the experience of the display followed by copulation. This interpretation is also supported by the behavior of the males that were socially isolated for six to nine months. These animals did not show waltzing (or the other displays) before mating with the female, but they often showed displays before attacking her. Thus, copulation seems to be the reinforcer that causes the motor mechanism for waltzing to become attached to the central coordinating mechanism for sex.

Tidbitting is a display that consists in ground pecking directed to edible or inedible objects and/or ground scratching, accompanied with high, rhythmically repeated calls. It develops out of the pecking and calling that accompany 'food running' (Kruijt, 1964), which can be seen in young chicks as early as two days. Tidbitting is especially interesting because it serves a courtship function in males, but a parental function in females. In all three contexts, it serves to attract conspecifics from a distance: food-running chicks attract other chicks and the mother hen, tidbitting males attract females, and food-calling (tidbitting) mother hens attract their chicks. As with waltzing, the form of the tidbitting display does not depend on social experience because it is seen in both chicks and adults that have been raised in social isolation. The causal factors controlling food running are complex and include escape, hunger, and possibly aggression (Hogan, 1966). Andrew (1966) reported that testosterone injections did *not* increase the occurrence of 'juvenile tidbitting', whereas they did increase copulatory behavior. Nonetheless, in adult males sexual factors play a primary role in the occurrence of tidbitting (Kruijt, 1964; van Kampen, 1997; van Kampen & Hogan, 2000), and in adult females, parental factors play a primary role (Sherry, 1977). Somewhat surprisingly, sexual factors are not implicated in the response of females to a tidbitting male (van Kampen, 1994). All these results imply that the motor mechanism for tidbitting develops new connections with central mechanisms in the course of development. Unfortunately, there have been no experiments to determine what kind of experience is necessary for the switch in causal factors to occur (but see Moffatt & Hogan, 1992).

The development of the perceptual mechanisms of the sex system and their connections to the central sex mechanism seem to be much more susceptible to the effects of experience than the development of the motor mechanisms. For example, junglefowl chicks become sexually dimorphic at about one month of age. By about two months, young males begin to show incomplete sexual behavior toward other animals, but such behavior is directed equally toward males and females. Only gradually, as a result of specifically sexual experience, does sexual behavior become directed exclusively to females (Kruijt, 1964). As seen above, the development of the partner recognition mechanism has been studied intensively for many years in the context of sexual imprinting, and there is extensive evidence documenting the influence of both prefunctional and functional factors (Bolhuis, 1991; Bischof, 1994).

Much of the work of Harlow and his students and of Hinde and his students on the development of social behavior in rhesus monkeys, *Macaca mulatta*, is also relevant to this discussion (see, for example, Harlow & Harlow, 1965; Sackett, 1970; Hinde, 1977). The parallels between the development of chicken behavior and monkey behavior are remarkable, and many of the points made in the previous discussion could have been illustrated just as easily by reference to the monkey results. There are also important parallels between the work discussed above and the development of human social behavior (see, for example, Rutter, 1991, 2002), but a discussion of these is beyond the scope of this chapter.

Play

The topic of play has been discussed extensively in the context of development. Here, I briefly present some ethological ideas about the causation of play, and I show how they complement the behavior system framework. A more general treatment of play that includes a discussion of problems caused by the confusion of cause and function is given by Martin & Caro (1985) and Burghardt (2001).

The first important idea was expressed by Lorenz (1956): "It seems characteristic of 'play' that instinctive movements are thus performed independently of the higher patterns into which they are integrated when functioning 'in serious'" (p. 635). In other words, the motor mechanisms are activated independently of an activation of the central mechanisms. Insofar as Kruijt's (1964) analysis of junglefowl development is correct, this is precisely the case in newly hatched chicks. As the animal grows older and causal factors for the central mechanisms grow stronger, the independence of motor and central mechanisms decreases, and one might expect the frequency of play to decline, which generally happens. Nonetheless, the analysis of the hunger system suggests that, even when particular motor patterns such as pecking and/or ground scratching become integrated into the system, these same movements can occur independently, especially when the causal factors that activate the central mechanism (i.e. the level of hunger) are weak.

Similar results have also been seen in other species. Lorenz (1956) described the behavior of a young raven, *Corvus corax*, that showed a wide array of 'playful' movements toward a strange object when not hungry, but that immediately tried to eat such an object if it was hungry. Likewise, Schiller's chimpanzees showed a playful manipulation of objects, especially when not hungry. The motor patterns of the raven and the chimps under these circumstances could be recognized as being similar to motor patterns belonging to various adult behavior systems.

Once various behavior systems have developed, it may be that play ceases. This, of course, is not true in many species. Morris (1956) suggested that play occurs when central mechanisms are switched off: "The mechanisms of mutual inhibition and sequential ordering mechanisms are not switched on and as a result there is no control over the types and sequences of motor patterns in the usual sense" (p. 643). Switching off central mechanisms would effectively return the animals to a very early stage of development, in which the appearance of play would again be expected. A more elaborate version of this idea was suggested by Baerends-van Roon & Baerends (1979) and was based on their observations of kittens. They proposed that, in cats at least, a central play mechanism exists that, when activated, inhibits other central mechanisms, allowing 'play' to appear. Thus, species-typical patterns of play can be understood as being due to a differential inhibition of central mechanisms. Further, when play occurs, its causation remains the same as Lorenz originally suggested: the independent activation of motor mechanisms.

DEVELOPMENT OF INTERACTIONS AMONG BEHAVIOR SYSTEMS

A basic tenet of ethological theory is that various behaviors of an animal – and often the most interesting ones – are the expression of the activation of not just a single behavior system, but of the interaction of two or more systems that are activated simultaneously. This conflict hypothesis was proposed by Tinbergen (1952) and has been discussed by Kruijt (1964), Baerends (1975), and Groothuis (1994). Two major studies have directly addressed the development of interactions among systems – those of Kruijt (1964) in chickens and Groothuis (1994) in black-headed gulls. I will restrict my discussion here to the behavior of chickens.

Kruijt's (1964) results show that the major behavior systems of escape, aggression, and sex develop in chickens in that order. Further, activation of a system already developed inhibits the expression of systems that are just beginning to develop. Thus, a young chick that shows frontal threatening and jumping to another chick may immediately stop this early aggressive behavior if it bumps into the other too hard. As the chicks grow older and the causal factors for aggression become stronger, however, such escape stimuli no longer stop aggressive behavior. Rather, attack and escape begin to occur in rapid alternation, and various irrelevant movements start to appear during fighting. Likewise, early sexual behavior is immediately interrupted if either the attack or escape system is activated, but later, behavior containing components of attack, escape, and sex can be seen simultaneously. As seen above, there is evidence that the basic organization of these major systems is formed prefunctionally and that their expression merely requires a sufficiently high level of causal factors. The gradual appearance of more complex interactions can be interpreted as reflecting changes in the strength of causal factors (i.e. motivational changes) rather than changes in the connections among central mechanisms (i.e. developmental changes).

The fact that another member of the species is the adequate stimulus for activating the escape, aggression, and sex systems means that all these systems must normally be activated when a conspecific is present. Kruijt (1964) pointed out that the precise state of activation of these systems at any moment depends on the previous history of the male and on the appearance, distance, and behavior of the other bird. He suggested that the appearance of smooth, typical adult courtship behavior depends on an increasing activation and mutual inhibition of the attack and escape systems, and that the relationship between attack and escape is stabilized by the activation of the sexual system. He posited a stabilizing factor in order to explain why the adult animals do not constantly switch quickly from performing one type of behavior to performing another.

The stabilizing influence of sex on the agonistic systems of escape and aggression is not merely a consequence of increasing hormone levels as the animals grow older. Experience also plays a major role. Kruijt (1964) found that junglefowl males reared from hatching in social isolation for more than nine months showed serious and apparently irreversible abnormalities in their courtship and sexual behavior. To a large extent, these abnormalities could be characterized as switching too quickly among escape, aggressive, and sexual behavior. In other words, the stabilizing influence of sex was present only after the experience of the sort that would occur during normal early development. Kruijt also found that as little as two and a half months of normal social experience immediately after hatching was sufficient to obviate the effects of subsequent social isolation for periods of at least 16 months. These results are difficult to interpret because during the first two and a half months of life, chicks show essentially no sexual behavior. Thus, they could not be learning anything specific about sexual behavior. Instead, it would seem that the experience a chick gains during normal encounters early in life provides the information necessary for it to stabilize its agonistic systems, and that normal sexual behavior can only occur if the agonistic systems are already stabilized.

Baerends' results, mentioned above, also support this interpretation. Their kittens that were raised in isolation from peers showed either unrestrained attack or complete avoidance when confronted with a normal kitten, and this pattern was also seen later in a sexual situation. Rhesus monkeys that were raised in isolation from peers also showed inadequate sexual behavior in adulthood (Harlow & Harlow, 1962). However, in both the cats and the monkeys, a particularly 'good' partner was able to compensate for the behavioral deficiency in the isolation-reared animals (Harlow & Suomi, 1971; Novak & Harlow, 1975). The description of these encounters suggests that the sexual behavior system itself had not developed abnormally, but that abnormal fear or aggression interfered with the performance of sexual behavior. The conclusion that can be drawn from these studies is that well-integrated interactions among behavior systems are necessary for the normal, well-coordinated behavior we see in adult animals, and that functional experience at various ages is necessary for such integration to occur. How a stabilizing influence develops has not been studied. We have seen that some of the experiences of a normally raised young chick, such as bumping into other chicks or being pecked at as a result of pecking another chick, are not necessary for normal aggressive behavior to develop. Such early social experiences, however, might be crucial for developing a normal attack-escape relationship.

TWO SPECIAL BEHAVIOR SYSTEMS: HUMAN LANGUAGE
AND CIRCADIAN SYSTEM

Human Language

In this section, I show how it is possible to consider human language to be a behavior system that is similar in many respects to the behavior systems I have already considered. Human language, of course, is vastly more complex than dustbathing or feeding in chickens, but, as a biological system, both the organization and development of language should share many of the principles governing these simpler systems. There is an enormous literature on language and its development, and only very restricted aspects are considered here.

To begin, it is necessary to identify the building blocks of the language system. What are the perceptual, motor, and central mechanisms comprising the system? For my present purposes, I will start with the perceptual and motor mechanisms that recognize and produce the sounds in a language. I will also restrict my discussion to specific speech sounds (i.e. phonemes) as opposed to other vocal aspects of language such as prosody (Locke, 1993, 1994; Locke & Snow, 1997; Kuhl, 2004). It has been known for some time that human infants as young as one month are able to perceive phonetic distinctions categorically in a similar way to normal adults (Eimas et al., 1971; Eimas et al., 1997). More recent evidence has demonstrated that these perceptual categories can be altered by linguistic experience. For example, in a cross-cultural study of six-month-old American and Swedish infants, Kuhl et al. (1992) found the two groups exhibited a language-specific pattern of phonetic perception to native- and foreign-language vowel sounds. Of particular interest is that these effects of experience are seen by six months of age, that is, before the infant itself begins producing speech sounds. Further, by one year of age, infants no longer respond to speech contrasts that are not used in their native language, even those that they did discriminate at earlier ages (Werker & Tees, 1992). Thus, the perceptual mechanisms responsible for speech perception in infants are both highly structured at birth and highly malleable in that they are shaped, instructively and selectively, by exposure to the linguistic environment (Kuhl, 1994, 2004, 2015; Wijnen, 2013).

Normal infants begin to babble between six and ten months (for review, see Locke, 1993). The initial sounds produced by the infant are species-specific (i.e. are similar in infants raised in different linguistic environments), and include phonemes not found in its native language. As the child grows older, the distribution of sounds comes more nearly to approximate the distribution in its linguistic environment, and the nonnative sounds drop out. The mechanism by which these changes occur involves a process of matching vocal output to the previously developed perceptual mechanisms (templates) by auditory feedback (Marler, 1976; Kuhl, 2004). Some of Kuhl's experiments with human infants are discussed in Chapter 7.

These results for the development of the perceptual and motor mechanisms that recognize and produce speech sounds involve the same problems of modularity, constraints, and processes that we have seen before, especially with respect to the changes that occur in the development of bird song. These parallels have been noted for many years (e.g. Lenneberg, 1967; Marler, 1970b), and continue to provide mutual insights into the development of both systems at both a behavioral and neural level (Hauser, 1996; Snowdon & Hausberger, 1997; Doupe & Kuhl, 1999; Bolhuis & Everaert, 2013).

Speech sounds, however, are only one aspect of normal spoken language. Sounds become combined into words (morphemes: units of meaning), and one can ask whether the phonemes or the morphemes are the basic units of the language system. Words can always be broken down into their constituent sounds, but there is now considerable evidence that infants learn utterances (words or short phrases) as a whole during the first two years with respect to both perception and production (Jusczyk, 1997; Locke & Snow, 1997). It is only later that children are able to break utterances down into smaller sound packets. For these and other reasons, Locke (1994) argues specifically that morphemes, and not phonemes, are the basic building blocks of human language. So, what do words represent?

Birds do not sing randomly. They sing when the appropriate internal and external factors are active. In most cases, this is when the sex and/or aggression behavior systems are activated. Humans speak in comparable circumstances, but the range of circumstances in which humans speak is very much broader. In fact, cognitive psychologists (Shelton & Caramazza, 1999) have proposed that humans possess a semantic system that receives input from spoken and written words (phonological and orthographic input lexicons) and responds with output of spoken and written words (phonological and orthographic output lexicons). Our speech (or writing) thus expresses the state of our semantic system. Of course, our semantic system is much more complex than the sex and aggression systems of songbirds (and much of the field of cognitive psychology is devoted to understanding the organization of the semantic system and the mechanisms of lexical access to it), but it seems certain that many of the principles of organization and development are similar. In the present context, some of Shelton & Caramazza's (1999) conclusions are particularly interesting. They reviewed studies of language processing following brain damage and found results that "broadly support a componential organization of lexical knowledge – the semantic component is independent of phonological and orthographic form knowledge, and the latter are independent of each other" (p. 5). In my terms, their language system can be considered to have a central semantic mechanism with perceptual mechanisms for recognizing words and motor mechanisms for producing words (cf. Berwick et al., 2013).

Here, I want to make a few comments about what I have just called the central semantic mechanism. With respect to the concepts I developed in Chapter 2, the central semantic mechanism would consist of a collection of representations that can be activated by the perception of spoken or written words; and activation of particular representations could lead to the production of spoken or written words. However, the central semantic mechanism is also a more general-purpose entity. It can be accessed by perceptual mechanisms and can activate motor mechanisms of most of an animal's other behavior systems. Further, the semantic system is much more than a collection of representations of words. It is the place where many memories are stored and also the place where many 'cognitive' tasks such as thinking and reasoning take place. As such, it is a much more complex behavior mechanism than the hunger or sex or aggression behavior mechanisms in humans, and studying it is basically the province of cognitive psychologists. Such a semantic mechanism may, however, exist in other animals, and it is important to realize that 'meaning' can exist without language. But, can language exist without meaning? Depending on one's definition of language, the answer is yes.

I have still not discussed what some consider an important component of language and others consider the essence of language itself: syntax and grammar. Words can be combined into sentences, and it is to this level of organization that the concept of 'language instinct' (Pinker, 1994), or language itself has been applied (Berwick *et al.*, 2013). The operation of grammatical structures is not normally apparent until sometime after the age of two years, when words become recombined into novel utterances that follow particular rules (Locke & Snow, 1997). One set of rules, called generative grammar, or universal grammar, was proposed by Chomsky (1965; 2012). These rules have been reasonably successful in describing the types of sentences produced by native speakers of English, as well as of many other languages. However, Chomsky's ideas have been extremely controversial, particularly with respect to how these rules develop in the child, and especially whether specific kinds of linguistic experience are necessary (e.g. Tomasello, 1995; Lieberman, 2016; Ibbotson & Tomasello, 2016). The details of this controversy need not concern us here, except to say that many of the issues are the same as we have already met in describing the development of grooming (Berridge, 1994) and dustbathing (Larsen *et al.*, 2000) sequences, some of which are considered further in Chapter 6.

These considerations imply that the human language system comprises three basic sets of components at two major levels of organization, and that these components develop largely independently. The sensory-motor components correspond to the perceptual and motor mechanisms depicted in Figure 1.1 (with additional connections between them and the central mechanisms), whereas the semantic (meaning) and syntax components correspond to two separate central mechanisms. This conception is basically the same as the conception of Berwick *et al.* (2013) depicted in Figure 2.11.

This general conception is also supported by the results of studies of deaf children. For example, deaf children born to deaf parents who communicate using sign language do not babble vocally; rather, such children babble with their hands (Petito & Marentette, 1991). Manual babbling occurs at about the same age that vocal babbling occurs in children with normal hearing who have been raised in a vocal environment. Further, the development of sign language proceeds in much the same way as the development of vocal language with respect to both structure and use. Goldin-Meadow (1997) found that the same general rules apply and that they appear at the same age. These results all suggest that the language system can use auditory-vocal units or visual-manual units equally well. Studies of the neural organization of language are also consistent with this interpretation. Newman *et al.* (2015), for example, using functional MRI, showed that the brain systems engaged by sign language are the same as those engaged by normal spoken language. Similar gestural sequences by nonsigners engaged other brain structures.

One can ask, finally, whether this conception of human language as a behavior system actually furthers our understanding of language and its development. I think it does in at least two important ways. First, by breaking the system up into its components, the study of the pieces becomes more tractable. There has already been considerable success in comparing the development of bird song and human speech (Doupe & Kuhl, 1999; Bolhuis & Everaert, 2013), and the development of grooming sequences may provide a useful model for some aspects of the development of syntax (but see p. 284). Further, insofar as these components are the 'natural' pieces of the system, it becomes easier to understand how the system could have evolved (Pinker, 1994; Hauser, 1996; Bolhuis *et al.*, 2014). The evolution of language is discussed in Chapter 8. A second important reason is that development of all three sets of components requires both functional and nonfunctional experience and involves the same problems of modularity, constraints, and processes that have appeared before. Solutions to these problems in one system should easily generalize to other systems.

Circadian System

The circadian systems in mammals and birds appear to develop prefunctionally, though there are very few studies addressing this question. Davis & Reppert (2001) provide a review of the development of the circadian system in mammals. In baboon (*Papio* spp.) and human fetuses, the suprachiasmatic nuclei (SCN) in the forebrain are responsive to visual stimulation by mid-gestation, but any physiological or behavioral circadian rhythms observed before birth have been shown to be due to the mother's circadian system. However, maternal entrainment of rhythms in the fetus and the neonate appears to be important for the development of autonomous circadian rhythms in the infant as it matures. Preterm baboon and human births allow investigators to study developing rhythms without the influence of the mother's rhythms. In both preterm and term infants, circadian rhythms are not apparent until sometime after birth, but preterm infants exposed to daily light cycles tend to develop circadian rest-activity and temperature rhythms earlier, with respect to post-conception age, than term infants (reviews in Rivkees, 2003 and Mirmiran *et al.*, 2003). These results imply that non-endogenous factors can influence the normal development of the circadian system in baboons and humans.

Once the circadian system is functional, the question arises of how it gains control over other developing or developed systems. To begin, the only method of determining that the circadian system is functional is to observe its effects. The earliest physiological rhythms to appear are usually body temperature and heart rate, and the activity-rest rhythm is usually the earliest behavioral rhythm. Temperature and/or activity are often used as markers for a circadian rhythm, but whether the SCN and the systems regulating temperature and activity are connected prefunctionally is not known.

Sleep is another system that develops a daily rhythm very early postnatally, and at least in humans, sleep itself develops both pre- and postnatally and interacts with the development of the circadian system (Mirmiran *et al.*, 2003). It is important to realize, however, that sleep and rest are not synonymous, and a rest-activity daily rhythm can appear independently of a sleep daily rhythm. Light-dark cycles, feeding schedules, and other disturbances all affect the synchronization of sleep and the circadian cycle, but, except for the effects of light-dark cycles, there are few systematic studies of the effects of these variables. Nonetheless, sleep and the circadian system are independent systems as can be seen in the forced desynchrony protocol discussed in Chapter 4: the 24-hour circadian rhythm of temperature continues at the same time as the sleep-wake cycle is forced to 28 hours.

In many ways, the central circadian pacemaker has properties similar to the central coordinating mechanisms (CCMs) of behavior systems: there are various inputs that affect it (entrainment factors) and it controls (i.e. is one of the causal factors influencing) a number of other CCMs (activity, feeding, dustbathing, etc.). It also affects cognitive performance and performance in instrumental learning situations (Terman, 1983). Light is the normal *Zeitgeber* (entraining stimulus) for the SCN, but other factors can affect entrainment as well. There has been controversy over whether a conditioned stimulus can act as a *Zeitgeber* (see Daan, 2000), but there is considerable evidence that food and feeding schedules can (Stephan, 2001;

Mistlberger, 2011). The feeding studies raise a number of important issues including the question of whether there is a separate food-entrainable oscillator (FEO – see Figure 4.4). Studies on Norway rats suggest that the light-entrainable SCN and a putative FEO control activity and feeding independently, while similar studies on Syrian hamsters suggest a single oscillator is sufficient (see Mistlberger, 2011, Fig. 2). Such species differences make structural and causal generalizations difficult.

The output of the central circadian pacemaker generates daily rhythms in a host of physiological and behavioral functions, some of which have been mentioned in Chapters 3 and 4. Whether, and how much, individual experience plays a role in the expression of these rhythms is a question seldom asked. It is usually assumed that the clock 'automatically' controls its various functions. However, we saw in Chapter 4 that rhythmicity in feeding and sleep in chicks declined to almost random over the course of a week under constant conditions, even though a robust daily rhythm of dustbathing continued under the same conditions. So, the clock is still running even though it no longer influences much observable behavior. In humans, the sleep-wake cycle is independent of the circadian cycle of temperature. In all these cases, if a normal light-dark cycle is reintroduced, circadian rhythmicity is reinstated. Factors responsible for connecting or detaching motor mechanisms to/from the clock remain to be investigated.

One area in which experimental evidence is available is in studies of time-place learning (review in Mulder *et al.*, 2013). Early studies of orientation in honeybees and starlings found evidence for a 'sun compass', which implied an internal timing mechanism that could be continuously consulted and that allowed the animal to keep a constant geographical bearing. More recent studies show that circadian oscillators can provide phase information to other behavior mechanisms, which allows an animal to remember when (time of day) an event occurred. Experiments with rats and mice have shown that the animals can learn that particular places can have different values at different times. For example, in one experiment, using a three-arm maze, a mouse learned that at 9 o'clock, it would get a mild foot shock in goalbox 1, but food without shock would be available in goalboxes 2 and 3. At noon, goalboxes 1 and 3 would have food available without shock, but goalbox 2 would give it a shock, and at 3 o'clock, it would get a shock in goalbox 3. A session was considered to be performed correctly if the mouse first entered the two non-shocked goalboxes and either avoided the shocked goalbox or entered it last. The mice readily learned this task, reaching an average performance of about 80% correct in just five days. With proper control procedures in place, the authors were able to conclude that the mice were using a circadian strategy to learn (Van der Zee *et al.*, 2008). All these results show that the circadian system has a ubiquitous influence on behavior, and many of its effects are yet to be discovered.

6

Ontogeny of Structure
Some Principles of Development

Conrad Waddington (1905–1975). From *Genetical Research*, 1976, **27**, 1.

The process of development is extremely complex, to a large extent because so many interdependent events occur simultaneously (Kuo, 1967; Hogan, 1978). Unfortunately, it is not possible to comprehend all the important variables at the same time, so that various sorts of distinctions and simplifications must be made in order to further our understanding (cf. Bateson, 1999). The basic simplification that I have made in this book is that of describing behavior in terms of motor, central, and perceptual mechanisms and the connections among them. These mechanisms are conceived of as structural units of behavior of a particular magnitude and complexity. Development is viewed as changes in these underlying behavior mechanisms and their connections during the lifetime of an individual. This conception makes it possible to see a clear analogy between behavior development and the development of specialized cells and tissues in the embryo (Waddington, 1966; Slack, 1991). I begin this chapter with a brief presentation of some basic embryology and genetics as background for understanding the concepts of innate and prefunctional. I then outline some aspects of the development of the nervous system and of perceptual and motor mechanisms, and discuss the problems of modularity, constraints, and developmental processes. I follow with consideration of the special role of early social experience on the development of later behavior, and finally discuss some functional aspects of development.

GENETIC AND ENVIRONMENTAL INTERACTION IN DEVELOPMENT

Much of the information in this section has been derived from Waddington (1966). Although written more than 50 years ago, Waddington was far ahead of his time in realizing the importance of molecular genetics for understanding developmental processes. His framework still reflects that of contemporary studies of embryology. His analysis of gene activities, while greatly oversimplified in light of more recent research, is generally correct. I am using it here because it makes very clear the problems with most people's concept of innate. A very readable exposition of recent developments is provided in the book by Jablonka & Lamb (2014).

Regulation of Gene Activities

Some genes, the structural genes, are concerned with the production of enzymes, proteins that are the basic building blocks of all cells. When the structural genes are active, molecules called messenger RNA (ribonucleic acid) are formed. The messenger RNA is transported to ribosomes that are located in another part of the cell, and, in the ribosomes, the messenger RNA is responsible for the production of enzymes (see Figure 6.1).

It is important to realize that early in development, all cells contain all the genetic information. Nonetheless, even from the very earliest stages of development, some cells are destined to become nerve cells, others to become muscle cells or liver cells, and still others to become part of the intestine. This must mean that what a particular cell becomes depends on which structural genes are active. Thus, an essential problem in understanding development is to determine how and when particular genes are switched on and off. In some cases, as in the bacteria, it is known that the structural genes are controlled by an operator gene that in turn is controlled by a regulator substance. The regulator substance, by interacting with the operator gene, can switch the structural genes on or off. The story of how this system works belongs in a course on molecular genetics, but it is sufficient for my purposes to know that structural genes can be turned on or off when various substances called inducers or repressors are present in the nucleus of the cell. Many of the inducer and repressor substances are produced in the cell itself, but others are produced in adjacent cells and these can have a determining effect on what a particular cell becomes. Early in development, this process is called 'embryonic induction'.

From the point of view of the genes, the repressors and inducers are part of the environment. Thus, in a very real sense, it is the environment that determines whether a cell becomes a nerve cell or a liver cell. For the embryologist, this is not a surprising conclusion because the phenomenon of embryonic induction has been known for many years. Embryonic induction is a process in which one part of an embryo influences another part and makes it develop into a tissue or organ different from what it would otherwise have become. For example,

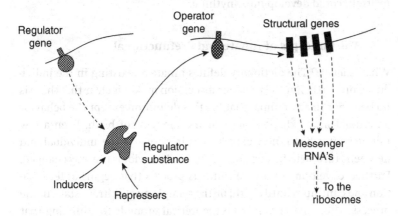

Figure 6.1 Regulation of gene activities in bacteria. Explanation in text. From Waddington, 1966.

a cell that would have become a piece of skin if it had been left undisturbed can be made to become part of the brain if it is moved to a particular position in the developing embryo. In its new position, the cell is being influenced by different inducers and repressors and, thus, different sets of structural genes are switched on and off.

From these and similar considerations, the only conclusion one can make is that *every* developmental process is influenced by environmental factors. It is true that many of these 'environmental' factors originate within the developing organism itself, but many other controlling factors originate in the external world of the organism. An especially tragic example is provided by the deformed children who were born without arms and legs to mothers that had taken the drug thalidomide during pregnancy (Speirs, 1962). A similar recent example is the microcephalic children born to mothers infected with the Zika virus. In recent years, much attention has been paid to such non-genomic effects on development, including incubation temperature and maternally derived hormones (see Crews & Groothuis, 2009).

Finally, although environmental (i.e. non-genetic) factors influence every developmental process, it would be wrong to conclude that the adult form of an organism is determined exclusively by environmental factors. Inducers and repressors are only able to switch on or off structural genes that actually exist. No environmental factors can make an elephant out of a fertilized mouse egg. Development is a continuous process of genetic and environmental (non-genetic) factors interacting with the developing organism. Neither set of factors, by itself, could develop into anything.

The Concepts of Innate and Prefunctional

What is *innate*? The dictionary defines innate as existing in the individual from birth. But the previous discussion makes it clear that there is no behavior in any animal that fits this definition except the behavior an animal may exhibit exactly at the moment of birth. Even a few minutes after birth, new experiences impinge on the individual and new genes are switched on and off, and its behavior is thereby changed. Further, development is a continuous process that begins at fertilization and continues until death; birth is simply one arbitrary stage in the process. There are certainly no theoretical grounds for thinking that developmental processes are any different before or after birth. Nonetheless, the 'nature/nurture' controversy continues to thrive.

In his early papers, Lorenz (1937) postulated that behavior could be considered a mixture of innate and acquired elements (*Instinkt-Dressur-Verschrankung*: intercalation of fixed action patterns and learning), and that analysis of the development of the innate elements (fixed-action patterns) was a matter for embryologists. In reaction to Lehrman's (1953) critique of ethological theory, Lorenz (1965) changed his formulation of the problem, and argued that the information necessary for a behavior element to be adapted to its species' environment can only come from two sources: from information stored in the genes or from an interaction between the individual and its environment. This formulation also met with considerable criticism from many who insisted that development consisted of a more complex dynamic. Gottlieb (1997) discusses many aspects of this debate, as does Bolhuis (2009); Bolhuis & Hogan (1999) have republished some of the original papers. Here, I will mention only two important issues.

First, Lehrman (1970) pointed out that he and Lorenz were really interested in two different problems. Lehrman was interested in studying the effects of all types of experience on all types of behavior at all stages of development, very much from a causal perspective, whereas Lorenz was interested only in studying the effects of functional experience on behavior mechanisms at the stage of development at which they begin to function as modes of adaptation to the environment. Thus, Lehrman used a causal criterion to determine what was interesting to study, while Lorenz used a functional criterion. Both these viewpoints are equally legitimate, but Lorenz' functional criterion corresponds to the way most people think about development (Hogan, 1994a).

The second issue is that even behavior patterns that owe their adaptedness to genetic information require interaction with the environment in order to develop in the individual. As Lehrman (1953, p. 345) stated: "The interaction out of which the organism develops is not one, as it is so often said, between heredity and environment. It is between *organism* and environment! And the organism is different at each stage of its development." This view has been developed by Oyama (1985) into what is now called Developmental Systems Theory (see Oyama et al., 2001), and has gradually been adopted by most students of behavior development. But, perhaps a better formulation of these ideas would be: *the organism develops in interaction with its genes and its environment, and the organism is different at each stage of its development*. This means that the same genetic activity or the same environmental input can have different effects at different stages of an organism's development. It is also necessary to remember that genes, in combination, only 'code' for the production of enzymes, and that there is no such thing as a gene *for* a particular behavioral trait. Development is a complicated, multi-layered process.

These considerations, however, have not dispensed with the idea of innate. Some kinds of behavior seem to have a different relation to experience than other kinds. Why do all herring gulls (and many other bird species) retrieve eggs in the same way? Why are green eggs preferred to brown? Why do broody hens sit on eggs? If we ask the question 'why' in a developmental context, we can show that environmental factors are essential for all these behaviors to develop. But, it is also true that the essential environmental factors are not functionally relevant. A gull does not need to practice egg retrieval or have experience with green eggs before it makes its response. Such behaviors can be considered to develop 'prefunctionally' (Schiller, 1949/1957). In fact, one could define innate behavior as behavior that develops without functional experience. I prefer the word *prefunctional* to the word *innate* because the latter has too many additional meanings. It is logically consistent to talk about behavior development that is prefunctional (or innate) versus behavior development that is learned when the criterion is the absence or presence of functional experience. And, as we have seen (p. 134), prefunctional behavior still presents interesting developmental problems that can be investigated in a causal framework. Labeling a behavior as prefunctional or innate by no means answers most of the questions concerning its development. And, as we shall see, the same is true of learned behavior.

It is important, however, to see some of the difficulties in using a functional definition. Perhaps the most important of these is that the function of a behavior is not always obvious. For example, if the function of pecking is viewed as being to provide nutrition, pecking becomes integrated into the hunger system of the chick prefunctionally; if the function of pecking is viewed as being to bring about ingestion, then pecking becomes integrated into the hunger system through functional experience. In either case, the causal process is the same. Similar problems arise when there are alternative routes to reaching the same end, as in the development of the oblique posture in the black-headed gull. A related problem is that the function of behavior can change over the course of time. Sometimes this change is due to changes in the environment, and sometimes to changes in the behavior mechanisms themselves. This means that, at best, the concept of prefunctional is only relative. It can only describe situations with respect to the particular function that the investigator has in mind.

DEVELOPMENT OF THE NERVOUS SYSTEM

Brown *et al.* (1991) divide brain development at the cellular level into four major stages: (1) genesis of nerve cells (proliferation, specification, and migration), (2) establishing connections (axon and dendritic growth, and synapse formation), (3) modifying connections (nerve cell death – apoptosis – and reorganization of initial inputs), and (4) adult plasticity (learning and nerve growth after injury). Stages 3 and 4 are the most relevant to behavioral development. Stage 4 also includes adult neurogenesis, which is now known to be widespread in many animals, including humans (Ming & Song, 2011).

During fetal development, many more nerve cells are formed than will be found in the adult brain. These nerve cells all send out axons and establish connections with target cells (other neurons and muscle cells), but a large proportion of them die before the synapses become functional. The mechanisms underlying this process involve electrical activity in the nerve cells and their targets, but they are still not fully understood (Oppenheim, 1991; see also Kristiansen & Ham, 2014). It is thought that neuronal death may serve to eliminate errors in the initial pattern of connections (Buss *et al.*, 2006). The axons of the cells that remain are often found to have more extensive branches and to contact more postsynaptic cells than they will in the adult. The mechanisms that bring about axonal remodeling – that is, the elimination and reorganization of these terminal branches – involve activity in the neurons. In brief, it has been shown that specific spatial and temporal patterns of electrical activity in both the nerve cells and their target cells are necessary for functional connections to form between them: "cells that fire together wire together" (Shatz, 1992, p. 64).

The process of axonal remodeling occurs both pre- and postnatally, and it is essentially irreversible. Once the axons have established functional connections with other neurons or muscles, those connections appear to be a permanent part of neural organization. The mechanisms that are responsible for adult plasticity include facilitation or inhibition of synaptic transmission, the growth of dendritic spines, which presumably correlates with the formation of new synapses, as well as formation of new neurons.

The work of Hubel and Wiesel established that visual stimulation plays a vital role in the development of the mammalian visual system (for reviews of the early work, see Blakemore, 1973 Wiesel, 1982). They showed, for example, that normal development of the connections between cells of the lateral geniculate nucleus (a relay body between the optic nerve and the cortex) and the visual cortex in the cat requires binocular visual stimulation soon after the kitten's eyes open. Allowing a kitten to see with only one eye at a time during the critical period results in most cortical cells being responsive to stimulation from one eye only, whereas binocular stimulation results in most cortical cells being responsive to stimulation from both eyes. These results were interpreted in terms of the eyes competing for control of cells in the cortex and are an example of axonal remodeling. These results were important because they showed that the organization of a sensory system was actually driven by stimulation from the environment. They also provide a model for how the perceptual mechanisms under-lying bird song learning and filial and sexual imprinting might develop (Bateson, 1987; Bischof, 1994; Bolhuis & Eda-Fujiwara, 2003).

The neural activity responsible for axonal remodeling in the visual cortex is triggered by stimuli originating in the environment after the kitten is born and has opened its eyes. More recently, other investigators have asked whether neural activity is also necessary for neural connections to form *in utero*, and, if so, how this activity is instigated. Shatz (1992) and her collaborators, for example, have looked at axonal remodeling in the lateral geniculate nucleus of the cat, which occurs before birth. They found that the same kind of action-potential activity is necessary for developing normal connections from the retina to the lateral geniculate, as is later necessary for normal connections to form in the cortex. Rather than being instigated by stimulation from the external world, however, the neural activity was caused by patterns of spontaneous neural firing. How these waves of activity are generated is currently being investigated (e.g. Blankenship & Feller, 2010).

These two cases of axonal remodeling illustrate the difference between development based on functional experience (organization of the visual cortex) and development that occurs prefunctionally (orga-nization of the lateral geniculate nucleus). What is important in the present context, however, is that the mechanisms for synaptic change are the same before and after birth, and it is irrelevant for the connec-tion being formed whether the neural activity arises from exogenous or endogenous sources. In fact, the same connection can be formed in either way. Some behavioral examples will be used to illustrate this point in the section on developmental processes.

The development of perceptual and motor mechanisms illustrates most of the problems encountered in the development of behavior systems in general: modularity, constraints, processes, and species differences. With respect to perceptual mechanisms, the postulation of a template or schema implies a kind of modularity in the brain in that a certain part of the brain is preassigned a specific function. There are constraints on the kinds of experience that can affect development and on the age or stage of development at which this experience can be effective. There is also the problem of developmental processes. Are the effects of experience direct or indirect? Is mere exposure sufficient, or is some sort of reinforcement necessary? How stable are an animal's preferences? Finally, there is much variability among species in the role played by experience and the types of constraints encountered. These problems are all interrelated and I discuss them below and again in Chapter 7 in the context of learning.

The development of motor mechanisms illustrates the same problems. The emphases are somewhat different in that most motor mechanisms (with the notable exceptions of bird song and human language) and even many motor sequences (such as dustbathing in chicks and grooming in rats) develop prefunctionally, whereas the development of almost all perceptual mechanisms is directly influenced by functional experience. Further, when functional experience is relevant, mere exposure is often sufficient for the development of most perceptual mechanisms, whereas some sort of reinforcement is usually necessary for the development of motor mechanisms and motor sequences. These topics are also considered again in Chapter 7.

Another similarity between perceptual and motor mechanisms is the existence of different functional levels of organization. In the case of perceptual mechanisms, the evidence supports their existence in at least three levels: feature recognition, object recognition, and function recognition (see Figure 2.6). Feature recognition mechanisms discriminate among various sizes, shapes, colors, smells, tastes, and so on. This is presumably the level at which the gustofacial reflex is organized in human infants. The reason for distinguishing between object recognition and function recognition is that objects with similar properties, such as food crumbs and aquarium gravel, mealworms and cinnabar caterpillars, or mice and shrews, are easily recognized (after appropriate experience) as being food or nonfood, whereas other objects with greatly disparate properties, such as grain, insects, fish, and the leaves of various plants are easily included in the food category. Similarly, a mockingbird mimics very accurately the songs of many different species (Baylis, 1982). Therefore, it must have a number of perceptual mechanisms for recognizing each different song. Further, the various songs that the mockingbird has learned are combined into an overall song that has species-specific characteristics (Logan, 1983), so there must also be an additional perceptual mechanism at a higher level of organization. Ten Cate's (1994) imprinting results with zebra finches tell the same story.

In the case of motor mechanisms, there is the level of motor primitives (see p. 220), the level of the individual motor pattern, and the level of motor pattern coordination. The results reviewed above for dustbathing, grooming, and prey catching all provided evidence that the motor pattern and motor pattern coordination levels are independent. In many ways, the level of motor pattern coordination is especially interesting because it provides the basis for the temporal patterning, or syntax, of functionally related behaviors. As Lashley (1951) pointed out, all skilled acts, including human language, seem to involve the same problems of serial ordering, and this topic is also discussed below and again in Chapter 8.

In one other discussion of motor development, Thelen (1995) proposes a dynamic theory in which "repeated cycles of perception and action can give rise to emergent new forms of behavior without preexisting mental or genetic structures" (p. 93). She opposes her theory to one in which the brain structures and movements appear when the appropriate level of maturity is reached. Although I would maintain that there are always preexisting structures, in some ways, her approach is similar to the one taken in this book. For example, we have seen that feedback from the performance of a song or display can affect the form of future performances. What is very different is that her basic units of action lie at a lower functional level of organization ('motor primitives') than I have considered here, and she explicitly considers factors that I call prefunctional and that I do not analyze further. With respect to the latter, her analysis is similar to Kuo's analysis of walking in chicks discussed earlier.

Modularity

An important assumption made in Chapter 5 is that particular parts of the central nervous system subserve particular functions, and that, by the time behaviorally interesting events are occurring, these parts, or modules, are preassigned. This means that, at the particular stage of development under consideration, the range of possibilities for further development of a particular behavior mechanism is so restricted that only special (i.e. already determined) kinds of experience can have a developmental effect on that mechanism. In practice, this means that, by the time of birth (or hatching), the central nervous system is already highly differentiated, with the general organization of pathways and connections already determined. By this stage of development, reversing the functions of major parts of the brain is generally impossible in the sense just discussed. Under these circumstances, it seems justified to speak of the song-recognition perceptual mechanism or the ground-scratching motor mechanism or the aggression central mechanism as prefunctionally developed units of behavioral structure subject to further (but quite restricted) differentiation on the basis of subsequent experience.

It should be realized, however, that, if we follow the development of any behavior mechanism backward in time, we can always find a stage in which the nerve cells making up the behavior mechanism could have subserved a different behavior mechanism under somewhat different conditions. If we go back still further, we will find a stage where the cells could have become something other than nerve cells, and so on. At the time of birth – an arbitrary time I have chosen for convenience – a particular set of nerve cells may have differentiated to the point where, if they survive, they will be the cells that mediate mate recognition, and in this sense, they are preassigned that function. But they are preassigned only from the point of view of future development.

Constraints: Irreversibility and Critical Periods

Constraints on development actually arise as an interaction between the structures (modules) available at any given time and the processes that can lead to changes in those structures. Two such constraints that are ubiquitous in discussions of development will be considered here: irreversibility and critical periods.

Insofar as behavior mechanisms can be regarded as preassigned, they illustrate the problem of the irreversibility of development. When Waddington (1966) discussed the question of whether the differentiation of cells is irreversible, his answer was, "it depends". It depends on what cell, in what animal, at what stage of development, and so on. This is already an important point because similar reasoning shows that it is nonsense to ask a question such as, 'is imprinting irreversible?' One can only begin to answer such a question after specifying the species, the particular imprinting procedures, the state of development, and so on.

More importantly, Waddington specified some of the processes that are responsible for the irreversibility of cell differentiation. For example, some or all of the genetic material may have been 'used up' or may have otherwise disappeared in the course of the development of the cell, or the genetic material may still be present, but, for various reasons, cannot be accessed. The most frequent reason for irreversibility, however, seems to be that

> development involves such a complicated network of processes that it would be an extremely long and tricky process to unravel them. One could, in theory, take an automobile, dismantle it, and build the pieces up again with a little modification into two motorcycles, but it wouldn't be easy; and it is something like this that we are asking a differentiated cell to do when we try to persuade it to lose its present differentiation and develop into something else (Waddington, 1966, pp. 54–55).

Processes with similar characteristics seem certain to underlie cases of behavior irreversibility.

The best-documented behavioral cases of apparent total irreversibility involve motor mechanisms for bird song, as exemplified by the 'crystalization' of song in the chaffinch (Thorpe, 1961) and the white-crowned sparrow (Marler, 1970a). Even here, however, more recent evidence indicates an important role for auditory feedback in maintaining stable adult song (Brainard & Doupe, 2000). The perceptual mechanisms, or templates, on which these songs are based are probably also fixed irreversibly once they have developed, although here the evidence is somewhat controversial (Baptista & Gaunt, 1997; Nelson, 1997). Many of the courtship and agonistic displays seen especially in birds, such as waltzing in chickens (Kruijt, 1964) or the oblique posture in the black-headed gull (Groothuis, 1994), are probably also fixed irreversibly once they have developed. These cases all involve axonal remodeling and are analogous to the case of cell differentiation, in which the genetic material either disappears or becomes inaccessible during the course of development.

Here, it is useful to emphasize the distinction between perceptual and motor mechanisms themselves, and the various connections that may exist between them. Even though a perceptual or motor mechanism has crystalized, there are still possibilities for alternative pathways among them. The concept of imprinting, for example, implies a change in a perceptual mechanism as a result of experience. In some species, such a change may be irreversible, but subsequent experience may lead to additional pathways being formed between other perceptual mechanisms and the sexual behavior system, and these new connections may mask the original imprinting. A rather difficult experimental analysis would be necessary to investigate this possibility. We have seen, however, a case such as this on the motor side of the hunger system in chickens. An original connection between pecking and ground scratching was masked, but not destroyed, by later experience.

The most common reason that behavior changes are apparently irreversible is probably the same reason that most cell differentiation is irreversible; so many events would have to be undone (or compensated for) that change becomes almost impossible. A very simple case where changes could still be made was training a kitten to catch fish after it had already learned to catch mice (Baerends-van Roon & Baerends, 1979). Here, there were two problems. One was an indirect, motivational problem; fear of water inhibited any attempt to catch the fish. Once the fear of water could be overcome, the kitten faced a direct, developmental problem: rearranging motor mechanisms in a different sequence. In this case, rearrangement was possible, although with some interference from the original learning.

A more complex case is the sexual behavior of male junglefowl raised in social isolation. Here, subtle aspects of the integration of the attack and escape systems seem to be permanently missing. Because this integration plays a determining role in permitting sexual behavior to occur, these effects of social isolation are effectively irreversible, even though the copulatory motor patterns remain intact. The fact that some consequences of normal social experience during the first few weeks of life are sufficient for the development of relatively normal adult behavior implies that axonal remodeling-type processes are involved. In effect, it could be that various perceptual neurons are competing for connections with the attack and escape systems, and that a stable attack-escape balance depends on the pattern of connections that finally develops. This is a speculative suggestion, but it does fit in well with what is known about the development of neural connections at earlier stages. Such a suggestion also implies that no new principles of development are required to understand the development of behavior system interaction. It remains to be seen whether some sort of 'therapy' could be devised to cope with this problem – as was possible in the cats and monkeys raised in social isolation – but this is an empirical matter.

The fact that development is not reversible (except as discussed above) means that constraints of various sorts are inherent in developing systems. The most commonly discussed constraint is a *critical* or *sensitive period* that corresponds to the embryological concept of *competence* (Waddington, 1966). In essence, these concepts refer to the fact that the developing system is especially susceptible to particular external influences at particular stages of development. This topic has often been a matter of controversy, especially with respect to the factors responsible for the beginning and the end of the period (see Bateson & Hinde, 1987, for an excellent discussion of sensitive periods). Nonetheless, the previous discussion should make clear that probably all aspects of development are associated with critical periods. At each stage of development, the animal is different from what it was; it is only to be expected that the effects of the 'same' experience will be different in the different stages (Schneirla, 1956; Schneirla *et al.*, 1963). The factors that are responsible for beginning and ending these periods are probably different in every case. Some of these are discussed in the next section.

Processes

What are the processes of behavior development? There is not yet a complete answer to this question, but I think several points are worth making. To begin, it seems very unlikely that the biochemical processes responsible for altering the structure of behavioral mechanisms and their connections are different before and after a particular behavior begins to function. If this is indeed the case, a number of results I have discussed become more easily understandable.

A first example is provided by the results of Groothuis (1992, 1994). He found that the oblique posture in the black-headed gull developed normally when a gull was reared either in social isolation or in large social groups, but that it sometimes developed abnormally when a gull was raised with only two or three peers. One can suppose that under circumstances of social isolation, endogenously produced patterns of neural firing provide the information necessary to develop the normal connections in the motor mechanism responsible for the form of the display, prefunctionally. When peers are present, functional social experience provides the information. Performance of precursors of the display often leads to reactions by the other gulls. These reactions, in turn, provide additional neural stimulation that could interfere with endogenously produced patterns and thus lead to different (abnormal) connections being formed in the motor mechanism. If these connections require repeated stimulation to form, the probability that the average experience of the young gull will be 'correct' is greater in a large group than in a small group, where the effects of the behavior of one abnormal individual companion would be relatively greater (see the results from groups of songbirds raised in isolation from adult song, Marler, 1976). This line of reasoning suggests that functional and prefunctional 'experience' provide alternative routes for the control of behavior system development, a suggestion that can also account for some of the results for the development of the aggression system in chickens reviewed above and for the results of play in several species (Martin & Caro, 1985).

As a second example, one of the interesting aspects of the perceptual phase of song learning in birds are the very large differences among species with respect to what kind of experience is needed for an adequate template to develop. At one extreme, a male cowbird raised in social isolation will develop a normal species' song (King & West, 1977), whereas a chaffinch or white-crowned sparrow raised similarly will develop a song that at best contains only a few species-specific elements (Thorpe, 1961; Marler, 1976). On the other hand, the time at which hearing the species' song is effective for learning is much more restricted in the white-crowned sparrow than it is in the chaffinch. Likewise, if socially isolated males are played variants of the typical species' song, or indeed songs of other species or even pure tones, some species are able to learn only the song of their own species, whereas other species are able to learn a much wider range of sound patterns. Similar species differences are also characteristic of the range of stimuli to which young birds will imprint and the time at which these stimuli are effective (Lorenz, 1935). In all cases, however, a perceptual mechanism develops that serves a species- or *Kumpan*-recognition function.

One way to understand how so many apparently different ways can lead to a similar functional outcome is to suppose that once certain kinds of structural change have occurred in the development of a perceptual mechanism, further change is no longer possible (crystallization, irreversibility). It then follows that the timing of triggering events becomes crucial in determining which events will affect development. In a particular species of songbird, for example, one can imagine that, if genetically triggered events occur in the perceptual mechanism for song recognition before the young bird can hear, then the perceptual mechanism is fixed, prefunctionally, in that species, and posthatching experience can no longer have an effect. If the triggering events are delayed however, the posthatching experience of the bird can provide the trigger. In this way, the same type of perceptual mechanism can be used for either 'innate' or 'learned' song recognition.

The timing of events that trigger irreversible changes in developing behavior mechanisms can also explain some apparent differences between perceptual and motor mechanisms. It is noteworthy that, with the exception of bird song and human language, the motor mechanisms of the behavior systems discussed above all develop prefunctionally, whereas all the perceptual mechanisms require at least some functional experience in order to achieve the normal adult form. This fact might suggest that there are some fundamental differences in the causal factors responsible for the development of perceptual and motor mechanisms. Such a conclusion is unlikely to be true because in both cases, the organization of neural or neuromotor connections depends on particular spatiotemporal patterns of neural activity that can be generated either endogenously or exogenously. Prior to birth, most of the causal factors would be endogenous, although external stimulation may play a role in some cases (e.g. the auditory system in ducks, Gottlieb, 1978). After birth, both internal and external factors could be important. The fact that most of the motor mechanisms I have considered develop prefunctionally very likely reflects the fact that motor mechanisms generally become organized earlier in development than perceptual mechanisms (Hogan, 1994b).

Greenough et al. (1987) proposed that different neural mechanisms might have evolved for brain systems that serve different functions. They distinguish between experience-expectant and experience-dependent information storage based upon the functional requirements of particular brain systems:

> Experience-expectant information storage refers to incorporation of
> environmental information that is ubiquitous in the environment and
> common to all species members, such as the basic elements of pattern
> perception ... Experience-dependent information storage refers to
> incorporation of environmental information that is idiosyncratic, or
> unique to the individual, such as learning about one's specific physical
> environment or vocabulary (p. 539).

They suggest that experience-expectant processes depend on selection
or pruning of overproduced synaptic connections (i.e. axonal remodeling, as discussed above), whereas experience-dependent processes
depend on formation of new synaptic connections.

Greenough *et al.* also suggest that their categories offer a new
view of phenomena that have previously been labeled critical or
sensitive periods. Instead of viewing these phenomena as due "to
the brief opening of a window, with experience influencing development only while the window is open" (cf. Bateson, 1979), their
approach "allows consideration of the evolutionary origins of
a process, its adaptive value for the individual, the required timing
and character of experience, and the organism's potentially active
role in obtaining appropriate experience for itself" (p. 539). This
view proposes a functional explanation for a causal phenomenon,
which leads to all the problems discussed earlier. As Bolhuis (1994)
points out, the development of perceptual mechanisms during
imprinting is not restricted to information that is ubiquitous in
the environment of the developing bird, but the perceptual mechanism would definitely be classified as experience-expectant (cf. van
Kampen, 1996).

It is tempting to speculate that development of behavior mechanisms that involve the elimination and reorganization of terminal axon
branches (axonal remodeling) is essentially irreversible. The critical
period then becomes the time at which the axonal remodeling occurs;
it would depend on all the factors that can affect the timing of the
remodeling. The production of new synapses (and new neurons) continues to occur throughout life and could modulate the structure of
behavior mechanisms after the critical period has passed. This suggestion separates experience-expectant and experience-dependent processes on the basis of mechanism rather than Greenough *et al.*'s
separation on the basis of function. These ideas have some similarities
to proposals by Bateson (1987), but are considerably broader. They are
also congruent with the putative neural mechanism underlying the
critical period.

It is instructive to consider the concept of experience-expectant mechanisms in more detail. If all experience-expectant mechanisms involve axonal remodeling, are all cases of axonal remodeling experience-expectant? As usual, the answer depends on the definition of experience. All structural changes, including those that involve axonal remodeling, require some sort of 'experience' (neurons that fire together, wire together), but neural mechanisms that could generally be considered to be experience-expectant seem to require additional *external* environmental experience to develop normally: visual cortex in cats, aural templates in birds, visual schemas in birds, and aural mechanisms in human speech (discussed in Chapter 7).

A further important factor is that the animal needs to be in the proper motivational state and/or to interact with the source of the information in order to incorporate the environmental input. In his discussion of the critical period for imprinting in the greylag goose, Lorenz (1935/1970, p. 126) noted that "the greylag gosling obviously 'expects' this experience during a receptive period, i.e. *there is an innate drive to fill this gap in the instinctive framework*". Van Kampen (1996) presents extensive experimental evidence on this point with respect to imprinting in chicks, as does Kuhl (2015) with respect to discriminating speech sounds in human infants. The extra experience needed for 'normal' development of the neural mechanism is sometimes very constrained (e.g. songbird species that only learn their own species' song), but in all cases, without the 'proper' experience, the neural mechanism develops abnormally. From the point of view of mechanism, the metaphor of a window opening and closing does seem appropriate, even if the factors causing the window to open and close may be different in every case.

Finally, if it is true that the processes responsible for altering the structure of behavior mechanisms and their connections do not differ before and after particular behaviors begin to function, it follows that the processes responsible for learning are no different from the processes responsible for development in general. In other words, the same structural change can be triggered by different events: for example, by genes or by the experience of 'reinforcement'. The important point is that the change itself cannot be classified as genetic or learned because it could have been triggered either way. In Chapter 7, I discuss how a consideration of the structures that are changing can provide a good basis for classifying different types of learning.

THE ROLE OF EARLY SOCIAL EXPERIENCE

A social behavior system could be defined as one in which the motor patterns belonging to that system are normally directed toward another animal (usually of the same species) and/or in which another animal provides the adequate stimulus for activating the perceptual mechanism(s) belonging to that system. In a discussion comparing the development of social and non-social behavior systems (Hogan, 1994b), I concluded that both kinds of systems develop according to the same rules, and that there appear to be no systematic differences between them. The question then arises why topics such as imprinting and bonding have assumed such an important role in the developmental literature. One answer is that scientists (who are people) are more interested in social behavior than in non-social behavior. Another answer is that this interest is related to Lorenz' original conception of imprinting: "the acquisition of the object of the instinctive behavior patterns oriented towards conspecifics" (Lorenz, 1935/1970, p. 124). In terms of the concepts used in this book, I would say that imprinting refers to the development of a perceptual mechanism (schema) that is responsible for species recognition and that is connected to all (or many of) the social behavior systems in the animal. The reason imprinting is so important is that Lorenz' definition implies that a single perceptual mechanism serves a number of different behavior systems and that this perceptual mechanism develops irreversibly very early in life.

Current evidence from imprinting studies is usually interpreted to mean that the object-recognition mechanisms for filial and sexual behavior develop separately (Bolhuis, 1991, 1996). In fact, Lorenz himself showed in his studies of jackdaws that the objects of the various functional systems he discussed (parental, infant, sexual, social, sibling) might be different and might develop at different periods in the animal's life. Thus, the implication of Lorenz' definition may generally not be true. Nonetheless, the idea that early experience has far-reaching, general effects on later social behavior has remained influential and is supported by a wide variety of evidence (e.g. Bowlby, 1991; Hofer, 1996; Rutter, 1991, 2002). The question is, how do these effects come about if the perceptual mechanisms of the various social behavior systems develop independently?

One suggestion is likely to be widely applicable. Hofer (1987, 1996) and his colleagues studied the processes of early social attachment in young rats and their responses to separation from their mothers. Their results show that separation has extensive effects on the young rats' behavior, similar to (though not as dramatic as) the effects of maternal separation on the behavior of young rhesus monkeys (Harlow & Harlow, 1962; Hinde, 1977). Hofer analyzed these effects into two components. The first involves the formation of an attachment system, which has similarities to the one proposed by Bowlby (1991) for human infants and to the filial system implicated in imprinting studies in birds (van Kampen, 1996). This system develops as the young rat learns the characteristics of its mother; when the infant is separated from her, it shows distress reactions, and shows relief when it is later returned to her. If one substitutes an alternative 'caregiver' for the mother, such as an inanimate object or another rat pup, Hofer's results show that the attachment still seems to function normally.

The novel aspect of Hofer's analysis is the second component; the behavioral and physiological effects that occur during long-term separation from the mother are shown to depend on specific aspects of the mother-infant interaction. Hofer isolated a number of such regulators, including body warmth, tactile and olfactory stimulation, stimulation peculiar to the suckling situation, etc. Many of the specific effects of these factors have been described by Fleming & Blass (1994) and Blass (2015). A real mother provides all the necessary regulators, but alternative caregivers do not. Under such circumstances, various behavioral and physiological abnormalities will develop.

More recent studies have looked at maternal behavior in female rats that were raised artificially and never had contact with a real mother (Lovic & Fleming, 2015). In comparison with normally-reared animals, the artificially-reared animals "are stimulus-driven, impulsive, and have attentional impairments. In turn, these cognitive-behavioral impairments alter maternal behavior even though primary maternal motivational dynamics might not have changed" (p. 45). They suggest that these differences may reflect changes in dopamine levels. These results also support Hofer's analysis.

Kraemer (1992) also interpreted the development of primate social attachment in similar terms to those of Hofer. He pointed out that a young rhesus monkey may become attached to an abusive mother or to a peer, and that such young monkeys can be seen in many ways to have a normally developed attachment system. But such monkeys also develop abnormally in many other ways. Kraemer provided evidence that absence of an adequate caregiver leads to aberrant development of brain biogenic amine systems that are implicated in the control of sensorimotor integration and emotion: "If the attachment system process fails, or if the caregiver is incompetent as a member of the species, the developing infant will also fail to regulate its social behavior and may be dysfunctional in the social environment" (p. 493). It seems likely that similar processes determine the attack-escape relationship with respect to the development of sexual behavior in chickens as discussed above. A young chick can become imprinted on an inanimate object and develop a normal filial system, but the inanimate object does not provide the conditions for normal agonistic behavior to develop.

Lorenz' and Hofer's theories are similar in that both postulate that a representation of the imprinting object or caregiver (perceptual mechanism) is formed early in ontogeny. They differ in that, in Lorenz' theory, the representation controls a number of social behavior systems, and long-term effects are seen because each system matures at its own time in the life of the animal. In Hofer's theory, the representation controls only the attachment system; long-term effects are seen because the object to which the animal is attached provides the necessary conditions for various biochemical and neural changes that are indispensable for normal development of other systems.

CONCLUSIONS

The development of behavior systems is a very complex process, involving intricate interactions of external and internal causal factors with the genes and their products at every stage. Yet the principles involved in this process seem relatively simple. Specific patterns of neural activity are responsible for the formation of the basic behavioral mechanisms and many of the connections among them, probably through the mechanisms of synapse pruning or axonal remodeling. Later stimulation causes the formation of new synapses, which probably underlie the modification of behavioral mechanisms and the formation of new connections among them; new synapses are probably also important in the development of new representations (cognitive structures, see Chapter 7). These neural processes are, in fact, sufficient for understanding a wide range of developmental phenomena including critical periods and irreversibility. Yet understanding the neural mechanisms that determine development tells us nothing about how a particular system will develop in a particular animal. The development of any specific system and of its interactions with other systems will need to be studied in each case.

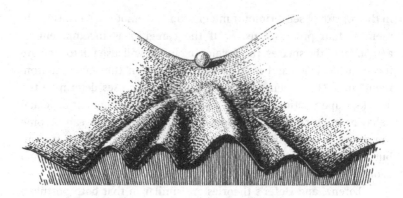

Figure 6.2 The epigenetic landscape. The various regions of a developing embryo have in front of them a number of possible pathways of development, and any particular part will be switched into one or other of these potential paths.
From Waddington, 1966.

One of the remarkable things about development is how normal most individuals become in spite of large variations in the experiences to which they are exposed. Waddington (1942) coined the term *canalization* to express this fact with respect to the morphology of the animal. He noted that the developing embryo has an inbuilt tendency to stick to the path of normal development even when exposed to unusual conditions such as an abnormal temperature or the presence of a few abnormal genes. And even cutting part of the embryo away often does not prevent it from finishing up as a normal adult. This is the result of systems of genes that interact in a self-stabilizing way to partly buffer out the effects of potentially harmful circumstances. Waddington visualized this process as an epigenetic landscape (see Figure 6.2). As with the concept of homeostasis, the concept of canalization is only a functional description of a process. The ways these systems of genes bring a normal organism about, of course, need to be investigated.

Canalization is an embryological concept, but there is a similar picture with respect to behavior. The basic structure of the perceptual, central, and motor mechanisms, as well as the basic interconnections among these units, develops, by and large, prefunctionally. The experience of the individual is, of course, important, often in very unexpected ways, but typically, the basic structure of behavior is extraordinarily stable. Nonetheless, development, especially of social behavior as we have seen, sometimes goes seriously wrong. Such disturbed development can often be traced to peculiarities in the social experience of the young animal, especially to periods of social deprivation. In general, the development of non-functional behavior is due to a combination of structural and motivational causes.

Structural causes for abnormal behavior include the development of aberrant behavior mechanisms. For example, a chick that is force-fed and is not allowed to peck in its first two weeks after hatching is later unable to peck at food when hungry, presumably because the motor mechanism for pecking remains independent of the central mechanism for hunger (Hogan, 1977); or, the partner recognition mechanism may develop with the image of the wrong species or of a member of the same sex, and interspecific courtship or homosexual behavior would be seen. However, structural aberrations probably account for only a small proportion of developmental problems. Most disturbed development probably results from motivational causes such as an abnormally high activation of particular behavior systems or atypical interaction among behavior systems. For example, excessively fearful animals have general difficulties in learning new tasks, like the older kittens learning to catch fish, and in expressing normal social behavior; and the inadequate integration of fear and aggression is probably the main reason for problems in the expression of sexual behavior, as seen in isolated roosters, cats, and monkeys. In all these cases, the basic behavioral structure is present, but the more subtle interactions among behavior systems are missing. It is, of course, sometimes difficult to distinguish between structural and motivational aberrations. Nonetheless, the causal analysis of behavior systems provides a framework within which to attack these problems.

SOME FUNCTIONAL CONSIDERATIONS

In this book, I have defined a behavior system in terms of its structure. As mentioned in Chapter 1, other investigators define a behavior system in terms of its functional characteristics (e.g. Timberlake, 1994). There may often be a close correspondence between systems defined in structural and functional terms, but this is by no means always the case; it is very easy for confusion to arise. For example, a structural definition of sexual behavior would include a description of the perceptual mechanisms that analyze stimuli and activate a central sexual coordinating mechanism plus a description of the motor patterns that occur when the central mechanism is activated. A functional definition of sexual behavior would emphasize reproduction – that is, those behaviors that lead to successful propagation of the species. It should be clear that many animals, including humans, engage in sexual behavior by the structural definition when that behavior will have no reproductive function. Further, courtship behaviors in many species are necessary for successful reproduction, even though the courtship behaviors themselves can be considered to belong to nonsexual behavior systems such as fear and aggression (Tinbergen, 1952; Baerends, 1975). Another example would be the language system that could be defined in terms of its communication function (Hauser, 1996; Lieberman, 2016), as opposed to its structure (Hogan, 2001; Berwick et al., 2013). Failure to make this distinction has led to heated debate (Bolhuis et al., 2014). A final example is the distinction between experience-expectant and experience-dependent neural systems that was discussed above. I should emphasize that one type of definition is not inherently better or worse than the other type; which type is most useful depends on the questions being asked.

Development implies changes in the structure of behavior, both changes in the organization of the behavior mechanisms themselves as well as changes in the connections among behavior mechanisms. To this point, I have only considered the causes of changes in behavioral structure. In this section, I briefly discuss some examples of functional questions. Since I have argued that there is no necessary relation between cause and function (Chapter 1), it might seem that there is nothing to be gained toward understanding causal mechanisms by asking functional questions. In theory, this should be true. In practice, however, the problems that an animal must solve in order to survive provide the selection pressures that are responsible for evolution by natural selection (see Chapters 8, 9, and 10), and it turns out that the evolutionary solutions to these problems sometimes use causal mechanisms that are related to the function that the behavior serves. I first consider some examples of functional questions that do not increase our understanding of development, and then some examples that do.

Is Development Selected?

It is almost a truism that natural selection should operate at all stages of development, and not only on the adult outcome, because any developmental process that reduces the probability of reaching adulthood will be strongly selected against, all other things being equal. Nonetheless, a genotype with advantageous consequences at a particular stage of development can be selected for only if its consequences in the adult do not reduce the fitness of the individual possessing it. What this means is that, at any particular stage of development, behavior may be far from optimal; it need only be good enough to bring the animal to adulthood.

This line of reasoning also leads to other conclusions. For example, it seems intuitively obvious that the best mechanism for regulating a particular outcome would be one that is directly sensitive to the outcome. Thus, the best mechanism for regulating nutrition, say, would be one that could directly sense the state of nutrition. This is another way of saying that an optimal mechanism should be based on a simple, direct relationship between cause and function. But as we have seen, this is often not the case (p. 114). Development is an extremely complex process and one in which optimal solutions may be the exception rather than the rule. It follows that development is opportunistic in the sense that any available means will be used to produce an acceptable end. Two examples should make this point clearer.

We have seen above that pecking in newly hatched chicks is not controlled by factors related to nutrition. When it became clear that experience was necessary for nutritional control to develop, it seemed reasonable to look at the effects of various kinds of nutritional experience on the occurrence of pecking. That approach turned out not to be the key to solving the puzzle because the necessary experience was not nutritional, but an association between the act of pecking and the effects of swallowing any solid object. These results were surprising (and took a long time to discover) because of our preconceptions about the relationships between the causes and functions of behavior. We intuitively feel that, when behavior changes in an adaptive direction, the cause of the change should be related to factors associated with the adaptation. Thus, when pecking changes in such a way that relatively more nutritive items are ingested, we infer that something about nutrition was responsible for the change. But in this case, our inference was wrong. Pecking behavior to food and sand during the test changes for reasons that are completely unrelated to nutrition.

A second example is provided by the analysis of Hall & Williams (1983) of the relationship between suckling and other ingestive behavior in rats. Suckling and eating are both behaviors that function to provide nutritive substances to rats, suckling normally for the first three weeks after birth and eating thereafter. In their search for the causal mechanisms underlying ingestion, Hall and his colleagues originally assumed that these mechanisms would be similar in both newborn and older animals. In fact, after many years of work, their results showed that the causal mechanisms controlling suckling are largely independent of the mechanisms controlling eating. Their analysis suggests that both systems coexist simultaneously, and that only one system is expressed at a time. Hall & Williams (1983, p. 250) concluded: "Such findings for suckling illustrate the general difficulty in determining the relationship between adaptive behavior of infancy and functionally similar representations in adulthood." Subsequently, Hall & Browde (1986) made similar studies of infant mice and discovered that the causal factors underlying eating are considerably different from those in rats. Thus, the study of the development of feeding behavior in chicks, rats, and mice shows that mechanisms for change have evolved that lead to an adaptive result, but that these mechanisms often bear little resemblance to our prior ideas of what they should be.

Adaptations for Development

The problems that an animal has to solve for survival put selection pressures on the causal mechanisms for the behavior that can evolve. It is for this reason that functional thinking can sometimes help us to understand causal mechanisms that we have discovered, and in some cases, it may direct our attention to seeking causal mechanisms that we would otherwise not have thought of (cf. Sherry, 2009; Daly, 2015). With respect to ontogeny, Oppenheim (1981) provides an excellent discussion of this issue. He pointed out that stages in development are often not merely a kind of immature preparation for the adult state – although they can be – but that each developmental phase involves adaptations to the environment of the developing animal. He called these "ontogenetic adaptations". As a consequence, certain early behavior patterns may disappear in the course of development. One example is some of the movements a chick uses when hatching from its egg. Another, as we have seen, is the finding of Hall & Williams (1983) that suckling is not a necessary antecedent for adult feeding in rats. The case of nutrition, which we have already considered from a causal perspective, provides other examples.

   In almost all species of animal, the method of acquiring nutrition changes – willy-nilly, at least once, and often two or more times – in the course of the animal's lifetime. In mammals, for example, nutrition is provided to the fetus via the placenta and, after birth, first by suckling and later by eating. Sucking, as a motor mechanism, exists before birth and after weaning, but it is not expressed then. Thus, at some stage in development, sucking must be 'switched in' to provide nutrition and later must be 'switched out'. Similarly, in birds, the yolk sac provides nutrition in the egg and for some time after hatching; then the young bird may receive food from its parents by gaping; and finally, it feeds itself using some sort of pecking movement. Here, too, something must regulate when gaping is used and when pecking is used. This, then, is the problem the animal must solve. How does it do it? The causal answer to this question is probably different for every species. We have seen at least a partial causal answer for chicks in the results we have obtained from pecking. But these results raise several obvious functional questions, two of which can be considered here. First, why should pecking not be controlled by nutritional factors at hatching? Second, why should experience be necessary for pecking to become integrated into the hunger system?

One can imagine that, if pecking were originally controlled primarily by the chick's nutritional state, pecking might not occur at all until the yolk reserves were exhausted. Such a chick would not have had as much experience with its world as a chick that had engaged in exploratory pecking during the first few days. Given that the control of pecking must shift sometime between hatching and the time when pecking is necessary for providing nutrients, there is no particular reason that experience should not provide the timing of the shift. On the other hand, there is one important reason that experience should provide the timing. Birds can hatch early or late with respect to their overall state of development and mammals can be born prematurely or past term. Endogenous timing of the switch in causal factors to or from pecking or sucking would be disastrous if, for example, a one-week premature baby could not suckle in its first week, or if a baby could not be weaned early if its mother's milk supply were interrupted. In general, it seems certain that experiential factors provide a more reliable timing cue than endogenous factors could provide in most cases where a switch between methods of acquiring nutrition occurs.

In this context, it is useful to return to the concept of a play system and to consider what function it may serve. We have seen that the essence of the concept is that motor mechanisms have a chance to be 'free' of influence from central mechanisms. Such freedom may give the motor mechanisms an opportunity to become incorporated into other central mechanisms. One can imagine that such a flexible system would be useful during development, especially in cases where something may have gone wrong, and in which the so-called normal connections would not function optimally. Similar functions for play have been suggested before, but one problem with such explanations is that adult behavior develops equally well in individuals that vary greatly in the amount of play they exhibit (Martin & Caro, 1985). Here, we can see a function for the developmental situation described by Groothuis (1994). Endogenous factors are sufficient to determine the development of particular behavior systems (or particular motor mechanisms), but during development, the possibility exists for experience to bring about a somewhat different outcome. Under normal conditions of development, either endogenous factors or play could provide alternative pathways to reach the same result. Only under special conditions (such as those provided to the gulls by Groothuis) would the different pathways lead to different results.

7

Causes and Consequences of Development
Reinforcement, Learning, Memory, Thinking

B. F. (Fred) Skinner (1904–1990). Photo by Bachrach/Getty Images.

Reinforcement is a consequence of activating certain behavior mechanisms, and also one of the causes of development. Learning is a consequence of development, and thinking can be considered as a process that is a precursor to learning. All three topics, plus memory, are considered in this chapter.

REINFORCEMENT

Reinforcement is traditionally thought to be a process that strengthens connections between representations of events, whether the representations are of stimuli (perceptual mechanisms) and responses (motor mechanisms), or of more central 'cognitive' mechanisms. This associationist view of learning has been challenged by some workers, and some of the issues involved will be discussed in the section on learning. But first I want to discuss some non-developmental aspects of reinforcement in the context of the associationist point of view. Much of this material properly belongs in Chapter 4, but I discuss it in this chapter because of its relevance to the way many people think about learning.

In an extensive review comparing the properties of different reinforcers, Hogan & Roper (1978) concluded "that with respect to learning, similar principles probably apply to all the reinforcers we have examined, but that with respect to performance, it is necessary to consider each reinforcer within the context of its own motivational system" (p. 157). There are three related issues raised by this conclusion that I will consider here. What is the relation of reinforcement to reinforcer? Why distinguish between learning and performance? What is the relation of a reinforcer to a behavior system? I will also consider the question of what aspect of a reinforcer is reinforcing.

A positive reinforcer is usually considered to be an event that, when made contingent on a response, increases the probability of that response occurring in a similar situation (Skinner, 1938). (Negative reinforcers raise additional issues that will be discussed below.) The 'event' is often a stimulus (e.g. food, water, a member of the opposite sex, a view of something interesting), but sometimes it is the opportunity to engage in particular behaviors (e.g. digging, running, fighting) or even having a pleasant thought. The situation is usually one in which the animal is deprived of food, water, sex, or beautiful views, or of the opportunity to dig, run, or fight. In other words, a positive reinforcer is something the animal wants, in the sense of 'is willing to work for'. However, what the animal is learning is not obvious: what connections are being strengthened?

What the animal may be learning will be discussed below in the section on learning, but in most cases, the reinforcing event has effects on behavior that are additional to its reinforcing properties. Presentation of food, for example, may reinforce a lever pressing response (a developmental consequence), but it also elicits eating and frequently drinking as well (motivational consequences). Further, many other aspects of the situation in which learning is usually studied also have motivational consequences that affect the performance of the subject. Shettleworth (1975), for example, studied the effects of food deprivation, feeding regime, exposure to a strange environment, and free food on the behavior of golden hamsters, *Mesocricetus auratus*. All of these factors, which are regularly present in learning experiments, had significant effects on a wide range of the hamster's behavior. Some of these effects are specific to the hamster's feeding system and such effects for feeding, drinking, nest building, and aggression systems will be discussed in the next sections. Further discussion of motivational effects on performance for heat, locomotor activity, sex, and electric stimulation of the brain as reinforcers can be found in the article by Hogan & Roper (1978).

Food and Water

Food and water are the prototypical reinforcers used in the large majority of experiments on learning, and results from these experiments are often considered to apply to learning using other reinforcers. Yet, as Hogan and Roper noted, even in the earliest literature, the reinforcing properties of food and water are not identical. Water, for example, appears to be a more effective priming stimulus in rats than food, and this affects performance in maze studies (Bruce, 1935).

The feeding and drinking systems also interact. On variable interval (Falk, 1961), fixed interval (Falk, 1966), and fixed time (Staddon & Ayers, 1976) schedules of reinforcement for food, hungry rats drink water after receiving a food pellet. The amount drunk in a typical experimental session is much in excess of normal water intake: polydipsia. Normal, free-feeding rats also drink water after eating dry food (post-prandial drinking), but the amount drunk is much more limited than seen in rats on interval reinforcement schedules (Kissileff, 1969). In a typical Skinner box experiment, there are not many things a rat can do except press the lever, drink, or move in circles. Staddon and Ayers, however, presented a food pellet every 30 sec (fixed-time schedule, no lever) in an apparatus that also had separate compartments in which the subject could drink, run in a wheel, enter a dark tunnel, or look at another rat. In all subjects, after a few trials, drinking was the response that immediately followed eating.

Falk (1971) termed polydipsia an *adjunctive* behavior because it is behavior induced as an adjunct to the behavior being reinforced. Falk also reviews examples of schedule-induced aggression, pica (eating nonfood material), and wheel running in a variety of species including pigeons, rats, and monkeys. As Staddon & Simmelhag (1971, p. 14) point out, these adjunctive activities are all related to motivational systems in that they are more or less interchangeable depending on the environmental support available. Adjunctive behaviors are the motivational consequences of the activation of various behavior systems by certain schedules of reinforcement and the presence of appropriate stimuli. The adjunctive behaviors themselves are not being controlled by the contingencies of reinforcement. Their occurrence must depend on the structure of the behavior systems involved.

Nest Material

Nest material is not usually considered in discussions of reinforcers, but for many species of birds and mammals, nest material can be a powerful motivating stimulus. Reinforcing effects of nest material have been demonstrated in several species, including the Mongolian gerbil, *Meriones unguiculatus* (see Glickman, 1973), but the most extensive data come from a series of experiments on mice by Roper (1973, 1975).

Nest building in non-reproducing rodents is primarily a thermoregulatory response. Nonetheless, even in the laboratory at normal room temperature, mice will build a nest if nest material is available. Roper observed the behavior of female mice while they were obtaining nest material from a paper strip dispenser. After the delivery of a few free reinforcements (shaping), obtaining a strip of paper was made contingent on pressing a key. This led to rapid acquisition of the key-pressing response. However, such rapid acquisition was only seen when the key was located very close to the paper dispenser. When the distance between the key and the dispenser was increased by only 2.5 cm, acquisition of the key-pressing response was significantly retarded, and usually required several sessions of manual shaping. In Roper's experiments, the distance between the key and paper dispenser affected many aspects of operant performance, including the rate (lower) and pattern (longer pauses after reinforcement) of responding. These differences are not seen to such an extent with food reinforcement. Analysis of the behavior of the mice showed that these effects occurred not because a paper reinforcer exerted weaker control over the operant, but because activities inherent in the nest-building sequence compete with key pressing.

The normal sequence of nest-building behavior in mice begins with gathering material and carrying it to the nest. When a certain amount of material has been brought to the nest area, the mouse then engages in nest building. In Roper's experiments, when nesting material was freely available, mice typically gathered and carried several pieces of paper to the nest area before engaging in a bout of building. When gathering was made contingent on making a key press, normal nest-building behavior was constrained. When the key was near the dispenser, a mouse could gather several pieces of material before carrying them to the nest and building. When the key was farther from the dispenser, a mouse typically gathered only one piece of material before carrying it to the nest; in this situation, the mouse also showed a much greater tendency to build with each paper strip. In fact, when the mouse was required to make ten responses before receiving a paper strip, building occurred after every reinforcement. This excessive nest-building behavior could account for the longer pauses between reinforcements. Roper compared the extra building behavior of the mouse to the polydipsia seen after food reinforcement on various interval schedules in rats.

Roper's experiments still leave unanswered questions, but it is clear that a piece of nest material has effects on the behavior of the mouse that are largely determined by the structure of the mouse's nest-building behavior system. A reinforcer not only increases the probability of responding, it also has many other motivational consequences depending on the structure of the animal's behavior systems.

Aggression

Fighting and aggressive display are actions, not stimuli. Of course, an action cannot occur without an appropriate stimulus – with respect to aggression, an opponent. But an opponent is not an opponent until the animal interacts with it. And it is the opportunity to engage in aggressive behavior that has been shown to be reinforcing. The opportunity to fight with, or display aggressively toward, a conspecific has been claimed to be a reinforcer for fighting cocks, mice, rats, and pigeons (see Hogan & Roper for review). But I will here confine myself to discussing aggression in fish, especially in the Siamese fighting fish (*Betta splendens*), the subject of some of my own work.

Fish have been popular subjects for studying aggressive behavior primarily because aggressive behavior can be elicited by the visual image of a conspecific behind glass, or by a mirror image of the subject. This simplifies experimental procedures and avoids actual fighting, which can be damaging or even lethal. Swimming through a ring or tunnel has been the most widely used response (Thompson, 1963), though some studies have used swimming down an alley (Hogan, 1961). In all these studies, the frequency of swimming through a ring or the speed of swimming down an alley increased when the response produced the image of an opponent. Rhoad *et al.* (1975) found a live fish behind glass, a mirror image, a moving model, and a stationary model to be increasingly less effective. Aggressive responses to these stimuli also wane increasingly fast, which suggests that the effectiveness of a stimulus as a reinforcer is proportional to the strength of the aggressive display that it evokes (see also Rasa, 1971).

Whether aggression could actually serve as a positive reinforcer has been a contentious matter, especially among social anthropologists (e.g. Montagu, 1968). Several behavioral biologists have also questioned this interpretation. Johnson & Johnson (1973), for example, provided evidence that the opportunity to view a large variety of objects, including another fish, a turtle, a marble, and an empty chamber supported swimming through a tunnel equally well. They suggested that curiosity could be the reinforcer. More extensive studies by Bols (1977), also using a variety of objects, found that swimming down an alley was supported by most of the objects, but only those that reliably evoked aggressive display gave results similar to those of most previous studies. Bols suggested that aggressive motivation plays an important role in studies of instrumental learning, but that in cases in which little or no aggressive display is seen, curiosity may be an additional factor. One other interpretation is that social motivation, and not aggression, may be the important factor (Bronstein, 1981). However, since fighting fish are solitary animals in nature, it is not clear what social motivation might mean.

We have also carried out a number of studies directly comparing the reinforcing properties of an opportunity to display with the reinforcing properties of food (see Hogan & Roper for review). In all cases, responses for aggressive display were different from responses for food, with food generally supporting higher levels of responding. One reason for this difference is that after a number of trials, stimuli in the runway come to elicit aggressive responses before the fish reaches the goal box; these responses interfere with swimming. In this situation, food does not seem to support such anticipatory feeding responses. Further, as discussed in Chapter 3, it is difficult to control levels of motivation with behavior such as aggression and sex. In one experiment, Hogan & Bols (1980) showed that swimming speed down an alley could be significantly increased using a priming procedure. As with food, water, and nest building, performance for aggression in a learning situation depends on the structure of the particular behavior system(s) involved.

What Is Reinforcing?

In reviewing the literature on reinforcers, Hogan & Roper (1978) noted that literally hundreds of studies had been published that claimed to show that some event in some species met Skinner's operational definition of a reinforcer. These included stimuli that fulfill some obvious physiological need in appropriately deprived animals, such as food, water, heat, cold, oxygen, and various drugs; the opportunity to engage in various species-specific activities such as pecking, gnawing, grooming, hoarding, sand digging, nest building, pup retrieving, courting, copulating, fighting, and running; the opportunity to explore; stimuli that seem biologically relevant to the species at hand such as sweet tastes, conspecific bird song, or an imprinting stimulus; electrical stimulation of the brain; and any stimulus change at all. Many of the studies were undertaken to prove or disprove some particular theory of reinforcement. Most of these theories have been concerned with identifying what particular aspect of the train of events that occurs in a reinforcing situation actually constitutes the reinforcer. In the context of behavior systems, these theories can be placed in a 2 x 3 classification: homeostatic (drive-reduction) and non-homeostatic (drive-induction) theories, on the one hand; and perceptual, central, and motor theories, on the other hand.

Homeostatic theories of reinforcement have been the most prevalent, and the ideas of Hull (1943) and Neal Miller (1959) have probably been the most influential in psychology. This theory could be characterized as a homeostatic central theory in that reinforcement is identified with the reduction of some central drive state (such as hunger or fear). Ethologists have proposed related theories, but have emphasized perceptual- or response-related reduction of drive. Lashley (1938) and Thorpe (1956), for example, have suggested that the perception of particular stimuli, such as a completed nest, can serve to reduce or terminate the drive state responsible for the nest-building behavior. Lorenz (1937), Tinbergen (1951, p. 106), and Hogan (1961) have suggested that the performance of particular responses, such as chewing and swallowing, can serve to reduce the 'action-specific energy' associated with those responses. Lorenz, Tinbergen, and Hogan imply that it is factors associated with performance of the responses that is reinforcing, and not the reduction of some central drive state such as hunger. Nonetheless, the common thread in all these theories is that the reduction in some drive state is presumed to be the reinforcer.

Non-homeostatic theories of reinforcement are more numerous than homeostatic theories, but they have generally been less influential since they stress the pleasurable aspects of reinforcement and identify an increase in simulation as the reinforcing event. Pleasure is often considered an unscientific concept, and seeking stimulation often seems 'unphysiological'. Nonetheless, these theories are widely held. Some non-homeostatic stimulus theories emphasize that the hedonic quality of the reinforcer (e.g. the taste of food) is crucial (e.g. Young, 1959; Pfaffmann, 1960), while others suggest that stimulus change of any kind is sufficient (e.g. Kish, 1966). Central theories ('pleasure centers') have been proposed to explain the results of electrical stimulation of the brain (e.g. Olds, 1958), and response theories have emphasized the arousal of 'consummatory' or 'species-specific' behavior patterns (e.g. Sheffield, 1966; Glickman & Schiff, 1967).

The most difficult problem in deciding among the various theories is that the performance of a response brings about stimulus changes, the perception of a stimulus brings about response changes, and all such changes probably have both drive-reducing and drive-inducing effects. Experimental efforts to separate these different effects usually produce results that are open to several interpretations. Our general conclusion was that all reinforcers share some characteristics of each theory, and that the aspect of a reinforcer that is responsible for reinforcement may well be different for each reinforcer (cf. Glickman, 1973).

LEARNING

Learning is a change in behavioral structure consequent on experience. But not all changes in behavioral structure are considered to be learning. As we have seen in Chapter 6, all structural changes (development) require some sort of experience. So what is special about learning? One view is that learning involves the association of representations. On this view, the study of learning becomes isomorphic with studying the laws of association. Some other views are discussed below.

This 'cognitive' view of learning is relatively new in experimental psychology, even though the study of associations has had a long history, especially in British philosophy (see, e.g. Boakes, 1984). For much of the twentieth century, however, most psychologists thought of learning in more peripheral terms as the association of stimuli with other stimuli or with responses; they were not concerned with the complexities introduced by thinking in terms of 'internal' events such as representations. But whether one has a more peripheral or a more central view of learning, there are two procedures that are typically used to study learning in psychology: classical conditioning and instrumental conditioning. (Operant conditioning is a form of instrumental conditioning that will be discussed separately below.) Classical conditioning involves pairing two stimuli, while instrumental conditioning involves pairing a response with a stimulus. An excellent introduction to the associationist view of learning is the chapter by Kirkpatrick & Hall (2005, in press). Mackintosh (1994) contains chapters by modern practitioners of animal learning, and Shettleworth (2010) gives an extensive review and discussion of the animal learning literature from a cognitive point of view.

Memory is generally considered to be a form of learning, but it has many features that are different from classical and instrumental conditioning. Some of these differences will be discussed below, as will the special features of perceptual and motor-skill learning.

Classical Conditioning

The basic facts of what happens when two stimuli are paired are clear. Early in the twentieth century, Pavlov (1937) paired the clicking of a metronome with presentation of food to a dog and measured the amount of saliva produced. (Actually, the clicking of the metronome was presented a few seconds before the food.) The food produced copious saliva (when the dog was hungry) with no training necessary. At first, the clicking of the metronome did not produce any saliva; but after several pairings of the clicks with food, the clicks did produce saliva before food arrived. Further, the amount of saliva produced increased with the number of pairings. The food was considered an unconditioned stimulus that produced an unconditioned response (salivation) and, after training, the clicking of the metronome became a conditioned stimulus; that is, it now produced salivation, a conditioned response. This procedure came to be called classical conditioning and is basically the response to an anticipated event.

Pavlov was hoping to elucidate brain mechanisms responsible for learning with his experimental procedure, but most psychologists were more interested in the effects of this procedure on behavior. Did the two stimuli merely become associated (S-S conditioning)? Or did the conditioned stimulus become directly connected to the conditioned response (S-R conditioning)? Were the unconditioned response and the conditioned response identical or not? The evidence from a century's work on these questions gave support to all these possibilities. In the framework presented in Chapter 1, I suggested (Hogan, 1994b) that in classical conditioning, the perceptual mechanism for the conditioned stimulus could become attached to an entire activated behavior system (the hunger system in the case of Pavlov's dog). In Chapter 5 (p. 156), I reviewed a number of studies that support this interpretation. In any specific case, of course, it is an empirical question what connections are being formed or strengthened, whether it be a simple stimulus-stimulus or stimulus-response connection or a connection between a stimulus (perceptual mechanism) and an entire behavior system (see also Domjan *et al.*, 2004).

Modern accounts of classical conditioning can probably be considered to begin with publication of two important papers by Rescorla (1967) and Rescorla & Wagner (1972). Reviews and discussion of subsequent work can be found in the book by Dickinson (1980), the chapter by Rescorla (1988), and several chapters in the book by Shettleworth (2010). In this more recent work, experimental results are interpreted in terms of associations between representations of events, which, in many ways, is very similar to the interpretations I have been presenting. There has also been further work investigating the properties of the representations involved in conditioning studies, some of which is discussed in the next section.

Instrumental Conditioning

Studies pairing a response with a stimulus began about the same time as Pavlov's work. Thorndike (1898, 1911) paired the response that led to escaping from a puzzle box (various, including pawing at a string or a button) with presentation of food to a hungry cat, and the cat quickly learned to escape more quickly. Some years later, Skinner (1938) paired pressing a lever with presentation of food to a hungry rat, and the frequency of lever pressing increased. Thorndike speculated that presentation of food (a reward) 'stamped in' the association of the stimulus (the puzzle box) with the response that led to escape – the 'law of effect'. Skinner stated that the reward (reinforcer) strengthens the response as measured by its frequency of occurrence. This procedure came to be called instrumental conditioning. With respect to the two procedures, the difference from the point of view of the animal is that in classical conditioning, the animal can be passive, whereas in instrumental conditioning, it must do something.

What is being learned when a response is followed by a stimulus? Let me take the iconic situation of a hungry rat pressing a lever for food as an example. The lever press, as a movement, is not being learned. The rat makes movements that will serve to depress a lever as part of its exploratory motor repertoire long before it ever meets a lever. The rat presses the lever 'spontaneously' the first time, food then appears, and its rate of lever pressing increases. What association is being strengthened? I suggested in Chapter 5 that the motor mechanism for lever pressing becomes attached to the central hunger mechanism. Because the rat is hungry, it will now press the lever again. It does not press the lever if it is thirsty.

However, the rat can only press the lever when the lever is present. In nature, the hungry rat might go looking for the equivalent of a lever to press. In the laboratory, such exploratory behavior is not necessary. Traditionally, the sight of the lever becomes a discriminative stimulus that 'sets the occasion' for lever pressing to occur when the rat is hungry. This implies that some kind of connection between the sight of the lever and the hunger system must also be forming because the rat only presses the lever when it is hungry. Further, the rat must also learn the contingency between the lever press and the reinforcer. Does it receive food every time it presses the lever, or only every few times it presses the lever? Or does it have to wait for some time interval before pressing the lever will deliver food? This analysis shows that what appears to be a simple situation actually presents a number of complex problems, not least of which is what connection is being strengthened.

Historically, the most controversial theoretical issue in studies of instrumental conditioning has been whether the association being strengthened is of the S-R or the R-S variety. Thorndike and his followers generally opted for a stimulus-response connection, but more recent studies have provided strong evidence for response-stimulus associations as well (Adams & Dickinson, 1981; Colwill & Rescorla, 1985; Rescorla, 1994; Dickinson & Balleine, 1994). These experiments use what is called a reinforcer-devaluation procedure. The subject (usually a rat) is trained to press a lever for food. An S-R connection

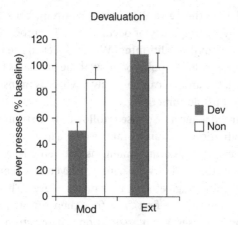

Figure 7.1 Two groups of rats were trained to press a lever for a 20% sucrose solution on random-interval schedules. Moderately trained rats (Mod) were allowed to earn 120 sucrose deliveries, and rats with extended training (Ext) were allowed to earn 360 sucrose deliveries. Half of the rats in each group were then sated on sucrose (devaluation – Dev), and the other half were sated on a different food (Non). All rats were then tested in extinction. Discussion of results in the text.
From Dezfouli & Balleine, 2012, with permission.

between the stimulus of the lever and the response of lever pressing presumably is being formed. The rat, in another situation, is now given the same food, but it becomes sick because of an injection of lithium chloride. The rat now avoids the food. When it is put back in the Skinner box, it no longer presses the lever. This must mean that the rat has also formed a connection between pressing the lever and the food, an R-S connection: it 'knows' that pressing the lever will give it food that made it sick. The reinforcer-devaluation experiments have made it clear that there is more to instrumental conditioning than merely strengthening an S-R connection. The example described in Figure 7.1 illustrates this procedure.

The first thing to notice about these results is that the moderately trained rats that had been sated on sucrose made fewer lever presses than those that had not been sated on sucrose. This is the typical result from reinforcer-devaluation experiments. However, rats that had originally received extended training failed to show this effect and responded at the same level as the non-devalued moderately trained rats. Such a result has led to the idea that instrumental conditioning reflects the interaction of two learning processes: goal-directed learning and habit formation (Dickinson, 1994). Goal-directed learning reflects the development of an association between a response (lever pressing) and its outcome (sucrose), an R-S association. Habit formation reflects the development of an association between a stimulus (the lever together with the conditioning apparatus) and a response (lever pressing), an S-R association. In anthropomorphic terms, the rat has learned how to press the lever (habit), and that pressing the lever produces the reinforcer (goal-directed learning). The value of the reinforcer (i.e. how much the animal 'wants' it) is important for both learning processes, but with extended training, the habit becomes stronger and the value of the original goal becomes less important. In fact, the habit can even be said to become a goal itself, an idea similar to the concept of functional autonomy of motives in personality theory (Allport, 1937). Habit formation has much in common with motor skill learning as discussed below.

Neurophysiological studies (Balleine & O'Doherty, 2010; Smith & Graybiel, 2013) have shown that early in its development, the habit is controlled by a particular part of the brain that is associated with reinforcement (dorsolateral striatum), but with continued training, control of the habit shifts to another part of the brain (infralimbic cortex). Graybiel & Smith (2014) propose that performing a mature habit is "acting without thinking". This is presumably the stage of habit formation reached by the rats with extended training in Figure 7.1. Goal-directed learning, on the other hand, is controlled by a different part of the brain (dorsomedial striatum), and remains sensitive to the value of the reinforcer.

Early accounts of the mechanisms of learning (formation of associations) relied on temporal contiguity of the behavior mechanisms being associated and on the contingencies between them. Later work showed that contiguity of the behavior mechanisms was not sufficient for an association to be formed: the stimulus to be associated had to supply information about the upcoming reinforcer (Kamin, 1969). In other words, learning only occurred when the animal did not expect the outcome, that is, when it was surprised. This idea has been formalized in the concept of *prediction error*. Current computational models of learning incorporate contiguity, contingency, and prediction error as variables (e.g. Rescorla & Wagner, 1972; Roesch et al., 2011; Berridge, 2012; Dezfouli & Balleine, 2012). The distinction between classical and instrumental conditioning as two procedures remains, but the same theoretical explanations are used to interpret the results of both, and the field has come to be called *reinforcement learning* (Sutton & Barto, 1998).

Neurophysiological evidence has provided considerable support for these behavioral theories (Schultz, 2006). Contiguity, of course, provides a situation in which neurons that fire together could wire together. Nonetheless, in the case of reinforcement learning, it appears that an additional factor, the neurotransmitter dopamine, is also necessary for new associations to form. Axons of neurons that secrete dopamine terminate in the area of the brain associated with reinforcement (striatum); and, it turns out that dopamine neurons fire at a higher rate when an unexpected stimulus is perceived. This provides a neural mechanism for learning in line with the theoretical models (Glimcher, 2011); it also provides a mechanism for the updating of expectancies during exploration (van Kampen, 2015).

Operant Conditioning

The preceding analyses of classical and instrumental conditioning are all in the tradition of associative analyses of learning. But there are competing traditions as well. The most ubiquitous of these is the operant conditioning tradition that can be defined as the study of behavior controlled by its consequences. Operationally, instrumental and operant conditioning are identical. However, in 1950, Skinner declared that learning theories were not necessary, and since then, operant psychology (now often called *behavior analysis*) has eschewed using theoretical concepts such as putative associations of representations. In this sense, it has become a descriptive science: presentation of a positive reinforcer increases the frequency of the response that preceded it; on a particular schedule of reinforcement responding changes in a particular way; this procedure produces that result; etc. Further, much of the work in operant psychology is not really about learning. Studies devoted to understanding the steady state of responding on various schedules of reinforcement, for example, are in most cases really studies of motivation, as are studies based on economic models. The analysis of adjunctive behavior above is an example. A review of current work in operant conditioning, especially with respect to problems of timing and choice, can be found in Staddon & Cerutti (2003).

Negative Reinforcers and Aversive Control of Behavior

Thorndike's (1911, p. 244) original formulation of the law of effect was:

> Of several responses made to the same situation, those which are
> accompanied or closely followed by satisfaction to the animal will, other
> things being equal, be more firmly connected with the situation, so that
> when it recurs, they will be more likely to recur; those which are
> accompanied or closely followed by discomfort to the animal will, other
> things being equal, have their connections with that situation weakened
> so that, when it recurs, they will be less likely to occur.

Since satisfaction to the animal has been replaced by the concept 'positive reinforcer', one might expect that discomfort to the animal would be replaced by the concept 'negative reinforcer'. That, however, is not the case. It became clear early on, as Thorndike himself was aware (e.g. 1931), that discomforting an animal, or *punishment* as it came to be called, did not affect behavior in a complementary way to satisfaction.

Punishment, or presentation of an aversive stimulus, often results in a decrease of the punished behavior, but sometimes has minimal effects or no effects at all on the strength of the punished behavior. Further, punishment often has motivational effects that obscure any effects it may have on response strength, so that the use of punishment as a method of controlling behavior is still controversial. In the operant conditioning tradition, a negative reinforcer has come to mean any stimulus from which an animal will learn to escape or avoid. In this sense, a negative reinforcer leads to an *increase* in escape or avoidance responses that remove the aversive stimulus. Punishment is the presentation of a stimulus that decreases the frequency of occurrence of the punished response (Skinner, 1953). This formulation avoids the use of concepts such as strengthening or weakening of connections. Whether a stimulus is aversive or effective as punishment becomes an empirical matter in each case.

One particular situation using an aversive stimulus, often called conditioned emotional response (CER) or fear conditioning, has been widely used as a tool in studies of memory (discussed below). The subject, usually a food-deprived rat or mouse, is trained to press a lever for food on a variable-interval schedule of reinforcement. In a typical experiment, after the subject has reached a stable level of responding, a tone is presented followed by a foot shock. The shock suppresses lever pressing, and after several pairings of tone and shock (classical conditioning), the presentation of the tone alone also suppresses lever pressing. The subject is tested later and/or after some manipulation, and the amount of suppression to the tone can be used as a measure of the subject's memory of the tone-shock pairing.

A variant of this procedure (contextual fear conditioning) involves shocking the subject in a specific environment, and measuring the duration of suppression of activity when it is placed in that environment at a later time.

Perceptual and Motor Skill Learning

Perceptual learning is the process of learning improved skills of perception (Kellman, 2002). Learning to read x-rays and learning to discriminate among wines are examples. Infants' learning of new auditory perceptual boundaries for phoneme recognition (see below) is also an example, as are imprinting and bird song template learning discussed in Chapter 5. Perceptual learning is different from classical and instrumental conditioning because no associations are being formed between a perceptual mechanism and other perceptual mechanisms or motor mechanisms: no pairing of behavior mechanisms (representations) is involved. In my terms, perceptual learning results in the *formation* of new object recognition mechanisms (stimulus recognition mechanisms might be a better term).

Recognition mechanisms form as a result of repeated exposures to particular stimulus complexes. Mere exposure is usually not sufficient for perceptual learning to occur: the animal must also be paying 'attention' to the situation (cf. van Kampen, 1996 and prediction error). In many cases, social interaction with another individual is also necessary for learning to occur (as some bird song – p. 137), and several further examples are given in the section on social learning. McLaren (1994) has proposed a theory of representation development based on the association of the elements ('primitives') of which the representation will be composed. It can be noted that these elements or primitives correspond to what I call features in Figure 2.6.

Motor skill learning is the process of learning to improve performance of a motor task such as learning to play the piano or to ride a bicycle. This process is illustrated in Figure 7.2. The representation of the task to be learned is activated at the selection level and each instruction (left to right in the figure) leads to execution of a motor primitive in turn (Figure 7.2a). With training, instructions can become 'chunked' at an intermediate level and a single instruction can lead to the same sequence of motor primitives (Figure 7.2b). With further training (not shown in the figure), the motor primitives can also become chunked, leading to a smoothly executed motor pattern ('practice makes perfect') (Diedrichsen & Kornysheva, 2015).

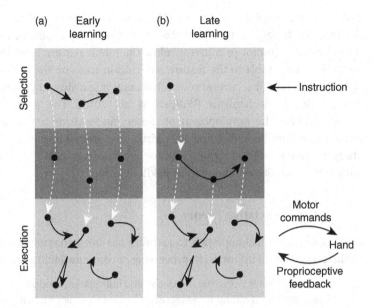

Figure 7.2 Levels of skill learning. The execution level encodes motor primitives. Early in learning, the appropriate primitives are activated (broken lines) from the selection level, and this involves explicit processing of task instruction. Later, skill learning may involve the formation of associations between the selected elements at an intermediate level and/or at the execution level (not shown). From Diedrichsen & Kornysheva, 2015, with permission.

As with perceptual learning, motor skill learning is different from classical and instrumental conditioning because no connections are being formed between perceptual and motor mechanisms. However, associations between various motor primitives are being formed, which presumably are being compared with the representation of the task. Those that match are strengthened, as was discussed with respect to the development of bird song. Wolpert *et al.* (2011) review behavioral mechanisms used in human motor skill learning.

In my terms, motor skill learning reflects the development of motor mechanisms. Since many motor mechanisms develop prefunctionally (locomotion, feeding movements, grooming movements), most investigators of behavioral development, including me, have not paid much attention to their development, except for the cases of bird song and human speech. The motor mechanism was taken as the unit of analysis. Two investigators who have considered lower-level units are Kuo (1967) and Thelen (1995), as was mentioned earlier (p. 146). With respect to Figure 2.6, it would be necessary to add a level of units (motor primitives) below the motor pattern level that would be comparable to the feature recognition units on the sensory side. In adult humans, current research on motor learning emphasizes computational mechanisms (Wolpert *et al.*, 2011). The processes responsible for the development of motor mechanisms are almost certain to be similar before and after birth, so it should be possible to study prefunctionally developed motor mechanisms using insights gained from studies of human motor skill learning.

A Non-associative Theory

One other theoretical approach to learning has been championed by Gallistel (1990b): an information-processing or computational analysis.

> The distinction between the associative and information-processing frameworks is of critical importance. By the first view, what is learned is a mapping from inputs to outputs. Thus, the learned behavior ... is always recapitulative of the input-output conditions during learning. An input that is part of the training input, or similar to it, evokes the trained output, or an output similar to it. By the second view, what is learned is a representation of important aspects of the experienced world. This representation supports input-output mappings that are in no way recapitulations of the mappings (if any) that occurred during the learning (Gallistel & Matzel, 2013, p. 179).

Traditional associative learning theory deals only with the strength of the association, and it is generally agreed that this is not sufficient to account for aspects of timing in learning studies and, for example, how animals represent and move about in space. Shettleworth (2010, pp. 123–126) reviews a number of studies bearing on this issue and compares the ability of the two theories to account for the findings. The theories, however, are not mutually exclusive and the behavior system approach presented in this book actually incorporates both ideas. Some of the navigation studies discussed in Chapter 4 provide examples that are best explained by the computational approach, as do studies on human infants' speech development, which I will discuss here.

As mentioned in Chapter 5, human infants up to the age of about six months are able to discriminate among all the phonetic units used in the world's languages, but by the age of one year, they only discriminate among the phonetic units used in their own language. How do they learn? Adult speakers utter speech sounds that vary in many dimensions. Experiments by Kuhl (2004, 2015) and her colleagues and by Maye *et al.* (2002) show that exposure to the distributional patterns of speech sounds in spoken language input results in new perceptual boundaries being formed that conform to the boundaries in the language to which the infants are exposed. Kuhl describes this as 'statistical learning' (see also Aslin, 2014). This happens within the first six months. Other experiments by Kuhl (2015) and colleagues and by Saffran (2002) and colleagues show that infants as young as eight months can discriminate word-like stimuli on the basis of transitional probabilities between adjacent syllables. Further, the frequency of hearing sounds, by itself, is not sufficient, in many cases, for learning perceptual boundaries. Especially in complex native-language learning situations, social interaction with another human is necessary (Kuhl, 2015).

Another question about these results is why infants by the age of one year are no longer able to discriminate between phonemes that are not in their native language. Kuhl (2004, p. 832) suggests "that the initial learning of native-language patterns eventually interferes with the learning of new patterns (such as those of a foreign language), because they do not conform to the established 'mental filter' [template, schema]". Here again, questions of modularity, constraints, and processes arise in the same way as we have seen in relation to imprinting and bird song development.

The development of both the perceptual mechanisms for phoneme- and for word-boundary recognition can be considered examples of computational learning of the sort postulated by Gallistel. Once formed, these representations can of course become associated with one another according to traditional association learning theory, as well as being used computationally. I should also mention that Gallistel's theoretical approach to learning includes modeling how the brain can deal with probabilistic inference and can form representations of uncertainty.

Constraints on Learning

I have already discussed problems of constraints on learning in Chapter 5, especially with respect to song learning and imprinting in birds, and in Chapter 6 with respect to irreversibility and critical periods. But this topic has also arisen in the instrumental conditioning literature. In the standard experimental situation, a response (e.g. key pecking) is followed by a stimulus (e.g. food), and the animal is hungry (the feeding system). Does it matter which response, which stimulus, and which behavior system, as well as which species? The associative theory assumes it does not matter (as long as the reinforcing stimulus is relevant to the animal's activated behavior system; that is, as long as the animal 'wants' or values the reinforcer), but in many cases, it turns out to matter a great deal. Shettleworth (2010, p. 98) points out that interest in biological constraints on learning has waned in recent years within the psychological learning community, but I will nonetheless discuss several examples here because of their relevance to a behavior systems approach to learning. Reviews and discussion of the earlier literature can be found in Shettleworth (1972) and in Hinde & Stevenson-Hinde (1973).

Hogan and Rozin (see Hogan, 1961) found that several days of hand shaping were insufficient to condition a Siamese fighting fish to press a target for food. The target was a small piece of black plastic attached to a thin steel rod. They finally succeeded after placing a small worm behind clear plastic on the target. As we have seen above, Siamese fighting fish learn very quickly to swim through a ring or down a runway for a food reward. When aggression was the reward rather than food, classically conditioned responses to situational stimuli came to interfere with instrumental performance, as we have also seen above.

In somewhat similar experiments, Sevenster (1968, 1973) observed male three-spined sticklebacks swimming through a ring or biting a thin rod when rewarded with either a ten-sec view of a courting female or of another male. He found typical learning results for both responses when the reward was fighting with another male, as were the results for swimming through a ring when the reward was courtship. However, the results for biting the rod were atypical when the reward was courtship. After a very detailed analysis of the situation, Sevenster was able to conclude that the rod became treated as a female, presumably through classical conditioning, and that courtship behavior to the rod inhibited the instrumental response of biting. In a later experiment, Sevenster & van Roosmalen (1985) reinforced the glueing movement (a nest-building activity) of the male stickleback with presentation of a courting female, but the rate of glueing did not increase, unlike the results for swimming through a ring.

In other experiments, Hogan (1964) reinforced preening in pigeons with food. He found that the rate of preening under continuous and variable interval schedules of reinforcement increased in much the same way as key pecking increases; however, stereotypy of responding developed differently. His results suggested that the preening system was being reinforced, and not a specific preening movement. Rice (1978) tried to affect the occurrence of shrill calls and twitters in young chicks by using food reinforcement, but he was unsuccessful.

Shettleworth (1975), using golden hamsters, also asked whether various behavior patterns could be influenced by food reinforcement. In one set of experiments, she reinforced animals with food when they performed various behavior patterns, including scrabbling, digging, rearing, face washing, scratching, and scent marking. She found that food reinforcement was effective in increasing the occurrence of scratching, digging, and rearing, but that it had very little effect on the occurrence of face washing, scratching, and scent marking. In separate experiments (described above), the first three patterns all increased in frequency in hungry hamsters, and the latter three decreased in frequency. Thus, behavior patterns that belonged to the hamster's hunger system – when the criterion used is a positive correlation with food deprivation – could be influenced by food reinforcement, whereas behavior patterns that belonged to other systems could not. See also Domjan et al. (2004) for further examples from sexual behavior systems.

These results all indicate a considerable degree of inflexibility with respect to which perceptual and motor mechanisms can become connected to which central mechanisms. Some responses are difficult to shape; others only work well with particular behavior systems. Some responses seem to be unreinforceable, others bring along a whole behavior system. A comparison of the results of the experiments of Hogan and of Sevenster using aggression as a reinforcer shows that the species also makes a difference: the organization of the aggression behavior system in fighting fish is different from the organization of the aggression behavior system in the stickleback. One can argue that these results all illustrate constraints on performance, and not constraints on learning. This formulation may, in fact, be true, but the problem remains because what we measure is performance. And usually, performance is what we are interested in. Once again, there are no easy causal generalizations.

Social Learning

Social learning refers to any learning that is influenced by interaction with, or observation of, another animal or its products (Heyes, 1994).

What makes it different from the types of learning I have already discussed? With respect to the mechanisms of learning, probably nothing! (cf. Heyes, 2012; Leadbeater, 2015) Nonetheless, it has been studied as a separate field for many years for at least two major reasons. The first is that many investigators have been interested in comparing human and non-human intelligence: the presence or absence of processes of social learning in various species might make it possible to differentiate humans from non-humans. The second reason is that social learning provides a mechanism for transmission of behavior between generations, and could thus provide insight into the evolution of behavior. I will discuss the mechanisms of social learning here, and some functions of social learning and its relation to culture and evolution in Chapters 8 and 9.

Examples of Social Learning

I discussed the development of food recognition in young chicks in Chapter 5, but one factor I did not mention there is the behavior of the mother hen. When a broody hen, in the presence of young chicks, finds a morsel of food, she picks it up and makes a 'food call'. This attracts the chicks. She then drops the food and one of them picks it up and swallows it (Kruijt, 1964). Subsequently, the chick will show a preference for similar foods. Suboski & Bartashunas (1984), using an arrow making pecking movements, found that one-day-old domestic chicks pecked selectively at stimuli (colored plastic pinheads) that matched stimuli placed on or near the arrow tip. In an analogous experiment, using broody hens, Wauters et al. (2002) found that chicks preferred feeding from food bowls that were preferred by the hen; these preferences were different from the food preferences of chicks that were raised without experience with a broody hen.

The role of social learning in the development of food preferences in Norway rats has been studied in a series of experiments by Galef and his colleagues (review in Galef, 1996). Weanling rat pups prefer foods eaten by the mother. Galef's experiments have shown that the development of these preferences depends on chemical traces left by these foods in the amniotic fluid, in the mother's milk, and on the mother's breath. Young kittens provide a somewhat different example. As we have seen (p. 145), naïve kittens recognize the smell of fish as food prefunctionally, but not the smell of mice. Berry (1908) found that a young kitten placed in a small enclosure with a live mouse would usually ignore the mouse until it moved. The kitten might then catch the mouse with its paw, but would then 'play' with it in much the same way as it would play with a small ball. After several minutes, it would once again ignore the mouse. When the mother cat was placed in the enclosure, she would immediately catch and eat the mouse. When the mother was removed, a new mouse introduced into the enclosure was treated as the original mouse had been. It was only after the mother cat caught the mouse, opened it, but did not eat it, that the kitten ate the mouse. Subsequent mice were treated as food.

Another well-studied example concerns foraging behavior in pigeons (*Columbia livia*). There have been several reports that pigeons (and doves) can perform relatively complex, novel food-finding behavior after observing the actions of an experienced conspecific. The most convincing evidence comes from a well-designed series of experiments by Palameta & Lefebvre (1985). These investigators were able to show that a pigeon, merely by observing another pigeon performing a learned response for food, can learn both where to direct its feeding behavior and what motor act to use. The experimental set-up used in their experiments is shown in Figure 7.3.

Social learning, of course, occurs in many contexts other than feeding. Song learning in birds is an obvious example, and social influences go beyond forming a template. Many bird species seem to

Figure 7.3 Observational learning in doves. The dove on the left is watching the dove on the right lift a cover off a source of food. Courtesy of Louis Lefebvre.

learn their song equally well from a tape tutor or from a live tutor. But other species, including the white-crowned sparrow, are able to learn their song outside the sensitive period, defined by a tape tutor, when exposed to a live tutor. And I have just discussed the results of Kuhl (2015) who found that social interaction with another human is necessary for learning auditory perceptual boundaries in human infants. A wide range of evidence for diverse social influences on vocal development can be found in the book edited by Snowdon & Hausberger (1997).

A final set of examples comes from studies of social learning in insects (Leadbeater & Chittka, 2007, p. R703):

> Well substantiated cases of social learning among the insects include learning about predation threat and floral rewards, the transfer of route information using a symbolic "language" (the honeybee dance) and the rapid spread of chemosensory preferences through honeybee colonies via classical conditioning procedures. More controversial examples include the acquisition of motor memories by observation, teaching in ants and behavioural traditions in honeybees. In many cases, simple mechanistic explanations can be identified for such complex behaviour patterns.

One example is an experiment by Leadbeater (2015) on observational learning in bumblebees (*Bombus terrestris*) that demonstrates second-order conditioning. In this experiment, 'subject' bumblebees were initially trained to associate conspecific presence with either a sucrose reward or an aversive substance. They were then permitted to watch conspecific 'demonstrators' choosing a particular flower color through a screen. When later allowed to forage alone, bees that had learned to associate conspecifics with sucrose preferred the flower color the demonstrators chose, while the aversive group actively avoided the same colors. A control group of naïve bees was not influenced by the demonstrators' choices.

Another example is provided by the work of Alem *et al.* (2016). These investigators trained bumblebees to obtain a sucrose reward using a string-pulling task. Only 11% of naïve bees spontaneously solved this task; after observing a trained demonstrator from a distance, 60% of the bees tested solved the task. The authors state that this is the first experimental demonstration of social learning of novel and non-ecological behavior in insects, and suggest that the miniature brains of bees might possess the essential cognitive prerequisites for culture. These examples illustrate the variety of studies of social learning, with respect to both the species and the types of social influences studied. Leadbeater (2015) has analyzed many of these examples using principles of associative learning and concludes: "it seems that social learning is really just learning" (p. 9).

The Problem of Imitation

At the end of the nineteenth century, animal intelligence was a much-discussed topic, and the ability to imitate was considered an important sign of intelligence. But as Thorndike (1898, p. 50) posed the problem:

> If one can from an act witnessed learn to do the act . . . he imitates . . . Now, as the writers of books about animal intelligence have not differentiated this meaning from the other possible ones, it is impossible to say surely that they have uniformly credited it to animals, and it is profitless to catalogue here their vague statements.

Differentiating the "associative processes in animals" was the purpose of Thorndike's experiments. And differentiating 'true' imitation from other sorts of imitation-like phenomena has remained a prime motive for studies of social learning to the present day.

Galef (1988) looked at the different definitions of imitation from the time of Darwin to the present and concluded that there has been "no resolution of the conflicting usages of the term imitation already evident at the turn of the century" (p. 10). He divided social learning terms into those that were descriptive [functional] and those that were explanatory [mechanistic]. The terms include imitation, local enhancement, social facilitation, observational conditioning, emulation, and several others. And, as Galef noted, all of these terms have been used in both senses by various authors. Since 1988, studies of social learning have increased dramatically (see Shettleworth, 2010, Ch. 13 for a review of many of these studies), and periodically, new classifications and discussions of social learning terms have been published. Although a certain amount of consensus vocabulary has emerged, Galef (2013, p. 123) has commented:

> Thorndike's . . . definition of imitation as "learning to do an act from seeing it done" has unduly restricted studies of the behavioral processes involved in the propagation of behavior . . . success in labeling social learning processes believed to be less cognitively demanding than imitation (e.g., local and stimulus enhancement, social facilitation, etc.) has been mistaken for understanding of those processes, although essentially nothing is known of their stimulus control, development, phylogeny or substrate either behavioral or physiological.

The more things change, the more they stay the same! I will return to social learning in an evolutionary context in Chapter 8.

MEMORY

Memory is the representation of past experience, and can be considered a process leading to the formation of behavior mechanisms (representations) as well as the outcome of the process. Psychologists have noted that there are different kinds of memories, though there is no consensus as to what those kinds are. One major proposed functional classification of human memory is that of Schacter & Tulving (1994). They have distinguished five major memory systems:

> (1) *working memory*, a short-term or transient form of retention that supports the online maintenance of internal representations for use in ongoing cognitive tasks, and also supports the controlled manipulation of these representations; (2) *episodic memory*, which supports the encoding and retrieval of personal experiences that occur in a particular time and place; (3) *semantic memory*, which refers to a person's general knowledge about the world, containing a complex web of associated information, such as facts, concepts, and vocabulary; (4) *the perceptual representation system*, a collection of domain-specific subsystems that operate on perceptual information about the form and structure of words and objects; and (5) *procedural memory*, which supports the acquisition and retention of perceptual, motor, and cognitive skills, and is involved in everyday tasks such as learning to ride a bicycle or becoming skilled in cognitive domains such as athletics (Schacter, 2009, p. 434).

As well, episodic and semantic memory are considered to be forms of *declarative* or *explicit memory* because they can be consciously recalled, whereas perceptual representations and procedural memory are considered forms of *implicit* memory.

Working memory is different from the other memory systems in that it is a temporary change (akin to motivation) that can be considered a representation of current experience (including retrieval of past experience), whereas the other memory systems involve permanent changes in the structure of behavior (akin to development). Working memory has been studied with respect to duration of memory, capacity (number of items that can be remembered), interference among items, and other properties as well. From a structural and causal point of view, however, generalizations are difficult to formulate because results can differ due to modality (vision, audition, olfaction, etc.), to the complexity of the items to be remembered, to the effects of previous experience, to the species, and for many other reasons. Functionally, working memory is the precursor to the formation of the representations characteristic of the other memory systems.

Procedural memory is probably congruent with perceptual and motor skill learning as discussed above, and the perceptual representation system would include the development of stimulus recognition mechanisms in an information-processing framework. But there is clearly overlap between these two systems with respect to the formation of perceptual mechanisms. Semantic memory is an important component of the human language behavior system, but it is likely that non-human animals also have non-linguistic semantic memory. Episodic memories are generally more complex than the other types because they include time and place information as well as stimulus information. There has been considerable controversy as to whether non-human animals also have episodic memory (see e.g. Clayton, 2015; Fugazza *et al.*, 2016).

The neural processes involved in the formation of memories in all these memory systems are probably similar, although the memories themselves can be encoded and stored in different parts of the brain. There are also developmental differences between declarative and procedural memory, in that children reach adult levels of performance for procedural memory tasks several years earlier than they reach adult levels for declarative memory tasks (Finn *et al.*, 2016). Nonetheless, all these types of memories are representations that can become associated in reinforcement learning.

"Memory is the capacity of an organism to acquire, store, and recover information based on experience. Memories are thought to be encoded as enduring physical changes in the brain, or engrams" (Josselyn *et al.*, 2015, p. 521). In other words, memories are a component of the structure of behavior. A cartoon of the development of a memory (engram) is shown in Figure 7.4. There has been considerable progress in recent years in unraveling the neural mechanisms involved in these structural changes and Josselyn *et al.* (2015, 2017) review the neurophysiological evidence for these mechanisms. Of special note is the fact that the neural ensembles that encode the memory (engram) are widely distributed in the brain, which is one reason why older studies, using relatively crude ablation methods (e.g. Lashley, 1950), were unable to find the location of the engram. A behavioral example of the consolidation and reconsolidation processes in engram development is the development of the partner recognition mechanism (sexual imprinting) in zebra finches discussed

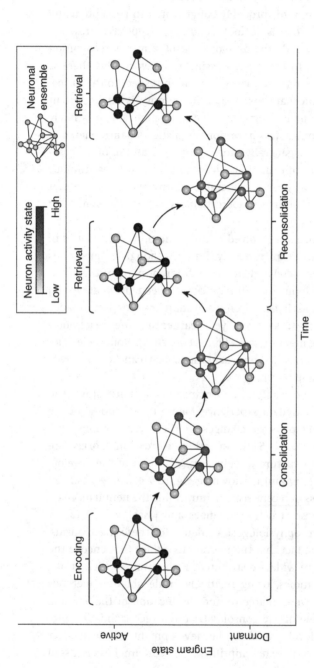

Figure 7.4 The lifetime of an engram. The formation of an engram (encoding) involves strengthening of connections between collections of neurons (neuronal ensembles) that are active (black) during an event. Consolidation further strengthens the connections between these neurons, which increases the likelihood that the same activity pattern can be recreated at a later time, allowing for successful memory retrieval. During consolidation, the engram enters a mainly dormant state. Memory retrieval returns the engram back to an active state and transiently destabilizes this pattern of connections. The engram may be restabilized through a process of reconsolidation and re-enter a more dormant state. Therefore, an engram may exist in a dormant state between the active process of encoding and retrieval required to form and recover the memory. In this way, an engram is not yet a memory, but provides the necessary conditions for a memory to emerge. From Josselyn et al., 2015, with permission.

in Chapter 5: the partner mechanism forms and consolidates in the young bird, but can be modified by sexual experience (retrieval) when the bird matures; it then reconsolidates.

Animals not only remember past experience, they also forget:

> The hippocampus is thought to automatically encode all experience, yet the vast majority of our experiences are not remembered later. Although psychological theories have postulated the existence of decay processes for declarative memory, the corresponding neurobiological mechanisms are unknown. Here, we develop the hypothesis that ongoing hippocampal neurogenesis represents a decay process that continually clears memories from the hippocampus (Frankland et al., 2013, p. 497).

That new neurons basically erase old neural connections is a novel idea, but there is now considerable evidence that this is the case (e.g. Akers *et al.*, 2014). To consolidate, most memories need repeated exposure or rehearsal (often during rest or sleep – Diekelmann & Born, 2010). Without repetition, neurogenesis provides a mechanism for forgetting. It should also be mentioned that new neurons in the hippocampus serve other functions as well, including enhancing the process of pattern separation, that is, making individual memories more distinct (Sahay *et al.*, 2011).

THINKING

Thinking is often considered to be the apogee of the human mind, but there is considerable evidence that many animals also engage in some forms of thinking. In his book on animal thought, Walker (1983, p. xiii) points out that

> ... human thought is intimately connected with the activities of the human brain; other vertebrate animals apart from ourselves have very complicated brains, and in some cases, brains which appear to be physically very much like our own; this suggests that what goes on in animal brains has a good deal in common with what goes on in human brains.

However, as with so many concepts, thinking has no accepted definition. What I will do in this section is begin with a series of quotations from well-known scholars that give their definitions of thinking in humans. I will then formulate my own definition and give some examples of behavior from spiders, bees, birds, and rats that might be construed as thinking, and finally draw some conclusions.

Thoughts about Thinking from Helmholtz to Kahneman

Hermann von Helmholtz (Lecture: The Facts of Perception, 1878).

> In a receptive, attentive observer, intuitive images of the
> characteristic aspects of things that interest him come to exist;
> afterward he knows no more about how these images arose than
> a child knows about the examples from which he learned the
> meanings of words ... [I have] called the connections of ideas which
> take place in these processes *unconscious inferences* ... Obviously we are
> concerned here with the elementary processes which are the real basis
> of all thought.

William James (*Principles of Psychology*, 1890).

> We talk of man being the rational animal; and the traditional
> intellectualist philosophy has always made a great point of treating the
> brutes as wholly irrational creatures. Nevertheless, it is by no means easy
> to decide just what is meant by reason, or how the peculiar thinking
> process called reasoning differs from other thought-sequences which
> may lead to similar results.
>
> Much of our thinking consists of trains of images suggested one by
> another ... In reasoning, we pick out essential qualities ... reasoning may
> then be very well defined as the substitution of parts and their
> implications or consequences for wholes. [James then quotes John Locke:]
> "The observer is not he who merely sees the thing which is before his
> eyes, but he who sees what parts that thing is composed of."
>
> The intellectual contrast between brute and man ... the mental process
> involved [in animal sagacity] may as a rule be perfectly accounted for by
> mere contiguous association, based on experience.

Lev Vygotsky (*Thought and Language*, 1934).

> We can trace the idea of identity of thought and speech from the
> speculation of psycho-logical linguistics that thought is 'speech minus
> sound' to the theories of modern American psychologists and
> reflexologists who consider thought a reflex inhibited in its motor part.
> The opposite view that thought and speech can be studied independently
> also precludes any study of the intrinsic relations between language and
> thought.
>
> Psychology, which aims at a study of complex holistic systems, must
> replace the method of analysis into elements with the method of
> analysis into units. What is the unit of verbal thought that is further
> unanalyzable and yet retains the properties of the whole? We believe
> that such a unit can be found in the internal aspect of the word, in *word
> meaning*.

Steven Pinker (*The Stuff of Thought*, 2007).

> In this book I have given you the view from language – what we can learn about human nature from the meanings of words and constructions and how they are used.
>
> Human characterizations of reality are built out of a recognizable inventory of thoughts. The inventory begins with some basic units, like events, states, things, substances, places, and goals. It specifies the basic ways in which these units can do things: acting, going, changing, being, having … We are constrained by this inventory, but we can expand our horizons with conceptual metaphor and the combinatorial power of language.

Daniel Kahneman (*Thinking Fast and Slow*, 2011).

> I describe mental life by the metaphor of two agents, called System 1 and System 2, which respectively produce fast and slow thinking. I speak of the features of intuitive and deliberate thought as if they were traits and dispositions of two characters in your mind. In the picture that emerges from recent research, the intuitive System 1 is more influential than your experience tells you, and it is the secret author of many of the choices and judgments you make.

What Is Thinking?

As you can see, reading through the list, there is no real consensus: like the blind men feeling the elephant – the physiologist, the philosopher, the psychologist, the linguist, the cognitive scientist – each has his view on what thinking is and whether non-human animals are capable of it. There are also many other views that I have not mentioned, including those of the behaviorists and those for whom consciousness is the key concept. Nonetheless, there are some themes running through the various views. All the authors agree that thinking (level 1) involves association of ideas (fast thinking); most would agree that thinking (level 2) involves deconstruction (deliberative thinking); and most would probably agree that thinking (level 3) involves language. Further, some authors believe non-human animals can only think at level 1 (though they might not all agree that association of ideas actually qualifies as thinking), and all would probably agree that thinking at level 3 is uniquely human.

Given that there is no consensus, I will make my own definition: thinking is consideration without action (although consideration often leads to action). And I will define consideration as the integration of information from two or more sources. This definition is extremely broad, but it makes thinking different from reflex action, in which a single source of information leads to a specific action. Further, (almost) all animals can learn (form associations), which is thinking at level 1, but several researchers have asked whether their thought processes can go beyond simple association. I will discuss several examples.

My first example deals with the hunting tactics of the jumping spider, *Portia fimbriata*, that eats other spiders (review in Jackson & Cross, 2011). When hunting other spiders, *Portia* uses mimicry, detours, or deception depending on the prey. When hunting web-spinning spiders, *Portia* uses aggressive mimicry. It will enter the web of the intended victim and produce vibrations that mimic vibrations made by the victim's prey. When the victim approaches, *Portia* attacks and kills it. (It should be noted that jumping spiders have excellent eyesight, unlike most other spiders.) *Portia* preys on many species of web-spinning spiders and the signals sent vary according to the species being preyed upon. Such signals can, of course, be part of the species' repertoire, but if *Portia* enters a web of a species for which it lacks an appropriate signal, it may vary its signals until it attracts its victim. Subsequently, it uses its new signal when it enters another web of the same species. Jackson considers this a case of trial-and-error learning. *Portia* also plans and uses detours:

> If *Portia* cannot get close to intended prey through deception, it might use a detour, as it often does in pursuit of *Argiope appensa*, a spider that builds orb webs on tree trunks. Although it might appear that *Portia* could simply walk straight from the tree trunk to the web, *A. appensa* is exceedingly sensitive to anything foreign touching the web and rarely lets *Portia* enter unchallenged ...
>
> *Portia* often walks up the tree trunk toward *A. appensa* and then stops, looks around, goes off in a different direction and reappears above the web. If there is a vine over the web, for example, *Portia* seems to look at the web, the vine and the neighboring vegetation before moving away, perhaps going to where the web is completely out of view, crossing the vegetation and coming out on the vine above the web. From above the web, *Portia* drops on a silk line alongside but not touching the intended victim's web. Then, when parallel with the spider in the web, *Portia* swings in to make a kill (Jackson & Wilcox, 1998, p. 353).

Another hunting tactic is used when the prey is *Zois genicularis*. *Portia f.* exploits opportunistically the times when *Zois's* attention is focused on its own prey. Relying on optical cues alone, *Portia* moves primarily during these intervals. One might even want to consider this an example of an early version of 'theory of mind' or mindreading in animals. Finally, detour behavior in *Portia labiata* has also been studied in the lab. Tarsitano & Andrew (1998) have shown that *Portia l.* uses visual scanning behavior to choose unblocked pathways to a spider-shaped artificial lure. All of these instances of hunting tactics would fit my definition of thinking.

My second example concerns concept formation by honeybees. There have been several experiments claiming that insects are able to learn concepts such as same or different (see Chittka & Niven, 2009), but Avarguès-Weber *et al.* (2012) have now provided evidence that honeybees are able to learn two abstract concepts simultaneously: a spatial relationship (above/below vs right/left) and a same/different relationship. The bees were trained in a Y-maze with stimuli presented vertically on the back wall of each arm. A small nipple in the center of each wall contained either a drop of sucrose solution (a reward) or a drop of quinine solution (an aversive substance). The correct stimulus always had two components that could be either the same or different (in color and pattern) and could be arranged either one above the other or next to each other. Bees were trained on the spatial arrangement and reached a level of 80% correct in 30 trials. During training, all the stimulus components were different. On transfer tests after training, the bees chose the stimulus with different components when the spatial arrangement no longer provided discriminant information. The authors conclude:

> These results demonstrate that the miniature brain of honey bees is capable of extracting at least two relational concepts from experience with complex stimuli and that these concepts can be combined to determine sophisticated performances, which are independent of the physical nature of the stimuli ... bees learned that they had to choose stimuli arranged in a specific spatial relationship and that all stimuli were composed of different visual elements (p. 7484).

My next example is two experiments purporting to demonstrate planning for the future by western scrub-jays (*Aphelocoma californica*). Jays, like the chickadees discussed earlier, cache excess food, which they can retrieve later. In the so-called 'planning for breakfast' experiments, Raby *et al.* (2007) found that the jays would preferentially cache food in a place where they had learned that they would be hungry the following morning. The birds also preferentially cached particular foods according to what their preferences would be the following morning. The authors concluded: "Jays can spontaneously plan for tomorrow without reference to their current motivational state, thereby challenging the idea that this is a uniquely human ability" (p. 919).

Another example is an experiment on causal reasoning in rats. In many ways, classical conditioning can be considered an example of acquiring causal knowledge: a formerly neutral stimulus comes to predict a change in the environment. The click of the metronome predicts food: from the dog's point of view, one could say that the click causes food to appear. Blaisdell *et al.* (2006) devised an experiment in which rats first learned (classical conditioning) that a certain stimulus (a light) sometimes produced a tone and sometimes produced food (sucrose solution). They were then tested by either being passively presented with the tone or by producing the tone themselves (by pressing a lever). When passively presented with the tone, the rats showed more food-directed behavior (nosing in the food compartment with no food present) than when they produced the tone with their own behavior. The authors argue that, when producing the tone themselves, the rats 'reasoned' that the light-tone connection was irrelevant to producing food, and that the "results contradict the view that causal learning in rats is solely driven by associative learning mechanisms" (p. 1022).

Two final examples come from studies investigating learning and cognitive processes in rats using advanced neurophysiological methods as well as behavioral observation. One example is two studies of expectancies in decision making in rats. In the first study, Johnson & Redish (2007) recorded neural activity in the hippocampus of rats that were making a right/left choice in a T-maze. Their data indicate that the rat is representing a location distant from its current location when making its decision to turn either left or right. In a subsequent study, van der Meer & Redish (2010) recorded neural activity in the ventral striatum (an area of the brain involved in reinforcement learning) of rats using the same apparatus and general methods of Johnson & Redish (2007). They found that neurons that fired at the feeder sites also fired at the choice point, but only during the early learning trials. They suggested "that ventral striatal reward representations contribute to model-based expectancies used in deliberative decision making" (2010, p. 29). Their results also support the distinction between goal-directed learning (early trials) and habit formation.

My last example is a study of 'regret' in rats (Steiner & Redish, 2014).

> ... regret entails recognition that an alternative ... action would have produced a more valued outcome. In humans, the orbitofrontal cortex [OFC] is active during expressions of regret, and humans with damage to the OFC do not express regret. In rats and nonhuman primates, both the OFC and the ventral striatum have been implicated in reward computations [reinforcement learning]. We recorded neural ensembles from OFC and ventral striatum in rats encountering wait or skip choices for delayed delivery of different flavors ... (p. 995).

Aspects of the experimental set-up and results are shown in Figure 7.5. This example provides evidence that in traversing the Restaurant Row, the rat is combining information from many sources (deliberative thinking) and is perhaps feeling 'regret'; but whether the rat 'feels' anything and how such a feeling compares to a human feeling of regret remain matters of conjecture.

Do Non-human Animals Think?

The studies I have just discussed demonstrate that spiders and insects are able to form strategies for action and to extract abstract concepts from a set of complex pictures and to combine them in a rule for subsequent choices. The bird and rat studies demonstrate instances of deliberative decision making, including planning for the future, engaging in causal reasoning, making decisions at a choice point, and behaving appropriately to opportunities lost. We can thus conclude that non-human animals are capable of deliberative thinking that goes beyond simple association. But of course, there are different forms of deliberative thinking (for example, imitation) and it is clear that not all animals are capable of highly complex (non-verbal) thinking. Wynne (2004) and Shettleworth (2010, Ch. 8) review many examples of the cognitive abilities of animals, ranging from ants and bees to birds, bats, dogs, porpoises, monkeys, and apes. Many of these animals can perform prodigious cognitive feats, though no species can match the (verbal) cognitive abilities of humans (cf. Bolhuis & Wynne, 2009). Further discussion of these issues can be found in the book edited by Menzel & Fischer (2011). My own conclusion reflects the comments of Walker quoted above, and echoes Shettleworth's conclusion about cognitive maps (p. 121): What goes on in animal brains must be similar to what goes on in human brains, and it is necessary to discover these similarities in each individual case.

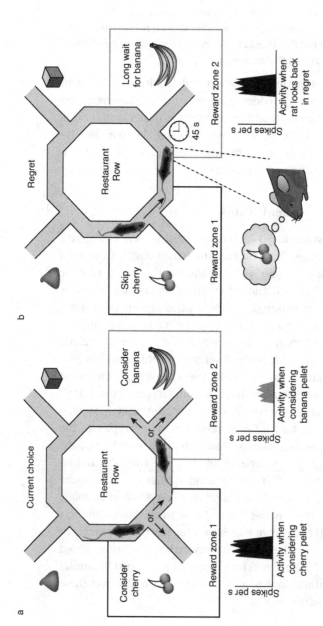

Figure 7.5 Neural encoding of regret in OFC and ventral striatum in rats performing the Restaurant Row task. (a) Rats run counterclockwise in an economic foraging task in which there were four reward zones (restaurants). If rats waited long enough in each zone, a flavored pellet was delivered at the end of the zone's arm. The length of the wait time was signaled by the frequency of a tone presented on entering that zone. Two zones, banana and cherry, are shown in this example. The rat likes both of these flavors. The plots at the bottom represent single cell activity seen in OFC during performance of this task. In this example, activity is high when the rat considers cherry and low when it considers banana. Thus, this neuron is selective for cherry. Other neurons are selective for other flavors. (b) The rat skips the cherry zone because the delay to reward exceeds the rat's threshold of willingness to wait for cherry. However, once the rat gets to the banana zone, the new delay for banana far exceeds the rat's willingness to wait for banana. At that instance the rat looks back with regret. During this look back, activity in OFC and ventral striatum is similar to that observed previously when the rat considered the cherry zone.

Re-drawn from Bissonette et al., 2014 (after Steiner & Redish, 2014), with permission.

Many students of behavior have the idea that behavioral and cognitive abilities and intelligence are positively correlated with brain size (e.g. Sol *et al.*, 2008). My invertebrate examples, however, show that even the miniature brains of spiders and insects can support cognitive processes such as planning and categorization. In a thoughtful and extensive review of insect behavioral capabilities, Chittka & Niven (2009) show that the brains of insects support highly differentiated motor repertoires, extensive social structures, as well as a surprising variety of cognitive processes. They point out:

> ... neural network analyses show that cognitive features found in insects, such as numerosity, attention, and categorisation-like processes, may require only very limited neuron numbers. [Larger brains] contain greater replication of neuronal circuits, adding precision to sensory processes, detail to perception, more parallel processing and enlarged storage capacity ... Yet, these advantages are unlikely to produce the qualitative shifts in behaviour that are often assumed to accompany increased brain size. Instead, modularity and interconnectivity may be more important (p. R9950).

A corollary of this point of view is that big brains do not necessarily provide new modes of cognitive processing. In fact, large brains do not necessarily provide even relatively 'primitive' modes of cognitive processing: some humans have only low levels of cognitive processing even though they have large brains. It is clearly the number of neurons *and* the connections among them that matter (cf. Olkowicz *et al.*, 2016).

I suspect that an important selection pressure for a big brain in humans was the need for storage. After the invention of speaking, but before the invention of writing, knowledge could only be passed between generations aurally. A large brain was necessary to remember and transmit ideas, general knowledge, and many cultural traditions, let alone to recite whole book-length poems. Writing reduced the need for a large memory, and computer storage has reduced it even more. Perhaps in several thousand years, human brains will be smaller!

8

Phylogeny of Structure
Evolution of Behavior

Konrad Lorenz (1903–1989). On a diving trip in Kaneohe Bay, Hawaii, in 1967. Photo by Ernst Reese.

The word 'evolution' is derived from the Latin *ēvolūtiō*, which means 'an unrolling'. It was originally used by eighteenth century biologists to describe the changes seen in the developing embryo, from egg to adult. It was later that it came to be used to describe changes in species' characteristics across generations (see, for example, Richards, 1992). Only after the publication of *The Origin of Species* (Darwin, 1859) has the word evolution been widely used in its modern sense. It is also noteworthy that Darwin himself did not use the word in the first edition of *The Origin*; he used the phrase "descent with modification". Modern studies of evolutionary biology investigate both the changes that have occurred in species over generations (phylogeny) and the mechanisms that are responsible for these changes. In this chapter, I begin by describing changes in behavior over generations, including some mention of mechanisms along the way, and discuss mechanisms of evolution in more detail at the end of the chapter.

EVOLUTION IS HISTORY

The study of the evolution of behavioral structure involves attempting to reconstruct the past: How have perceptual, central, and motor mechanisms, and the connections among them, changed over generations? A major problem, however, is that one of the most important methods used to study the evolution of morphological structures is to analyze evidence left by fossils, and behavior mechanisms leave no fossils. This has meant that most studies of behavioral evolution rely on evidence from living species. Of course, it is sometimes possible to infer behavior of extinct species from structure: Could this species walk on two legs? Could that species fly? Could this species eat meat? Morphological structures may suggest answers to these questions, but nevertheless, all such inferences remain speculations because, as we have seen, structure is never a perfect predictor of function. With this caveat, I will discuss some of the evidence that suggests how some behavior may have evolved.

Evolutionary studies of behavior are usually interested in discovering either the progression of changes in some behavioral character or trait of a species across generations or the historical relationships among closely related species. If these changes involve descent with modification (*homology*), it is possible to make strong inferences about the historical relationships within and among species. One problem, however, is that some traits may end up being similar even though the changes to reach the same endpoint have taken different paths (*homoplasy*); this is usually called similarity by *convergence*. Ryan (2005, in press) gives a survey of the methods used to disentangle homology from homoplasy in behavior, and I will mention several examples below.

EVOLUTION OF MOTOR MECHANISMS

This question was investigated in early studies on pigeons by Whitman (1911) and on ducks by Heinroth (1910), but it only became a question of general interest due to the influence of Lorenz (1937, 1950). Lorenz pointed out that an *Erbkoordination* (a motor mechanism) could be considered an 'organ' or taxonomic character of a species, and that its evolution could be traced in the same way as the evolution of morphological structures. A major focus of interest has been on the evolution of displays.

A *display* is a movement (motor mechanism) that is specifically adapted to serve as a signal to another individual, usually a member of the species, but sometimes a member of another species. It is equivalent, in most ways, to the older ethological concept *Auslöser* (Lorenz, 1935). The zigzag dance in sticklebacks, waltzing in chickens, and song in birds are examples of displays. A human example is the eyebrow flash shown in Figure 2.2. Other movements, such as pecking in chicks, may serve as signals, but these are not usually considered to be displays, because their primary function is not related to influencing the behavior of another individual (such movements are called 'cues' in the behavioral ecology literature). Displays have attracted much interest among behavioral biologists because they are so conspicuous and often bizarre, which raises many questions about how such strange behaviors could have evolved. In this section, I will discuss the evolutionary origins of displays and some other movements. The functions of displays are considered in Chapter 9 and the question of adaptation is considered in Chapter 10. A seminal paper on these topics is by Tinbergen (1952).

One display described earlier was the 'upright' posture of the herring gull (Figure 3.9). This was considered to be an ambivalent movement that contained elements of attack (bill held pointing down) and escape (head held high with sleeked neck feathers). Sometimes when this display is shown, attack components are stronger than escape components, and, at other times, the reverse is true. In general, one can see a continuous transition between an aggressive upright and an escape upright. Such a continuous transition is not seen, however, with other displays. Two other agonistic displays seen in many species of gulls are the 'oblique' and the 'forward'. Tinbergen (1959) analyzed the transition between these two displays in the black-headed gull (*Larus ridibundus*) using cinefilms. Frame-by-frame analysis showed that, in comparison with the time taken by each posture, the transition between them is extremely short (see Figure 8.1). If we assume that these postures also originated as a superposition of attack elements and escape elements and that the motivational states underlying attack and escape vary continuously in intensity, the abrupt transition between the oblique and the forward must imply that

OBLIQUE   TRANSITION   FORWARD

Figure 8.1 Transitions in postures in the black-headed gull. Average number of cinefilm frames for Oblique (32+), transition (5), and Forward (19+) in 3 film shots (24 frames per second).
From Tinbergen, 1959.

changes in behavior mechanisms have occurred in the course of evolution.

In the examples I have just discussed, evolutionary change has been inferred by comparing a particular movement – the display – with other movements in the same species. Another approach that has proved to be successful in suggesting evolutionary relationships among behavior patterns is to compare similar behavior patterns in closely related species. Similarity in form, in causation, and in development can all be considered (see Baerends, 1958). From these comparisons, a tentative picture of evolutionary changes can be constructed. Studies of courtship displays in ducks provide an example of the comparative approach. As discussed earlier, court-ship is a situation of conflict. Stimuli are present that release attack and escape behavior, as well as sexual behavior. One type of move-ment that often occurs in conflict situations is a displacement activ-ity; and displacement preening is a movement that is often seen during courtship in many species.

In several species of ducks (see Figure 8.2), displacement preen-ing has acquired a signal function during courtship. In the shelduck (*Tadorna tadorna*), courtship preening is quite similar to normal preen-ing, but is somewhat more vigorous and is always directed to the wing. The mallard (*Anas platyrhynchos*) raises its wing and preens a brightly colored patch on the wing. The garganey (*Querquedula querquedula*) makes very incomplete preening movements toward a blue patch on the back of its wing, while the mandarin (*Aix galericulata*) merely touches a single, large, bright orange feather that sticks out of the other wing feathers. It is likely that these courtship displays form a series from less-to-more ritualized (see p. 304), and that all of them originated from the same displacement activity (Lorenz, 1941; Tinbergen, 1965). Possible scenarios for the evolution of the

SHELDUCK

MALLARD

GARGANEY

MANDARIN

Figure 8.2 Ritualization. Explanation in text.
From Tinbergen, 1965 (after Lorenz, 1941)

mandarin's display are easier to imagine because of the presence of the intermediate forms.

One of the most extensive of the earlier comparative studies was that of Tinbergen (1959) and his colleagues on the behavior of gulls. They observed the behavior of 12 species of gulls from the genus *Larus*, plus the kittiwake *(Rissa tridactyla)* and the ivory gull *(Pagophila eburnea)*. One of their goals was to understand how the various displays seen during pair formation and courtship originated and diverged in the course of speciation. They analyzed ten displays that occurred in most of the species studied (including the upright, oblique, and forward discussed above), with respect to their form and causation, as well as by comparing the displays in the different species. Tinbergen tentatively concluded that all the displays were movements derived from components of attack, escape, feeding, and nest building behaviors. Interestingly, unlike the ducks, components of preening did not appear to be involved in any of the displays. He also concluded: "the individual displays are each the result of divergent evolution started from a common root ... They have each descended from either a display already possessed by the common ancestor, or have developed [in] parallel after the ancestral species had broken up" (p. 55).

Another of the early studies was that of Lorenz (1941) comparing displays in various species of ducks, as discussed above. More recently, Johnson *et al.* (2000) followed up on the Lorenz study using a phylogenetic perspective to investigate the evolution of postcopulatory displays in dabbling ducks (Anatini). [For a brief introduction to phylogenetic reconstruction see Davies *et al.* (2012, p. 37ff.) or Ryan (2005, in press).] They observed the postcopulatory displays of 48 species, which they divided into initial displays (immediately following copulation) and additional displays. The initial displays in most species were either 'wing-up-bill-down' or 'bridal' (see Figure 8.3). When the presence or absence of these displays was mapped onto a phylogeny

Figure 8.3 'Bridling' of the mallard drake following copulation. From Lorenz. 1941.

derived from mitochondrial DNA sequences, it appeared that the occurrence of these two displays was highly stereotyped and phylogenetically conservative. The wing-up-bill-down display was characteristic of the less derived species, while the bridal was characteristic of the more derived species. In contrast, the occurrence of the additional displays was less stereotyped and less phylogentically conservative. These results implied that few evolutionary changes were necessary for the evolution of the initial display sequence, but more were necessary for the additional displays. The evolution of the additional display sequences implies convergence. I should also point out that this study did not consider the evolutionary origin of the displays, but rather was concerned with the evolution of the sequence of displays after copulation. Evolution of the displays themselves is discussed by McKinney (1975) and Johnson (2000).

Not all studies of motor mechanism evolution involve birds and displays. For example, Langtimm & Dewsbury (1991) investigated the phylogeny and evolution of rodent copulatory behavior in 22 species of Sigmodontine murids (New World woodrats and deer mice) using a phylogenetic perspective. Five characters of copulatory behavior in these species were examined: locked copulation; pre-ejaculatory intromissions; intravaginal thrusting; stereotyped dismount; and post-ejaculatory intromissions. Their analysis revealed that "behavior was consistent with a phylogeny derived from morphological characters. Primitive species display a mechanical lock between partners, whereas more derived species show no lock but produce a copulatory plug. The more derived species also display increased complexity in the copulatory sequence" (p. 217). As with the dabbling duck study, this study was concerned with the evolution of the copulatory sequence and not with the evolution of the individual behavior mechanisms.

A final example is the evolution of song in the *Drosophila repleta* group of species of fruit flies. During courtship, *Drosophila* males produce 'songs' using low frequency vibrations of one or both wings. There are about 1500 species of *Drosophila*, and in those species that have been investigated, species' song is unique. Ewing & Miyan (1986) investigated the songs of 22 species of the *repleta* group. Songs in this group have either one or two components that the authors labeled A and B. Some species have only A songs or B songs and some species have both. A songs occur at the beginning of courtship and are thought to be involved in species recognition, while B songs occur later in courtship and are thought to stimulate females sexually. One might have hypothesized that the more complex two-component song evolved from the one-component songs. However, comparing the songs of these species with a phylogeny of the group based on cytological evidence led to the conclusion that the two-component song was present in the common ancestor, and that species with only one-component songs have lost one of the songs. This example shows that it is a mistake to assume that evolution always proceeds from the simple to the complex.

EVOLUTION OF PERCEPTUAL MECHANISMS

The definition of a display as a movement that is adapted to serve as a signal implies that the 'receiver' of the signal must be able to understand the signal; that is, it must have an appropriate recognition mechanism. If the receiver behaves appropriately to the signal, one can assume that it has some sort of recognition mechanism, but further investigation is required before concluding that the mechanism has been adapted for that function. Recall the egg recognition mechanism of the herring gull. The gull, in the appropriate situation and motivational state, does retrieve its egg, but supernormal eggs are greatly preferred. There are actually very few comparative studies that have specifically considered the evolution of perceptual mechanisms, but several studies of acoustic communication in frogs have considered the evolution of both motor and perceptual mechanisms.

The sensory structures responsible for auditory reception in frogs are the amphibian and basilar papillae located in the inner ear. In an early study, Frishkopf & Goldstein (1963) recorded responses to acoustic stimuli from single units in the eighth (auditory) nerve of the bullfrog, *Rana calesbeiana*. Their study revealed two strikingly different kinds of auditory units that exhibited sharply frequency-dependent sensitivity (tuning curves). They called these units 'simple' (lower-frequency units) and 'complex' (higher-frequency units) and identified them as derived from the basilar and amphibian papillae, respectively. Subsequent studies by Capranica (1965) showed that the tuning of the auditory units matched the frequencies with the greatest relative energy in the bullfrogs' mating call. He proposed that call recognition results from stimulation of the peak sensitivity regions of the papillae in characteristic ways. Further studies in other species of frogs (review in Gerhardt, 2001) found that mating calls of some species (now often called advertisement calls) may stimulate only one or the other structure, but the frequency peaks of the call still match the tuning curves of the papillae. From these and other data, it appears that the sensory structures are phylogenetically conservative, which implies that their characteristics constrain the evolution of the calls.

A major comparative program by Ryan and his colleagues (review in Ryan, 2009) has been investigating the evolution of calls and auditory tuning in the *Physalaemus* (=*Engystomops*) *pustulosus* species group of treefrogs. One aspect of the program was concerned with the evolution of the advertisement call of the túngara frog, *Physalaemus pustulosos* (Wilczynski et al., 2001). This call contains two components: a whine (similar to the advertisement calls of other members of the species group) and a chuck (unique to this species).

> Initially, the evolutionary interpretation was that females evolved a preference for complex calls and low-frequency chucks because it allows them to choose larger males and thus maximize reproductive success. Thus trait and preference, signal and receiver, must have coevolved to bring about such benefits. The addition of an evolutionary perspective [phylogenetics], however, changes this interpretation. Both the preference for chucks and the auditory tuning that guides females to lower frequency chucks are not restricted to species with chucks; both of these receiver traits are found in species that produce only simple calls. Thus it appears that there was a preexisting preference for chucks and that in evolving chucks males exploited this pre-existing preference (Ryan, 2009, p. 141).

This conclusion is similar to the interpretation of Enquist and Arak (p. 34) that the mechanisms concerned with signal recognition have inherent biases in responsiveness.

Ryan and his colleagues have also been interested in the evolution of the sensory (perceptual) mechanisms that subserve species recognition in the *pustulosus* group of species. All these species have a species-specific whine as an advertisement call. In discrimination tests between conspecific and heterospecific calls, túngara females showed a strong preference for their conspecific call. In recognition tests, however, females often responded to heterospecific and reconstructed ancestral calls (Ryan & Rand, 1999). Using a variety of analyses, including artificial neural networks (Phelps et al., 2001), they were able to conclude that both phylogenetic relatedness (history) and call similarity contribute to explaining the behavior of the females. For these and other reasons, they suggest that the perceptual mechanisms used to determine mate choice in these and many other species have evolved in other contexts. This means that the species-specific mating calls have evolved to match perceptual mechanisms that were already present (Ryan & Cummings, 2013). I will return to these ideas in Chapter 10.

EVOLUTION OF CENTRAL MECHANISMS

Representations are a simple form of central mechanism, yet it is not clear what evolution of representations would mean. Most representations are formed during the ontogeny of the individual and depend on specific interactions with the external environment. Such development can be strongly genetically biased, as is the development of many of an individual's perceptual and motor mechanisms, but in general, most representations are unique to an individual during its lifetime. The evolution of representations reduces then to a consideration of the evolution of developmental mechanisms, more specifically of the evolution of learning.

As we have seen in Chapter 7, however, learning is a somewhat nebulous concept, and the evolution of learning would involve determining when particular learning mechanisms first appeared and how they have changed with time. Considering all the ramifications of these issues would require a whole book (e.g. see Hodos & Campbell, 1969, for relevant discussion) but, for my purposes, it is sufficient to propose that each species has evolved an 'innate schoolmarm' (Lorenz, 1965, 1981) that filters or selects the information that is to be stored in a representation, as well as the learning mechanisms that build the representation. Presumably, these mechanisms evolved very early in evolutionary history, and may not have changed substantially since then. Of course, individual representations can become organized in various ways (e.g. into modules), but the type of individual experience that is necessary for the development of such organization has been a subject of much controversy (e.g. see Bolhuis, 2009, 2015), which has major implications for their evolution. See also Leadbeater, 2015.

Central coordinating mechanisms are more complex central mechanisms that must also have existed in one form or another for many millions of years, because they coordinate the perceptual and motor mechanisms that are necessary for the basic functions of life. Hunger (nutrition), sex (reproduction), attack (defense), and escape (defense) mechanisms must have been present in the earliest forms of life, but the structure of these mechanisms in modern species almost certainly represents both descent with modification and convergent evolution. As with motor and perceptual mechanisms, most studies compare these mechanisms in closely related species. Although few, if any, studies consider central coordinating mechanisms explicitly, the studies of postcopulatory displays in dabbling ducks and of rodent sexual behavior discussed above, for example, can actually be considered studies of central coordinating mechanism evolution.

Another example is provided by studies of feeding and parental behaviors of cichlid fishes found in the large East African lakes (Meyer, 1993; review in Henning & Meyer, 2014). These fish

> have evolved adaptations to eating every conceivable food source in their environment. Algae scrapers have flat teeth like human incisors that allow them to nibble the nutritious growths on rock surfaces; insect eaters have long, pointy teeth that help them to get into rock crevices; ambush predators possess huge extendable jaws with which they can suck in their unsuspecting prey in a matter of milliseconds (Meyer, 2015, p. 72).

These studies have been carried out in the context of adaptive radiation, which occurs when one lineage evolves into numerous species that each evolve specializations to a variety of ecological niches. Adaptive radiation is discussed in Chapter 10, but the results of these studies imply that the central coordinating mechanisms controlling feeding have also changed: they are activated by different stimuli and they control different motor mechanisms. Other vertebrates have also evolved specialized modes of feeding, and similarities in feeding techniques among these various species must have evolved independently and convergently.

The evolution of parental behavior provides examples of descent with modification as well as of convergent evolution. In Chapter 3, I pointed to the wide variety of parental behavior that exists: male only, female only, both parents, and no parental behavior. These forms all occur throughout the various classes of vertebrate species, and the complexity of the parental behavior varies similarly, which implies that parental behavior has evolved independently many times. Descent with modification, however, can be inferred from comparisons of closely related species. Among the cichlid fishes, for example, most species exhibit bi-parental care, in which females lay eggs on a suitable substrate, which are then fertilized by the male. Both parents then fan and guard the eggs and also guard the young for some time after they hatch and become free-swimming. There are many variations on this theme, however, and a number of cichlid species in the genera *Geophagus* and *Haplochromis* have evolved mouthbrooding behavior, in which the eggs are taken into the mouth immediately after laying and remain there until they hatch. It is usually the female that broods the eggs, although in a few species, it can be both parents or only the male. The substrate species often mouth the eggs to remove dead embryos or readjust their position, so it is easy to imagine how holding eggs in the mouth could have evolved by descent with modification. Of special interest, however, is the fact that *Geophagus* species live in South America, while *Haplochromis* species live in East Africa. This means that mouthbrooding behavior not only can be considered as descent with modification, but also represents convergent evolution.

My final example of the evolution of central coordinating mechanisms is the evolution of human language, a very controversial issue. A recent publication, authored by eight prominent experts in the field, concluded: "Based on the current state of evidence, we submit that the most fundamental questions about the origins and evolution of our linguistic capacity remain as mysterious as ever" (Hauser *et al.*, 2014, p. 1). A major part of the problem, however, is that there is little agreement on the definition of human language. I have suggested that human language can be considered a behavior system, like any other behavior system, and that it contains two major central coordinating mechanisms, syntax and semantics (p. 272). This is a structural definition of language, similar to the definition of Berwick *et al.* (2013), and opposed to a definition based on communication (function) – e.g. Hauser, 1996; Sterelny, 2012; Lieberman, 2015. The essence of language evolution then becomes the study of the evolution of semantics and of syntax. The semantics mechanism consists of representations, both concrete and abstract, and following my suggestion above, the evolution of semantics (meaning) is congruent with the evolution of learning and the evolution of semantic organization. The evolution of syntax is more problematic.

In Chapter 2, I discussed evidence supporting the existence of motor pattern coordinating mechanisms and suggested an alternative name of syntax generating mechanisms. Such mechanisms order the sequence of motor patterns (words), and in principle can follow particular 'inbuilt' rules. Insight into the evolution of these rules would normally be provided by comparing the rules of human syntax with the rules of closely related species. Unfortunately, there are no closely related species extant, though the rules in other species may not be so different (cf. Lashley, 1951). Berwick *et al.* (2013) propose that the syntax module, which they call 'merge', did not evolve under a selection pressure for communication, but under some other unspecified selection pressure; it was later co-opted for the communication function. If such a scenario is correct, then both sender and receiver would possess the same central coordinating mechanism, and mutual understanding would not require a co-evolutionary process. Rather, some mechanism such as a 'mirror neuron' (Rizzolatti & Craighero, 2004) could be sufficient. The rules of behavior sequences in many animals exemplify hierarchical structure, but whether the recursive hierarchical structure ('merge') found in human language could have evolved from the structure found in other animals is currently a hotly debated issue (Pulvermüller et al., 2010; Moro, 2013, 2014; Bolhuis et al., 2014).

DOMESTICATION (ARTIFICIAL SELECTION)

As a postscript to the section on 'evolution is history', a few words on domestication seem appropriate because, although domestication is also history, it is recent history that has occurred during the period when humans could talk and, sometime later, write. Verbal and written (or pictorial) descriptions can actually be considered behavior fossils. The process of domestication is an example of evolution in action, and has provided many insights into the mechanisms of evolution (see below). With respect to animals (as opposed to plants), domestication usually means changes in attack and escape central mechanisms: domesticated animals are more docile than their wild counterparts. The study of the process of domestication was pioneered by Darwin (1868), but recent advances in genomics have created new understanding of, and renewed interest in, the process (Wayne & vonHoldt, 2012; Larson & Fuller, 2014).

Archaeological evidence indicates that the dog was the first species to be domesticated. Genomic evidence suggests that dogs and gray wolves are sister taxa that are descended from an unknown ancestor that has gone extinct (Freedman et al., 2014). About 16,000 years ago, these two taxa began to diverge from each other. The branch that became dogs gradually lost its fear of humans, while the wolves remained wary. It is thought that dogs began to approach human camps and eat food that had been uneaten or left behind. Later, there is evidence that dogs were accompanying humans on hunting expeditions (J. Clutton-Brock, 1995). The actual progression of these changes is unknown, of course, but it is clear that by about 12,000 years ago, the escape and/or fear central mechanisms of dogs had changed significantly. Morphological features also changed during this period. Even greater changes in both morphology and behavior occurred during the period between 4,000 and 2,000 years ago, when dogs were selected for various functions such as hunting, guarding, shepherding, and decoration (lapdogs). Because these recent changes in behavior have occurred during historical times, it is possible to follow the progression of changes more accurately. Coppinger & Schneider (1995) provide an analysis of the evolution of behavioral changes in working dogs.

After many humans changed from a hunter-gatherer lifestyle to an agricultural lifestyle about 11,500 years ago, other species came to be domesticated. Cattle, sheep, goats, and pigs were among the earliest. A few thousand years later (about 4,000 to 3,000 years ago), chickens, horses, and camels were domesticated, and since then, many other species, including fish (see Larson & Fuller, 2014, for review). Zeder (2012) has proposed three pathways to domestication. The first she called the *commensal pathway*, in which animals begin to forage around human settlements, become less wary of humans, and are tolerated by humans because of the benefits they provide. Dogs, cats, chickens, and possibly pigs would be included in this category. The second she called the *prey pathway*, because these species were first hunted for food or fur and later captured and managed by their human captors. This category would include cattle, sheep, and goats, as well as various species raised for their fur such as foxes and chinchillas. Under conditions of captivity and directed breeding, these species also became less wary of humans and more easily managed. The third she called the *directed pathway*, because these species were originally deliberately captured with the goal of domesticating a free-living animal to obtain a specific resource of interest. This category includes horses, donkeys, and camels. For all three categories, the domestication process ends with directed breeding. Of special interest for the evolution of behavior, Zeder discusses characteristic pre-adaptations of behavior that are favorable or unfavorable for domestication that include social structure, sexual behavior, parent-young interactions, feeding behavior and habitat choice, and responses to humans. Differences in these characteristics can explain why dogs, but not wolves, and cattle, but not gazelles and zebras, have become domesticated. Such pre-adaptations also place limitations on possible further adaptations as is discussed in Chapter 10.

CULTURAL TRANSMISSION AND EVOLUTION

All the examples discussed above are changes in behavior observed over generations that were assumed to be caused by genetic changes. In order to prove that this is so, it would be necessary to do genetic experiments. Very few such experiments have been done, but the scant evidence that does exist supports a genetic hypothesis. An alternative mechanism for evolutionary change in behavior is cultural transmission. We saw in Chapter 7 that social learning provides a mechanism for passing information between generations. Whether socially transmitted information should be considered cultural transmission, however, has been, and still is, a very controversial issue. The problem, as usual, lies with the definition of the word 'culture'. Many social scientists believe that culture is a uniquely human phenomenon, and do not accept its use by behavioral biologists. Even behavioral biologists have differing definitions. Bonner (1980, p. 10), for example, defined culture as "the transfer of information by behavioral means", which makes culture synonymous with social learning. A more restrictive definition was proposed by Nishida (quoted by Whiten et al., 1999, p. 682): "A cultural behaviour is one that is transmitted repeatedly through social or observational learning to become a population-level characteristic." Pagnotta (2014) traces the history of different conceptions of culture among social and biological scientists, and discusses possible ways to reach consensus; for the time being, however, there is no consensus.

Non-human Cultural Transmission

Using Nishida's definition, are the examples of social learning I discussed in Chapter 7 also examples of cultural transmission? Most are probably not, even taking population-level to mean a restricted group of individuals within a species. One definite exception, however, is song learning in birds (Slater, 1986). Marler & Tamura (1964), in laboratory studies, showed that any of the three 'dialects' found in white-crowned sparrows in the San Francisco Bay area could be taught to any young white-crowned sparrow, regardless of its geographical origin. They considered these results implied cultural transmission. Since then, there have been many more studies of dialect learning in several species of birds, including starlings and zebra finches (e.g. Fehér & Tchernichovski, 2013). Most of these studies have been carried out in the laboratory, but in one study, Payne and his colleagues (1981) were able to follow the local song dialects of the indigo bunting (*Passerina cyanea*) for fifteen years in the wild. Song types outlived the birds themselves by about three to one. Another field study (Ince et al., 1980) found that several chaffinch songs lasted from at least 1960 to 1980 in a particular location in England. These results mean that the local songs are relatively long-lived behavior traditions that persist by social song learning.

Figure 8.4 A capuchin monkey cracking a cashew nut with a stone in Serra da Capivara National Park in northeastern Brazil.
Courtesy of Tiago Falótico.

There is now also considerable evidence that various primates pass on traditions from one generation to the next. For example, some, but not all, groups of chimpanzees (*Pan troglodytes*) push sticks into the nests of ants and termites. The insects crawl onto the stick, which is then withdrawn and the insects eaten. Observations of infants with their mothers suggested that the young chimps imitate the older animals in this behavior (van Lawick-Goodall, 1970). Later studies by several researchers studying groups of geographically separated chimps found even more differences in behavior repertoires among the groups that included tool usage, grooming, and courtship behaviors. These results were taken as strong evidence for cultural transmission in non-human animals (Whiten *et al.*, 1999, 2001). Similar evidence for cultural transmission has also been found in groups of orangutans, *Pongo pygmaeus* (van Schaik *et al.*, 2003).

Another example of primate cultural transmission is provided by studies of capuchin monkeys (*Sapajus* spp.) found in various regions of Brazil and neighboring countries (Ottoni & Izar, 2008). Many, but not all, groups of monkeys have been observed to use heavy stones as pounding tools to crack open hard palm nuts, and sometimes other hard-shelled nuts as well (see Figure 8.4). Of special interest are two groups of *Sapajus libidinosus* that have been observed in the wild in Serra da Capivara National Park in northeastern Brazil. These two groups manufacture tools to probe for lizards by trimming branches and thinning their tips, and use pounding stones for breaking and/or enlarging holes in tree trunks or rocks, as well as for cracking nuts (Falótico & Ottoni, 2014, 2016); the use of pounding stones to open cashew nuts goes back at least 600 to 700 years (Haslam *et al.*, 2016). These behaviors have not been observed in other groups of capuchins. The young monkeys have been seen observing older monkeys performing these behaviors, but it takes several years of practice before the young monkeys become proficient. Resende *et al.* (2014) have studied how the young monkeys learn to crack nuts. Other examples of cultural transmission in non-human animals can be found in Laland & Galef (2009).

Human Cultural Transmission

By definition, social information transmitted by humans must be cultural. There is one well-analyzed example that shows how cultural transmission can lead to the evolution of behavior in a very similar way to genetic transmission: the evolution of human languages. Using what they call the 'comparative method', linguists have been able to trace the origins of most modern-day languages and have constructed diagrams showing probable relationships among them (see Figure 8.5). For example, Swedish and Danish are more closely related to each other than either is to English or German. Nonetheless, all four of these languages originated from a common Germanic ancestor, and all are more closely related to each other than they are to Latin or Slavic languages; and all these and many others are descended from an extinct language called Proto-Indo-European (Watkins, 1969). More recently, Gray *et al.* (2011), using modern phylogenetic methods, have been able to determine the approximate dates at which these various languages diverged from each other, which allows the integration of evidence from linguistics, anthropology, archaeology, and genetics in making inferences about human history. (See also Levinson & Gray, 2012.)

In general, it is assumed that languages evolve because individuals make small mistakes in learning their language originally, or develop new expressions in their lifetime, and these changes get passed on to their children and neighbors. When small groups of people become isolated or migrate, these small changes can stabilize, and dialects are formed. Continuation of the process leads to new languages. Exactly this same process occurs in the evolution of new species or of new behaviors that are based on genetic changes that get passed on to

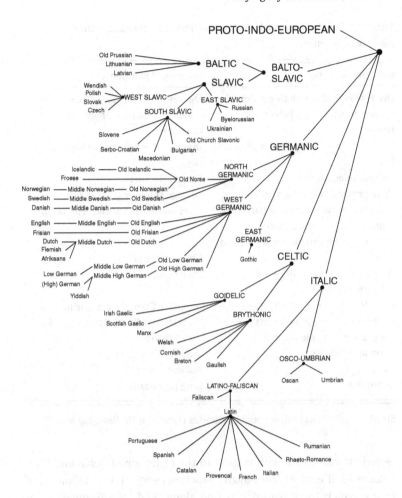

Figure 8.5 A partial phylogeny of human languages descended from an original Proto-Indo-European language.
Copyright 1981 by Houghton Mifflin Harcourt Publishing Company. Adapted and reproduced by permission from the *American Heritage Dictionary of the English Language*.

offspring and may, through a variety of mechanisms, become fixed in the population. Some of these mechanisms are discussed below.

Genes, Culture, and Evolution

Considering cultural transmission in the context of evolution raises two important issues: Is cultural transmission really different from

Table 8.1 *Parallels between biological and linguistic evolution. From Pagel et al., 2007, with permission.*

Biological Evolution	Language Evolution
Discrete heritable units – e.g. genetic code, morphology, behavior	Discrete heritable units – e.g. lexicon, syntax, and phonology
Homology	Cognates
Mutation – e.g. base-pair substitutions	Innovation – e.g. sound changes
Drift	Drift
Natural selection	Social selection
Cladogenesis – e.g. allopatric speciation (geographic separation) and sympatric speciation (ecological/reproductive separation)	Lineage splits – e.g. geographical separation and social separation
Anagenesis	Change without split
Horizontal gene transfer – e.g. hybridization	Borrowing
Plant Hybrids – e.g. wheat, strawberry	Language Creoles – e.g. Surinamese
Correlated genotypes/phenotypes – e.g. allometry, pleiotropy.	Correlated cultural terms – e.g. 'five' and 'hand'.
Geographic clines	Dialects/dialect chains
Fossils	Ancient texts
Extinction	Language death

Similar parallels were first noted by Charles Darwin in *The Descent of Man*.[39]

genetic transmission with respect to the evolution of behavior? Does culture itself evolve? Once again, much depends on the definition of terms, which I will consider as I go along. And, as a framework for discussing these issues, I have reproduced a table published by Pagel *et al.* (2007) in an article on lexical evolution. This table looks at parallels between 'biological' and linguistic evolution; linguistic evolution can be considered a prototype of cultural evolution.

Table 8.1 lists 13 characteristics of biological evolution and suggests 13 similar characteristics of linguistic (cultural) evolution. One of these characteristics is structural (heritable units), four refer to mechanisms (mutation or innovation, drift, natural or social selection, and horizontal transfer or borrowing), and the others refer to consequences of evolution. In what ways are these parallel characteristics the same or different? With respect to heritable units, the biology list contains genes, morphology, and behavior, while the culture list contains words and could, for example, contain behaviors that use tools (pounding with a stone or poking with a stick). Now, words and tool use (and cultural units in general) are behaviors; that is, they express the activation of specific behavior mechanisms. In that sense, they are no different from any other behaviors. Since all behavior mechanisms are biological, cultural units must also be considered biological. In an evolutionary context, these behavior mechanisms are the units that get passed from generation to generation, and there is no reason to think that the process of evolution is different for behaviors based on different modes of transmission.

In an article challenging the notion that biological and cultural evolution are different, Ingold (2004) compares walking with playing the cello, as a putative example of a biological versus a cultural behavior. He notes that walking in an individual develops during the first months of life and can take different final forms, depending on the culture into which the individual is born. For example, the Japanese, traditionally, were taught to walk from the knees, while Europeans are taught to walk from the hips. Ingold posits that being taught to walk is in principle no different from being taught to play the cello, both of which require individual skill learning.

> Now if, by evolution, we mean differentiation and change over time in
> the forms and capacities of organisms, then we must surely allow that
> a skill like cello-playing has evolved, as has the skill of walking in
> a certain way. No-one, however, would seriously suggest that people from
> different backgrounds walk in different ways, or play different musical
> instruments, because of differences in their genetic make-up. But nor
> does it make sense to suppose that these differences are due to
> something else, namely culture, that overwrites a generalized biological
> substrate (p. 218).

This line of argument is similar to that of Developmental Systems Theory (p. 181) with respect to the dichotomy of innate/learned, and of the role of epigenetic inheritance in developmental evolutionary biology (discussed below). See also Griesemer, 2000.

If cultural units of transmission, with respect to evolution, and genetic units of transmission are both biological, what differences remain? The mode of transmission (genes or social learning) has no differential consequences for evolution, nor does the agent of selection (nature, a member of the species, or humans). In all cases, heritable units change over time. In fact, as can be seen in Table 8.1, for each characteristic of biological (genetic) evolution listed, there is an equivalent characteristic of cultural evolution. It is often argued that genetic evolution is slow, while cultural evolution is fast, but there are many examples where the opposite is true (Laland & Brown, 2011, pp. 169–170; Pennisi, 2016a). I would point out, however, the conclusion that cultural units of change are also biological does not mean that the effects of genes and of culture on behavior cannot be studied independently; it only means that evolution, as a history of changes in behavior mechanisms over time, is agnostic as to the mechanisms by which these changes come about. In a remarkably prescient article, Waddington (1959a, p. 1637) wrote:

> When we can discern cultural differences between two human groups, does the main responsibility of these lie with the system of cultural transmission or with that of genetic transmission? Again, from an *a priori* point of view one must expect differences in both systems to be involved, and it is exceedingly difficult to assess the relative importance of the two contributing factors. It is quite clear that races such as the West Africans, the Maoris and the Chinese, differ genetically from Europeans. Some of these genetic differences are obviously expressed in skin colour; there must surely be others, more difficult to detect. But is there any reason, for example, to suppose that it is differences in the gene pool between the Chinese and the European populations which have caused the one to develop a social system based on such relatively unindividualistic systems as Confucianism and Buddhism and the other a civilization inspired by such a different system of thought and feeling as Christianity? I see no *a priori* reason to suppose anything of the kind; nor is there sufficient factual evidence to establish such a conclusion.

With respect to the question of whether culture itself evolves, there is much less difference of opinion: whatever one's definition of culture, culture clearly changes over generations. However, the mechanisms of cultural evolution remain controversial. Much of Darwin's (1871) discussion in *The Descent of Man* can be considered a study of cultural evolution, and biologists have generally assumed cultural evolution to be a fact, as can be seen in Waddington's (1959a, p. 1636) statement: "Human evolution has been in the first place a cultural evolution." But recent interest in cultural evolution probably stems from the theoretical models of Cavalli-Sforza & Feldman (1981) and of Boyd & Richerson (1985), as well as from developments in the study of the evolution of human cognition (Rendell et al., 2011; Heyes, 2012). The study of cultural evolution has developed into a subfield of its own, and Laland & Brown (2011, Chs. 7 & 8) provide an excellent overview.

One focus of interest has been the evolution of 'cumulative culture', which is the accumulation of learned improvements via social learning from one generation to the next. Several theoretical investigations showed that cultural traits could evolve if they enable individuals to obtain adaptive information that is otherwise costly or difficult to learn (e.g. Richerson & Boyd, 2005). However, Enquist & Ghirlanda (2007) pointed out that social transmission can involve maladaptive information as well as adaptive information, and showed that cumulative culture can only evolve if individuals can identify and discard maladaptive culture. They suggest "that the evolution of such 'adaptive filtering' mechanisms [cf. 'innate schoolmarm', above] may have been crucial for the birth of human culture" (p. 129). All these considerations have spurred a number of investigators to do empirical studies to test some of these theoretical ideas (e.g. Caldwell & Millen, 2008; Dean *et al.*, 2012; Morgan *et al.*, 2012).

As an example, Dean *et al.* (2012) asked why social learning is so much more widespread than cumulative culture. They provided small groups of capuchin monkeys, chimpanzees, and three- to four-year-old children with an experimental puzzle box that could be solved in three stages to retrieve rewards of increasing desirability. The appropriately scaled cumulative culture puzzle box required individuals to first slide a door, then to depress a button, and finally to rotate a dial. Although two monkeys solved stage one and one chimp solved stage three, many (but not all) children solved stage three. Analysis of the results led the authors to conclude that "reaching higher-level solutions was strongly associated with a package of sociocognitive processes – including teaching through verbal instruction, imitation, and prosociality – that were observed only in the children and covaried with performance" (p. 1114). Although one could argue that the puzzle box task does not really reflect cumulative culture and that the actions required are more suited to children than to monkeys or chimps, the results do suggest that certain social behavior prerequisites (pre-adaptations) are necessary before cumulative culture can evolve.

MECHANISMS OF EVOLUTION: A BRIEF EXCURSION INTO EVOLUTIONARY BIOLOGY

This is a book about behavior, but, as we have seen in this chapter, understanding how behavior has evolved requires some knowledge about evolutionary mechanisms. Many of the issues discussed in the next two chapters on function and adaptation also require such knowledge. This section on evolution gives a brief outline of the development of evolutionary ideas, and of current understanding of evolutionary mechanisms. There is actually a great deal of controversy in the field these days, especially between those who find the mainstream theory of evolution sufficient and those who feel evolutionary theory needs a major overhaul to accommodate new findings (see Laland *et al.*, 2014 and Wray *et al.*, 2014). It is not clear when the dust will settle, but the reader is advised to consult the most recent publications for further information and interpretations.

Toward the end of the eighteenth century and the beginning of the nineteenth, the long-held idea that species were a given and were immutable came under discussion. Many people of the time no longer believed the myth of creation as proclaimed in the Bible, although most still believed in a deity. Among these were two French naturalists, Jean-Baptiste Lamarck and Étienne Geoffroy Saint-Hilaire, professors at the *Muséum National d'Histoire Naturelle* in Paris at the turn of the century. Both men proposed that species were not immutable, but could be transformed with time. Geoffroy's idea was that the environment directly induced organic (heritable) change in organisms, and these changes could occur in big steps (saltation). He also had a theory of the 'unity of plan in organic composition', which is discussed below. Lamarck had a more complete theory of evolution, 'the transmutation of species', in which organisms changed from simple to complex over time, and on the way developed adaptations to their environment by the use or disuse of certain of their characteristics (inheritance of acquired characters). These and several other ideas were circulating in the mid-nineteenth century, but it was Darwin's theory, published in 1859, that became the basis of modern evolutionary theory, primarily because Darwin had an abundance of well-organized evidence to support his theory.

Darwinian Selection

Darwin's (1859) theory is encapsulated in the title of his book: *On the Origin of Species by Means of Natural Selection or the Preservation of Favoured Races in the Struggle for Life*. Organisms vary in their characteristics and more organisms are born than the environment can support, so in the struggle for life, it is the "favoured races" (the most fit individuals) that are preserved. This *is* natural selection or 'survival of the fittest' as Herbert Spencer (1864) described it. As mentioned above, Darwin did not use the word evolution in *The Origin*, possibly because it connoted the idea of continuous progress, or an unfolding of species from simple organisms to man, an idea he disagreed with. (He did use the word 'evolved' once: it is the last word in the book!) Darwin used the phrase "descent with modification" and envisaged a branching structure: when taxa diverged, each branch would continue evolving independently. In later publications, he did use the word evolution, and in *The Descent of Man* (1871), man does emerge as the pinnacle of evolution. These days, however, biologists no longer refer to different species as higher or lower, but rather compare species as more or less derived from a common ancestor (see Hodos & Campbell, 1969).

For species to evolve, there must be continuous production of variation, and the progeny of the successful individuals must inherit their parents' favorable characteristics. In Darwin's time, the laws of inheritance were not known beyond the folk wisdom that 'like begets like', nor were the many sources of variation known. Based on his observations of the process of domestication, Darwin assumed that variation was random, although he did believe that the Lamarckian mechanism of use/disuse might play a role in addition to selection; and most of his speculations about heredity were wrong. Nonetheless, as many authors have pointed out, the logic of his argument is correct. Lewontin (1970, p. 1) wrote:

> As seen by present-day evolutionists, Darwin's scheme embodies three principles:
> 1. Different individuals in a population have different morphologies, physiologies, and behaviors (phenotypic variation).
> 2. Different phenotypes have different rates of survival and reproduction in different environments (differential fitness).
> 3. There is a correlation between parents and offspring in the contribution of each to future generations (fitness is heritable).
> These three principles embody the principle of evolution by natural selection. While they hold, a population will undergo evolutionary change.
> It is important to note a certain generality in the principles. No particular mechanism of inheritance is specified, but only a correlation in fitness between parent and offspring. The population would evolve whether the correlation between parent and offspring arose from Mendelian, cytoplasmic, or cultural inheritance. Conversely, when a population is at equilibrium under selection (for example, a stable polymorphism due to heterozygous superiority), there is *no* correlation in fitness between parent and offspring, no matter what the mechanism of inheritance. Nor does Principle 2

specify the reason for the differential rate of contribution to future generations of the different phenotypes. It is not necessary, for example, that resources be in short supply for organisms to struggle for existence. Darwin himself pointed out that "a plant at the edge of a desert is said to struggle for life against the drought." Thus, although Darwin came to the idea of natural selection from consideration of Malthus' essay on overpopulation, the element of competition between organisms for a resource in short supply is not integral to the argument. Natural selection occurs even when two bacterial strains are growing logarithmically in an excess of nutrient broth if they have different division times.

The generality of the principles of natural selection means that any entities in nature that have variation, reproduction, and heritability may evolve.

Note that in this formulation, natural selection is the *outcome* of variation, reproduction, and heritability (cf. p. 16).

The 'Modern Synthesis'

From Darwin's time to the present, many aspects of his theory have been the subject of debate, especially those concerning the factors causing the variation on which selection acts, and the nature of heredity. Soon after the publication of *The Origin*, new discoveries required changes to Darwin's ideas or additions to the theory. In the 1880s, Weismann's insights into the mechanisms of heredity ('continuity of the germ plasm'), derived from the developing field of cell biology, led to a rejection of the idea of the inheritance of acquired characters. The rediscovery of Mendel's work in 1900 showed that inheritance was particulate, and not a blend of characteristics of the parents; the mathematical theorizing of Fisher, Haldane, and Wright in the 1920s and 1930s demonstrated how evolution could be characterized as a change in the genetic composition of populations. These and other advances forged an expanded theory of evolution variously called the 'synthetic theory' or 'neo-Darwinism' or, finally, the 'Modern Synthesis' (Huxley, 1942). Extensive histories of these changes are provided in the books by Mayr (1982) and Gottlieb (1992). Jablonka & Lamb (2014, Ch. 1) also present a history of these changes emphasizing changing fashions in evolutionary studies. This is how they characterize the Modern Synthesis (p. 29):

> Heredity is through the transmission of germ-line genes, which are discrete units located on chromosomes in the nucleus. Genes carry information about characters.

Variation is the consequence of the many random combinations of alleles that are generated by the sexual process, with each allele usually having a small phenotypic effect. New variations in genes – mutations – are the result of accidental changes; genes are not affected by the developmental history of the individual.

Selection occurs among individuals. Gradually, through the selection of individuals with phenotypes that make them more adapted to their environment than others, some alleles become more numerous in the population.

In this formulation, evolution becomes synonymous with genetic change, and mutation and selection are its instruments. After the discovery of the role of DNA in reproduction in the 1950s, there were some tweaks to the theory; in particular, the theory became more 'gene-centric', the *selfish gene* (R. Dawkins, 1976b), rather than focusing on selection of individuals. Nonetheless, the Modern Synthesis is/was the prevailing theory of evolution well into the twenty-first century. However, not everyone was convinced.

The Role of Development in Evolution

One of the tenets of the Modern Synthesis is that genes are not affected by the developmental history of the individual. Waddington (1959a, p. 1635) proposed an alternative account:

> Natural selective pressures impinge not on the hereditary factors themselves, but on the *organisms* [my emphasis] as they develop from fertilized eggs to reproductive adults. We need to bring into the picture not only the genetic system by which hereditary information is passed on from one generation to the next, but also the 'epigenetic system' by which the information contained in the fertilized egg is expanded into the functioning structure of the reproducing individual.

In a series of experiments on *Drosophila*, Waddington (e.g. 1959b) exposed eggs or larvae to stressors such as heat shock, ether vapor, or salt, and selected individuals with particular effects of the stressor as breeding stock for the next generation. After as few as thirteen generations, the effect of the stressor appeared *before* the stressor was applied. Waddington termed this 'genetic assimilation' of an acquired character. This effect could be traced to genetic changes, but the point was that the changes were not random: they were responsible for an adaptation to the stressor. Waddington also pointed out that the behavior of the animal can modify its environment, which can change the selective

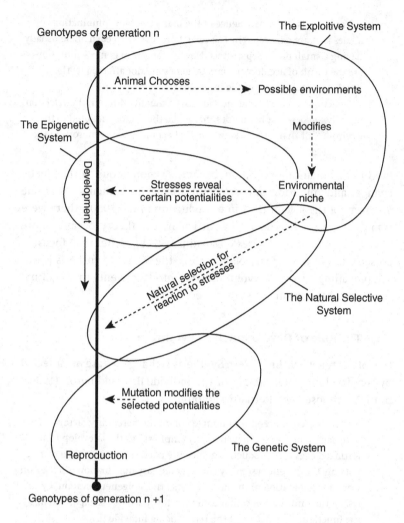

Figure 8.6 The logical structure of the biological evolutionary system.
From Waddington, 1959a, with permission.

pressures to which it is exposed. He called this the exploitive system.
Figure 8.6 shows his conception of how the exploitive, epigenetic,
selective, and genetic systems are involved in biological evolution.

Waddington's ideas were mostly ignored for many years, but
since the 1990s, similar ideas have been championed by the developmental psychologist Gilbert Gottlieb (1992), the theoretical biologists
Mary Jane West-Eberhard and Eva Jablonka, and the evolutionists John
Odling-Smee, Kevin Laland, and Marcus Feldman, among others. West-

Eberhard's (2003) book, *Developmental Plasticity and Evolution*, presents evidence for 'phenotypic accommodation', which is adaptive adjustment, without genetic change, of variable aspects of the phenotype following a novel input during development. It sets the stage for evolution by natural selection in the same way as Waddington's genetic assimilation: "Genes are followers, not leaders, in adaptive evolution." Odling-Smee *et al.* (1996) point to the role the animal plays in determining its own selection pressures, 'niche construction'. Jablonka's book, *Evolution in Four Dimensions: Genetic, Epigenetic, Behavioral, and Symbolic Variation in the History of Life* (Jablonka & Lamb, 2014), marshals a wide range of evidence for the role of various kinds of epigenetic and nongenetic factors in evolution. She sometimes refers to herself as a neo-Lamarckian. (See also Bohacek & Mansuy, 2015.)

A somewhat different role for development in the process of evolution is given by biologists who study the evolution of developmental mechanisms, a field called 'evo-devo' (Raff, 2000; Moczek *et al.*, 2015). Most of the work in evo-devo involves molecular genetics, detailed coverage of which is outside the purview of this book (and of this author's competence). Nonetheless, some of the findings in the field have many implications for the evolution of behavior. For example: "The formation and differentiation of many structures such as eyes, limbs, and hearts – so morphologically divergent among different phyla that they were long thought to have evolved independently – are governed by similar sets of genes and some deeply conserved genetic regulatory circuits" (Carroll, 2008, p. 28). This has been called 'deep homology' by Shubin *et al.* (2009). The regulation of the development of these structures in each species is controlled by regulator genes, which is where most of the genetic diversity among species is found, and where most changes (mutations) occur in the course of evolution. Since the brain is also a morphological structure, much behavior must also have evolved in this way. Interestingly, Geoffroy's theory of the unity of composition, mentioned above, is very similar in logic to the evo-devo account: All animals are formed of the same elements, in the same number; and with the same connections: homologous parts, however much they differ in form and size, must remain associated in the same invariable order. Development is a working-out of existing potential in a given type, and transmutation occurs as a direct effect of the environment (paraphrased from Wikipedia).

No one disputes that genetic mutations cause changes in organisms that affect their fitness. And no one disputes that development can also influence an organism's fitness. But whether the Modern

Synthesis version of evolution needs a major overhaul remains a very contentious issue. The developmentalists (Waddington, Gottlieb, West-Eberhard, Jablonka, Carroll, Odling-Smee, Laland, Feldman) are quite vocal in their call for change, whereas the traditionalists (e.g. Wray *et al.*, 2014) are quite content with tweaks to the theory. Time will tell the outcome.

Causes, Mechanisms, and Processes of Evolution

What are the causes of evolution? As usual, the answer depends on one's conception of cause and of evolution. In a recent article comparing evolutionary 'causes' with evolutionary 'processes', Laland (2015, p. 97) argues that many of the conventions within the evolutionary sciences have become counterproductive.

> These include the views (i) that evolutionary processes are restricted to those phenomena that directly change gene frequencies, (ii) that understanding the causes of both ecological change and ontogeny is beyond the remit of evolutionary biology, and (iii) that biological causation can be understood by a dichotomous proximate – ultimate distinction, with developmental processes perceived as solely relevant to proximate causation. I argue that the notion of evolutionary process needs to be broadened to accommodate phenomena ... that bias the course of evolution, but do not directly change gene frequencies, and that causation in biological systems is fundamentally reciprocal in nature.

I generally agree with these comments, though I would add that the concept of cause also needs some discussion (cf. p. 20).

Mechanism is a structural concept. It actually corresponds to Wouter's (2003) first meaning of function: what an item does by itself or, more simply, how it works. Mechanism can be concrete or abstract: it can be an organization of neurons (e.g. behavior mechanism) or a description of a process (evolution is the outcome of variation, differential fitness, and inheritance). Accepting these arguments means the number of mechanisms of evolution at the various levels of organization becomes very large indeed, and even listing them would be a difficult task. I do not try, but I will mention a few examples of behavior evolution that illustrate the variety of mechanisms involved and will return to this issue in Chapter 10.

When speculating on the genetic mechanisms responsible for canalization and genetic assimilation, Waddington (1959a) noted that there are many genes that are not expressed during normal development (genes 'hidden' from natural selection). However, if the developing organism is submitted to a 'stressor', these genes can become activated, and become exposed to natural selection (see Figure 8.5). With respect to canalization, these hidden genes are a type of back-up mechanism for stability. With respect to genetic assimilation, the hidden genes actually lead to a specific (and often adaptive) response to the stressor.

The process of domestication is also very stressful for an animal, and studies of domestication have uncovered many unexpected evolutionary mechanisms. A famous example is the farm-fox experiment. In 1959, the Russian geneticist Dmitri Belyaev began a breeding experiment with silver foxes (*Vulpes vulpes*), in which animals were selected for tameness, i.e. lack of fear of humans, and absence of aggressive-avoidance responses (see Trut, 1999; Trut *et al.*, 2009). The tamest 10% of each generation were used for breeding. Over the course of more than 35 generations, most foxes did become tame, indicating changes in the central coordinating mechanisms for aggression and fear; but the results of most interest are the various characteristics that appeared along with tameness that were definitely not selected for. These included physiological changes (reproductive cycle altered), morphological changes (droopy ears, curly tail, coat color), and behavioral changes (wagging tails, whimpering, seeking contact with humans). Such changes appeared in about 2% of the animals by the sixth generation, but in almost 50% by the thirtieth. A number of these same changes have been noted in other domesticated animals, including dogs, pigs, horses, and cattle, which suggests the possibility of a 'domesticated phenotype' and the presence of 'deep homology'.

A very recent study of domestication of red junglefowl, the wild ancestor of the domestic chicken, by Per Jensen and his colleagues, is finding some similar results (Agnvall *et al.*, 2015). By the sixth generation, the chickens in the line selected for lack of fear to humans showed a significantly reduced reaction to a novel object while feeding, and higher metabolism compared to the unselected line. A final example from the domestication literature involves genomic studies of dogs, wolves, and jackals by Robert Wayne and his colleagues mentioned above (Freeman *et al.*, 2014). These studies have provided strong evidence for gene flow between species, which implies successive hybridization events along the way and a more web-like course of evolution than the branching image of Darwin. Such a web-like course of evolution was also found by Lorenz (1941) in his studies of dabbling ducks.

My final example returns to the studies of the cichlid fishes in the East African Lakes, also discussed above. These fishes have evolved into numerous species in a much shorter time than would be expected according to the Modern Synthesis.

> The recent sequencing and analysis of genomes has revealed a number of mechanisms that may have spurred cichlids to quickly diversify into myriad forms [these include abundant mutations, gene duplication, jumping genes, mutations in DNA that typically does not change, and novel microRNAs]. Some of these mechanisms could also help explain another mysterious aspect of this group of fishes: namely the extreme degree of parallel evolution, in which the same highly specialized traits emerged again and again" (Meyer, 2015, p. 74).

Think mouthbrooding. Once again, the concept of 'deep homology' comes to mind.

The mechanisms of evolution are many indeed: genetic, developmental, and cultural. Darwin's theory, as formulated by Lewontin (1970), is still correct, but it is only one piece of a much larger puzzle.

9

Phylogenetic Consequences
Survival Value (Current Fitness)

Karl von Frisch (1886–1982). Photo © Bettmann.

In an essay dedicated to Lorenz at the occasion of his 60th birthday, Tinbergen (1963, pp. 417–418) noted:

> In the post-Darwinian era, a reaction against uncritical acceptance of the selection theory set in, which reached its climax in the great days of Comparative Anatomy, but which still affects many physiologically inclined biologists. It was a reaction against the habit of making uncritical guesses about the survival value, the function, of life processes and structures . . . I am convinced that this is due to a confusion of the study of natural selection with that of survival value. While I agree that the selection pressures which must be assumed to have moulded a species' past evolution can never be subjected to experimental proof, and must be

traced indirectly, I think we have to keep emphasising that the survival value of the attributes of present-day species is just as much open to experimental inquiry as is the causation of behaviour or any other life process.

Tinbergen suggested that Lorenz' work had revived interest in the study of survival value, and urged ethologists to divert more attention to studies of function. How much effect Tinbergen's exhortations had on the scientific community is moot, but it is certainly the case that the publication of Wilson's (1975) book, *Sociobiology*, was the major impetus for the switch of attention from causal to functional questions in behavior studies, which is still a major focus of behavioral biology studies today.

In the 1950s, when Tinbergen was promoting the study of function, evolutionary biology had still not defined its major functional concepts. For example, Tinbergen (1951, pp. 151, 157–158) could write:

> Many different terms pivot around the concept of self-maintenance. 'Biological significance', 'adaptiveness', 'directiveness', 'purposiveness', 'survival value', 'ecological function', &c., are all related concepts. They are all intended to indicate the fact that the mechanisms and/or structures considered contribute to the maintenance of the organism ... I will present the facts in the following arrangement: (i) Activities of direct advantage to the individual; (ii) Activities of advantage to the group.

He placed feeding behavior, escape behavior, and learning mechanisms in the first category, and cooperation, mating, fighting, and care for offspring in the second category. After the gene-centric view of the Modern Synthesis took hold in the 1960s, most biologists began to believe that all adaptations could be considered to be of direct advantage to the individual or its genes (Williams, 1966), and the idea of advantage to the group or 'group selection' was generally discarded. There are still some biologists who argue that group selection has a role to play in evolution (e.g. Wilson & Wilson, 2007), but their arguments will not be considered here.

In this chapter, I will be considering function as biological advantage, that is, the value for the organism (survival value) of particular behavior mechanisms having certain characteristics rather than other characteristics (Gould & Vrba, 1982; Wouters, 2003, see p. 15 above). The function of much behavior is fairly obvious (eating, drinking, sex, escape, care for young), but even with such obvious cases, the possible function of aspects of these behaviors needs investigation. I begin by discussing the concept of fitness and then continue with examples of studies of the primary functions of behavior: feeding, defense, and reproduction. I conclude with a discussion of situations in which the primary functions of behavior interfere with each other or are conditional on the behavior of other individuals. I will be emphasizing empirical studies and the methods used. The nature of adaptation and selection are considered in Chapter 10.

THE CONCEPT OF FITNESS

Fitness is an important concept for evolution, because, without differential fitness of entities in a population, evolution cannot occur! But fitness is a slippery concept. It is a concept like homeostasis and canalization: it is defined by its effects. It is one of those concepts that everyone thinks he/she understands, but then finds it difficult to specify exactly in a particular situation. A typical textbook definition of fitness is the measure of an individual's success in passing on copies of its genes to future generations. But fitness can be defined with respect to the genotype or phenotype of a trait or an individual or even a group; it can have two or three components (viability, mating success, fecundity); it can be considered a property of an individual or of a class of individuals (average value of the group); it can be absolute or relative; it can be direct or indirect or inclusive; and individuals with identical genotypes may have different fitnesses depending on the environments with which they have interacted during development. Also, how many generations need to be considered? The next? Two? Many? Sober (2001) and Orr (2009) discuss most of these issues. I will generally not be too concerned with these problems of definition, but they do become important in interpreting the results of some of my examples, and will be discussed there.

In any particular situation, an acceptable definition of fitness can usually be determined, but measuring fitness can still be a difficult task. For that reason, as we shall see, most studies use a surrogate for fitness that is more easily measured. Further, there are two general approaches to studying the fitness or survival value of behavior. The first begins by observing an animal's behavior and asking how that behavior contributes to its fitness; this was the method used by Tinbergen and might be called the ethological tradition. The second approach is more theoretical and starts at the other end and is sometimes called 'reverse engineering'. It is generally used by behavioral ecologists and is described below. I give examples of both approaches.

FORAGING AND FEEDING

Finding and ingesting food are obviously functional behaviors: they provide nutrients for the development and maintenance of the individual and also provide the energy needed to sustain its activities. Honeybees live in colonies and require a large amount of pollen and nectar in order to provide for themselves and their numerous offspring, and successful colonies have large numbers of workers searching for supplies. Pollen and nectar may be locally abundant (as when an apple tree is in blossom, or a field of clover is in bloom), but the location of these resources varies greatly from time to time. It would clearly be most efficient if a few bees could scout around for rich sources of nourishment and then return to the hive and inform their fellow workers where to forage. Karl von Frisch, a Nobel Prize recipient in 1973, discovered that the bees do exactly that.

Von Frisch (1955) combined accurate observations of bee behavior with elegant field experiments and showed that returning worker bees perform two types of 'dances' at the hive. These dances provide information about the location of a source of food. The 'round dance' is an excited series of circular movements that informs the other bees of a source of food near the hive. If the food is more than about 100 meters away, the returning bee performs a 'wagging dance':

> The bee runs along a narrow semicircle, makes a sharp turn, and then runs back in a straight line to her starting point. Next she describes another semicircle, this time in the opposite direction, thus completing a full circle, once more returning to her starting point in a straight line . . . The characteristic feature which distinguishes this 'wagging dance' from the 'round dance' is a very striking wagging of the bee's abdomen performed only during her straight run (1955, p. 116).

The remarkable feature of the wagging dance is that the speed of dancing is inversely related to the distance of the food, and the angle between the straight part of the dance and the vertical is the same as the angle between the direction of the sun and the feeding place. Worker bees in the hive follow the returning worker while she is dancing and thus become informed of both the distance and direction of the food.

The accuracy of the information transmitted by the returning bee to the other bees is extraordinary. In one experiment, von Frisch allowed several numbered bees to feed on sugar water to which lavender scent had been added, at a distance of 750 meters from the hive. Nine scented boards without food were then placed at distances between 75 and 2500 meters from the hive in the same direction as the feeding place. An observer sat at each board and counted the number of new bees that arrived in the next 1½ hours. More than 60% of the bees that were seen appeared at the two boards 700 and 800 meters from the hive. In a similar kind of experiment, scented boards without food were arranged in the shape of a fan at angles of 15° from the hive at a distance of 200 meters. Here too, 60% of the bees appeared at the board located in the direction in which a food source had been originally found. There are many other facets to the bees' language, but even this description of the two dances indicates how unexpected some of the functions of behavior can be. Wynne (2004, Ch. 2) gives a very readable account of honeybee behavior.

A different kind of example is a classic set of studies on self-selection of foods that investigated growth rate as a measure of the nutritional fitness of a young rat's food choices (Richter, 1942, pp. 78–79). In one of these so-called 'cafeteria' studies,

> rats were given access to an assortment of 11 substances, one source of each of the substances known to play an important part in nutrition. [Each rat's cage] contained several food cups for solids and from 8–20 graduated inverted water bottles ... for fluids ... On the selections made from these substances the rats grew at a normal rate [see Figure 9.1], were normally active, reproduced, and showed no signs of deficiency, actually thriving for as long as 500 days, at which time the experiment was discontinued ... That the rats actually made very efficient choices from this assortment is shown by the fact, although they grew at the same rate, their total food intake as measured in grams was 36 per cent less than that of the control rats kept on our stock diet. The difference in total caloric intake was less great, 46 and 56 calories on the self selection and stock diets respectively, by virtue of the higher fat content of the self selection diet.

Although the studies of Richter and his colleagues were designed to elucidate the mechanisms of food choice, the results clearly indicate the survival value of the rat's choices. Rozin (1976) provides an excellent review of many of the earlier studies of food selection by a variety of animals, including humans; it is still relevant today.

In recent years, most studies of foraging have taken the reverse engineering approach to studying function and have been conducted by behavioral ecologists. Such studies often begin by constructing an optimality model, which is a hypothesis about how an animal should behave if it is maximizing its fitness (McNamara et al., 2001). The model contains a hypothesis of some measure of fitness (the 'currency') and

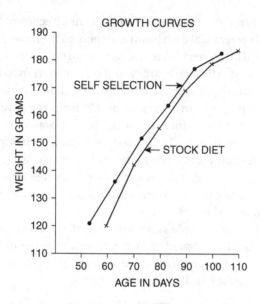

Figure 9.1 Average growth curves for eight rats on self-selection diet and 19 rats on stock diet.
From Richter, 1942, with permission from the Harvey Society.

takes account of the 'constraints' assumed to be influencing the animal when it is making its choice ('decision'). Predictions from the model are then formulated and tested experimentally. If the experimental results confirm the predictions, the currency and constraints are considered to be factors of importance influencing the animal's decisions. If the results do not confirm the predictions, a new currency or new constraints need to be investigated. It should be noted that optimality models actually *assume* that the behavior being investigated is optimal (due to natural selection), and are really testing whether the hypothesized currency and constraints are reasonable assumptions.

An example is the optimal 'prey choice' model. Giraldeau (2005, in press) gives a step-by-step description of how such a model is built and tested. The decision the animal must make is whether to take or leave prey when encountered (is it a generalist or specialist forager?). However, an individual's behavior does not necessarily satisfy both energy and nutritional needs at the same time, so that many studies choose one or the other to measure fitness. In most foraging models, rate of energy intake is taken as the currency of fitness; nutritive value is normally not considered. The constraints include the time taken to deal with the prey (handling costs), the rate of encountering prey (search costs), the absolute and relative abundance of the different prey types, plus a host of other assumptions that depend on the particular case. Davies *et al.* (2012, pp. 79–81; 432–436) discuss some of the advantages and disadvantages of optimality models.

An experiment by Krebs *et al.* (1977) is an early example of a test of an optimal foraging model involving sequential prey encounter. These investigators presented individual captive great tits (*Parus major*)

> with two prey types, profitable and unprofitable, on a moving belt. Both prey types were made out of mealworms [and had equal handling times]. When the encounter rate with both prey types was low, the birds were non-selective, but at a higher encounter rate with profitable prey, the birds selectively ignored the less profitable type and did so irrespective of the encounter rate with them. These results are as predicted by the model, but the birds did not as predicted change from no selection in a single step. We suggest that this is because the birds invest time in sampling to determine the availability and profitability of different prey types (p. 30).

These results illustrate a typical outcome of most tests of optimality models: some of the results are as predicted, but further factors (constraints) also need to be included to make the model more realistic.

An animal must choose its diet, but it should also do so efficiently. In many cases, food is located in patches in the environment. How long should the animal forage in one patch before it moves on to another patch? This question has given rise to 'patch residence' foraging models. For such models, rate of energy intake is usually also the currency chosen, but the constraints are different from the optimal prey choice models. The most relevant new constraints are the rate of patch depletion as a consequence of foraging and the amount of effort it takes to find or travel to another patch. One prediction of the model is that optimal time in a patch increases as the average travel time between patches increases. The patch residence model also applies to the behavior of 'central place' foragers, animals that forage in a patch, but then return with their prey to a burrow or nest (a central place) either for storage or to feed their mate or young. An extra constraint for central place foragers is the cost of carrying the prey ('load' factor). An example is a field experiment by Giraldeau & Kramer (1982) using eastern chipmunks (*Tamias striatus*). Chipmunks were offered trays of sunflower seeds at various distances from their burrow. They made repeated trips to the same tray, in different locations, collecting seeds in their cheek pouches and carrying them to their burrow, where they were stored. The chipmunks collected larger loads and spent more time doing this as distance, and hence travel time, to the burrow increased, as predicted. However, the rate of food delivery achieved was lower than the rate they would have achieved had they taken larger loads. Here, too, qualitative results confirmed the predictions of the model, but quantitative results did not, indicating that additional constraints needed to be considered (see also McAleer & Giraldeau, 2006).

Starlings have been a favorite subject for many patch residence foraging studies because the birds are abundant and relatively easy to observe in the field, can be trained, and can also be kept in the laboratory (J. Tinbergen, 1981; Kacelnik, 1984). These studies have varied a number of constraints (e.g. nest with young or no nest; number of young in nest; distance to food source) to determine which currency best fits the data. When all things are considered, rate of energy intake turned out to be the best. Interestingly, experiments with honeybees gave a different result. Honeybees have a nectar crop that can be filled while the bee is foraging. Schmid-Hemple *et al.* (1985) found that the bees did not fill up their crops as full as would be expected if they were maximizing rate of energy intake. Instead, it turned out the bees were maximizing energy efficiency. The difference between the birds and the bees in this case is that, on the return flight to the hive or nest, the weight of the nectar affects energy consumption much more in the bee than the load of insects affects energy consumption in the starling. The results of these and related experiments are reviewed and discussed by Piersma & van Gils (2011, Chs. 6 and 7).

Theunis Piersma and his colleagues have carried out a series of field and laboratory studies on red knots, *Calidris canutus*, over a period of more than 20 years that give further insight into functional aspects of foraging. Knots are medium-sized shorebirds that breed in the Arctic tundra and over-winter in various places further south, including the Dutch Wadden Sea. There, they forage during low tide on bivalve molluscs and various crustaceans that live buried in mudflats along the coast (see Figure 9.2). They swallow their prey whole and crush the hard shells in their gizzard, a strong muscular stomach. Early studies on knot foraging found that the birds were feeding at a slower pace than expected and not maximizing rate of energy intake as predicted by the patch residence model. But foraging knots face a special problem due to the composition of their prey, hard-shelled molluscs. Their gizzard, necessary for crushing the shell and extracting the energy-rich meat, can only crush shells at a certain rate, which is slower than the foraging rate. So, when the stomach is full, the birds need to wait for

Figure 9.2 Red knots (*Calidris canutus*) foraging in the Wadden Sea.
Courtesy of Jan van de Kam.

some digestion to take place before they can recommence eating. When the constraint of digestion time is added to the model, the birds are found to be maximizing rate of energy intake.

Foraging knots also posed a problem for the patch choice model. As well as eating hard-shelled cockles (*Cerastoderma edule*), they also eat soft-shelled shrimp (*Crangon crangon*). The cockles and shrimp generally occur in separate patches, and energy calculations show that cockles are more profitable than shrimp with respect to rate of energy intake and inter-capture interval. But some birds choose shrimp patches over cockle patches, and captive birds in the laboratory always prefer shrimp. It turns out that the gizzard is the problem again: it changes size over the course of the year, and a small gizzard digests shrimp more efficiently than cockles. One more constraint then, gizzard size, needed to be added to the model. How and why gizzard size changes is a story in itself, but that story is especially relevant for ecologists, and will not be discussed here. A review and discussion of all these studies can be found in the excellent book by Piersma & van Gils (2011).

In nature, knots are social birds and usually forage in groups of thousands of individuals, which raises a host of additional functional questions. Here, I will concentrate on why and how group foraging contributes to individual fitness. The main cost of group foraging is competition for resources. In the Wadden Sea, cockles are abundant, so competition for prey is not very important for knots. However, the cockles occur in patches that are spread out over a large area of more than 1000 km^2 and their density in particular areas is continually changing, so an important problem for the birds is to find a good patch in which to forage. 'Public information' (in this case, observing the behavior of other birds) can be used to locate food as well as to estimate its quality, which allows more effective foraging.

Bijleveld *et al.* (2015) carried out two experiments with captive knots studying the knots' use of public information. In the first experiment, a focal bird was allowed to observe 'demonstrator' birds foraging on two separated patches, two demonstrators on each patch. One patch had blue mussels hidden in the sand, the other patch had none. The demonstrators were all hungry and did not know which patch they occupied, so they all foraged. The only difference between the demonstrators was that one set found and could eat mussels, while the other could only forage. Focal birds joined the patch with food more than 76% of the time. In the second experiment, knots were allowed to forage in groups of one, two, three, or four in an arena with 48 small patches, only one of which was baited with mussels. As the number of foragers increased, the time to find the baited patch by all birds decreased proportionally. Neither dominance status nor sex appeared to be important for rate of energy intake in either experiment.

One further result in the group foraging experiment was that 'personality' differences emerged among the knots. Some birds were more active in searching for food, the producers, while others appeared to wait for the food to be found, the scroungers. In this case, since all birds found food faster when in a group, group foraging benefitted all the individual birds (though maybe not as much benefit as could have been achieved if no scroungers had been present). Producer-scrounger scenarios present interesting problems for evolutionists and will be discussed further below. Individual differences ('personalities') will be discussed in Chapter 10.

Group foraging can have costs as well as benefits; in addition to prey depletion, these costs can include stealing food from each other (kleptoparasitism) and time spent in agonistic encounters among members of the group. If these costs increase, it would pay an individual to leave the patch in which it is foraging and move to a different patch. Knots generally exhibit neither kleptoparasitism nor agonistic behavior while foraging in groups, and prey depletion is also not an important issue. Nonetheless, in nature, knots are seldom observed foraging in tightly knit groups. In a laboratory experiment, similar to the two just described, Bijleveld *et al.* (2012) let groups of two to eight birds forage in a single patch, 0.7 x 0.7 m, that moved slowly, exposing new prey as it moved. In spite of the fact that essentially none of the conventional interference mechanisms were seen, both rate of food intake and searching efficiency declined as a function of group size. The authors suggest 'cryptic interference' as the reason: the birds spend time trying to avoid each other and are thus less efficient at foraging. There is much more to foraging than food availability and ingestion.

It is instructive to compare the ethological studies with the ecological studies. Von Frisch saw some strange behavior in the hive whenever a worker bee returned from a foraging trip. He wondered what function that behavior might have. He did much careful observation, constructed a hypothesis, and carried out some experiments. The result was the story of the bee dances. Piersma observed knots foraging in the Wadden Sea. He wondered if they were maximizing their energy intake. He made some calculations and saw that they were not, and wondered why. After several hypotheses were tested and found to be at best partial answers, he realized the gizzard might be the key to the puzzle. Many experiments later, the gizzard story was the result. Is one approach better than the other? In both cases, interesting and important facts about the function of behavior were discovered even though the investigators were asking different kinds of questions. For the ecologist, the fact that bees maximize energy efficiency rather than rate of energy intake might be more meaningful than the bees' dances, and for the ethologist, the behavioral mechanisms used to space out the knots might be more interesting than the fact that they are maximizing rate of energy intake; but answers to one question may facilitate finding answers to the other question, and answers to both questions give greater understanding of the animals' behavior.

DEFENSE

When an animal is approached by a predator, it can engage with the predator, hide, or flee. Animals use all three strategies, and which kind of response is best depends on many factors. For example, a mother hen with chicks will spread her wings, make an alarm call, and viciously attack an approaching potential predator, even a human. At the same time, the chicks may run and hide motionless in the nearby vegetation. A hen without chicks may give an alarm call and flee from the same potential predator. In this case, which response occurs depends on the hen's hormonal state.

Engaging with a predator is very common throughout the animal kingdom, though unlike the hen with chicks, it seldom involves direct combat. Many kinds of engagement can be considered distractive behavior that serves to protect the individual and/or its offspring. A classic example is the behavior of the Eurasian peacock butterfly (*Aglais io*). At rest, it sits quietly with its wings folded together, resembling a leaf. When quickly approached or touched by a predator (usually a bird), it suddenly opens and closes its wings rapidly, exposing brightly colored 'eyespots' on the front of its wings. Early experiments by Blest (1957) used a simple apparatus to study the effectiveness of this behavior in reducing predation. Yellow buntings, great tits, and chaffinches were trained to 'capture' a mealworm placed between two translucent windows that hid the stimuli (models) being tested. When the birds were comfortable with the situation, the models were suddenly illuminated as the bird came to pick up the mealworm (presumably mimicking the wing opening of the butterfly). Most birds tested were clearly frightened by the sudden appearance of the model and flew away without picking up the mealworm. The escape response of the birds quickly died out, but Blest was able to show that the models that most resembled an eye were the most effective in releasing escape. More recent experiments (reviewed by Stevens, 2005; see also Stevens *et al.*, 2008) suggest that conspicuousness, rather than resemblance to an eye, may be the stimulus for the bird's response. Nonetheless, whatever the effective stimulus, the butterfly's behavior clearly distracts the predator, and in many cases would allow it to escape and avoid being eaten.

Another example of distractive behavior is 'mobbing', a form of harassment. In colonial breeding birds, mobbing can be a communal response to a potential predator (e.g. a raptor or a carrion crow) that can include flying about the predator with loud alarm calls, dive bombing, and defecating on the intruder. Numerous experiments have shown that this behavior does reduce predation on eggs and young (e.g. Kruuk, 1964). The term mobbing is also used for similar behavior shown by other animals even when no mob and only individuals are involved. Hinde (1970) reviews and analyzes many of the early studies of mobbing behavior.

Many species have evolved quite specialized displays that serve a defense function (Armstrong, 1954). A well-known, striking example is the 'broken-wing' display of the killdeer (*Charadrius vociferous*), a medium-sized shorebird. When a potential predator is detected, the incubating bird leaves its nest and loudly limps off holding its wing in a position that makes it appear broken. The predator follows its 'debilitated' prey, but the bird keeps a safe distance from the predator until it finally flies off when it is far from the nest; it only returns to the nest when the predator is gone. A more recent example is the behavior of the California ground squirrel (*Otospermophilus beecheyi*) toward northern Pacific rattlesnakes (*Crotalus viridis oreganus*). When a snake is encountered, these squirrels throw substrate at the snake and show a specific tail flagging display (Coss & Biardi, 1997). Experiments showed that these behaviors deterred snakes from striking the squirrel, and that tail flagging by adult squirrels increased the likelihood that snakes would leave their ambush site (Barbour & Clark, 2012).

While mobbing and other distractive behaviors clearly function to reduce predation, a less obvious function of mobbing is cultural transmission of enemy recognition. Curio *et al.* (1978) designed a clever conditioning experiment in which an 'observer' blackbird (*Turdus merula*) could see a stuffed Australian honeyeater (*Philemon corniculatus*), a novel non-predator of blackbirds, while a 'teacher' blackbird was mobbing a stuffed owl (*Athene noctua*) that the observer could not see. Subsequently, the observer blackbird showed strong mobbing behavior when only the stuffed honeyeater was presented. Alarm calls of many species serve the same function of cultural transmission of enemy recognition, usually to the caller's own young, but alarm calls present special problems for the assessment of fitness because it is not always clear how the individual making the call benefits. I return to this problem in Chapter 10.

A second strategy in response to an approaching predator is to hide. Avoiding detection by a predator has obvious survival value, but the means animals use to avoid detection are not always so obvious. There is a large literature on the use of camouflage by animals, but I will only discuss two examples. The first is a classic study by Tinbergen and his students (1962) on egg shell removal by the black-headed gull (*Larus ridibundus*). Within a few hours after a chick has hatched, the parent gull removes the empty egg shell from the nest and, either walking or flying, deposits it anywhere from a few centimeters to 100 meters from the nest. The parent returns immediately to the nest. Based on his general knowledge of gull biology, Tinbergen surmised that neither the avoidance of injury, nor of parasitic infection, nor of interference with brooding is the main function of egg shell removal. Because the inside of the empty egg shell is conspicuously white, the most likely function of removal seemed to be maintenance of the camouflage of the brood. To test this hypothesis, he and his students carried out several experiments in which they placed painted or natural hens' or gulls' eggs, widely scattered, in the vegetation outside the colony. They also placed empty egg shells near some of the eggs. They then waited to see which eggs were taken by the local carrion crows (*Corvus corone*) and herring gulls. In one experiment, 450 half-concealed gulls' eggs were laid out in 15 tests with empty egg shells placed at either 15, 100, or 200 cm from the eggs. The predators took 42%, 32%, and 21%, respectively, of the offered eggs. The authors concluded that proximity of an egg shell endangers the brood and that this effect decreases with increasing distance, which leaves "little room for doubt about the survival value of egg shell removal as an antipredator device" (p. 85).

My second example is of a different kind of camouflage: hiding in a group. There are many reasons for an animal to join a group, including increasing foraging success as we have seen above. Another important reason is that it can decrease risk of predation. Hamilton (1971) showed theoretically that individual fitness can be enhanced by joining a group, and that the value increases with the size of the group, as long as larger groups do not attract proportionally more predators. In essence, you are less likely to get caught the more of you there are and the closer you are together. But joining a group may have additional antipredator benefits as well, one of which is that groups of prey may confuse or distract a predator. Neill & M. Cullen (1974) designed experiments to see how groups of 1, 6, or 20 small schooling fish affected the hunting behavior of squid (*Loligo vulgaris*), cuttlefish (*Sepia oficinalis*), pike (*Esox lucius*), and perch (*Perca fluviatilis*). The first three species are ambush predators and perch chase their prey. Their results showed that capture per contact of prey decreased with increasing size of the groups in every case. Based on these results and on observations of hunting behavior, they concluded that a shoal (school) of prey fish provides additional protection to the prey by hampering the attack of a predator.

The third strategy for avoiding a predator is to flee. This is probably the best (fittest) response if you can outrun, outfly, or outswim your predator, though even then circumstances may indicate that some other response is better (e.g. Kramer & Bonenfant, 1997). The book by Caro (2005), *Antipredator Defenses in Birds and Mammals*, gives an excellent review of the wide variety of antipredator behaviors.

REPRODUCTION

The survival value of reproductive behavior is obvious, but reproductive behavior has many facets and calculating its fitness is not straightforward. For example, early work of Lack (1954) on the European swift (*Apus apus*) found that swifts that lay only two eggs raise more young on average than swifts that lay three eggs. Later studies have looked in more detail at the fitness of reproductive behavior in egg-laying birds. An example is a series of studies investigating 'family planning' in great tits and kestrels (*Falco tinnunculus*) carried out in the Netherlands over a period of more than 20 years (Daan et al., 1990; Dijkstra et al., 1990; J. Tinbergen & Daan, 1990). Determination of optimal clutch size depends on current reproductive behavior, as well as on future reproductive behavior including the reproductive behavior of current offspring. It also depends on the time the bird begins laying. In these studies, investigators observed free-living pairs of ring-banded birds breeding in nest boxes in nature, and recorded date of laying, clutch size, number of chicks hatched and surviving after various intervals, availability of food for the young, effort of parents in feeding the young, reproductive success of the offspring, as well as general environmental conditions, and used these data to calculate measures of fitness. They also carried out experiments in which the number of eggs or young in the nest were increased or decreased. As might be imagined, calculating fitness is a complex business, and even with all the data collected, there are still many unproved assumptions. Nonetheless, the results suggested that a majority of the great tits were maximizing their individual fitness, and 60% of the kestrel clutches observed obeyed the maximization criteria. As with the foraging studies described above, however, these departures from optimality imply that additional factors ('constraints') need to be included in the models. Such problems are studied in the context of life history strategies and evolution.

Drent & Daan (1980) proposed that the natural variation in the dates when a bird lays its eggs and the size of the clutch depends largely on phenotypic tuning to individual nutritional circumstances, rather than on genetic variability (see also Högstedt, 1980). This hypothesis was tested in other experiments on kestrels. The breeding cycle in kestrels begins in the spring when voles, their major source of food during breeding, appear in large numbers. At that time, the male increases his daily hunting time and the amount of food he delivers to the female. Egg laying begins two weeks later. In food experiments on free-living and captive kestrels, Meijer *et al.* (1990) showed that increased food delivery to the female led to a significant advance in laying date. They also found that females spent more time incubating initially-laid eggs as the breeding season progressed. When females spend about 50% of their time incubating, follicles in the ovary become resorbed, which determines the number of eggs the female can lay. This leads to smaller clutches in birds that begin to lay later in the season. The tendency to incubate is most likely controlled by day length, so that the natural variation in laying dates and clutch size is determined by the joint action of nutritional and endogenous factors. Like the gizzard story described above, this is another example of how causal and functional studies can complement each other.

If we accept the proposition that fitness can be measured by an individual's success in passing on copies of its genes to future generations, the studies of optimal clutch size just discussed show that even actual reproduction (laying eggs) is not always optimally fit. This conclusion is even more applicable to behaviors that only promote reproduction indirectly, such as territorial behavior, courtship, and care of offspring. Nonetheless, it is generally assumed that such behaviors also promote fitness, and I will discuss some examples of these behaviors in the remainder of this section.

Agonistic Behavior (Competition)

Aggressive behavior is ubiquitous among animals, but how does it promote fitness? One frequent consequence of aggression is the acquisition of a territory, an area defended by an individual or a pair, generally against intrusions by a conspecific. A territory can provide resources that may serve many functions: it may provide food, a secure place to live, an area where courtship and mating can take place, a location for building a nest, a burrow, or a den, etc. It may serve one or more of these functions, depending on the species. The territory of the male stickleback, discussed in Chapter 3, serves all these functions, whereas the territories of gull species do not provide food, but only a place for courtship, mating, and nesting. An interesting example is provided by the Saharan gerbil, *Psammomys obesus*, that lives in the Algerian desert. This species is a specialized feeder on the leaves of the Chenopod bush, and is highly aggressive. The female sets up a small territory around her food bush and burrow, about ten meters in diameter, and vigorously defends it from other females and even from males unless she is in estrus (Daly & Daly, 1974). Territories are often acquired and defended with actual attack behavior, but such behavior can be very costly, and in most cases, other behaviors have evolved that serve the same function but are less costly. The gull displays, 'upright', 'forward', and 'oblique', discussed in Chapter 8 are examples of aggressive behavior that functions in territory defense, but does not involve actual attack. Aggression in territory defense can also be expressed in other ways. Many mammals, for example, mark their territories with scent. Humans use fences and locks.

Aggressive behavior can promote fitness in other ways too. In many species, males compete with each other for access to females. This can range from two males competing for a single female to many males competing for groups of females (harems). A striking example of the latter is the behavior of northern elephant seals (*Mirounga angustirostris*) that breed on traditional islands off the California coast. The males arrive at their rookery first, in early December, and immediately begin fighting and threatening each other. They quickly establish a stable hierarchy in which the dominant males monopolize the space in the rookery as well as the females that find themselves in their territories. The pregnant females arrive later in the month, settle where space is available, deliver their pup six days after arrival, nurse their pup for 28 days, and then depart. Females are in estrus the last four days of nursing, and that is when copulations occur. In a long-term study over six seasons, LeBeouf (1974) found that fewer than one-third of the males copulated with a female, and that the five top ranking males each season (about 3%–4% of the males) made 50%–85% of the observed copulations. Most males never copulate. Females presumably benefit from being impregnated by high-quality males. Similar harem-type behavior is found in some baboons (*Papio* spp.) and red deer (*Cervus elaphus*), as well as in species of birds, fish, and insects. From a fitness point of view, it should be noted that a recent study by de Bruyn *et al.* (2011) provides evidence that some female elephant seals mate at sea, which means that mating opportunities for males also exist outside the rookery.

Female-female competition also occurs, though it tends to be studied less because female behavior is often less 'spectacular' than male behavior. The behavior of the female Saharan gerbil discussed above is an example. However, the classic example of female-female competition is the so-called 'pecking order' in chickens first described by Schjelderup-Ebbe (1922). Male chickens (roosters, cocks) are always higher in rank than female chickens (hens), but among the females, each has her place in a linear hierarchy. When a new hen (or one that was removed from the flock for 48 hours or more) is introduced to a flock of ten or twelve hens, small fights break out between the new hen and the others, which continue until the new hen has interacted with all the other hens: a new hierarchy has been established. Chase (1982) has analyzed how such linear dominance hierarchies, based on individual recognition, can develop. This organization gives the hens priority access to food, nesting, and roosting sites according to rank. Experiments comparing well-integrated flocks with flocks undergoing constant reorganization found that the hens of organized flocks pecked each other less, consumed more food, and maintained body weight better when placed on restricted access to food (review in Wood-Gush, 1971). In this way, the pecking-order system benefits everyone, not only the high-ranking individuals.

Another social system involving male-male competition is the lek system (Swedish *leka* means 'to play'). A prototypical example is the behavior of the black grouse, *Lyrurus* (= *Tetrao*) *tetrix*.

> Males assemble on a communal display ground, arena, or lek, especially during the early morning hours in spring. Each male defends its site from neighboring males by means of displays and, to a lesser extent, by actual fighting. Females visit the lek and are courted by the males. During these short visits the female chooses a male with which she mates. After mating the female departs and no further parental cooperation between the sexes takes place: nestbuilding, incubation, and care of the young are the exclusive task of the female (Kruijt & Hogan, 1967, p. 204).

In our study, the lek was located in a meadow used for grazing cattle, adjacent to a large area of heather moorland in which the females nest. In 1963, from early April, our lek was visited regularly by eight territorial males and irregularly by as many as five non-territorial males. Females visited the lek, either singly or more usually in small groups, between mid-April and early May.

The territories were staked out such that four of them were surrounded on all sides by other territories. These four were considered central and the other four were considered peripheral (marginal). Whether central or peripheral, all encounters at the borders of a territory never resulted in a change in boundary: boundaries were mutually respected. In this sense, all territorial males were considered equal in social rank. Black grouse are very sociable (except females when nesting), and since the density of birds is highest in the center of the lek, females mostly land on or near a central territory when arriving. She then walks about, often visiting several territories. In each territory, the resident male courts the visiting female; if and when a female crouches, copulation occurs. Although courtship can be intense, females give the impression of being relatively relaxed on the lek. From the point of view of fitness, peripheral males were seen to copulate only once (out of 26 observed copulations) and the two most successful central males (25% of the males) copulated sixteen times (62%). Further, in about one-third of the observed copulations, a neighboring male crossed the boundary and attempted to interfere. Only one such attempt was successful, presumably because mating only takes about two seconds and the territories are large enough that the intruder has no time to reach the copulating male. Similar lek-type systems have been studied in other species of grouse, other species of birds, and in mammals, fish, and insects, although the details are different in every case.

In both harem-type and lek-type social systems, a few males obtain most of the copulations, and most males do not copulate at all. But the dynamics of the two systems are very different. In the harem-type system, it is the male that controls the females and guards them from dalliances with other males. In the lek-type system, it is the female that makes the choice, and she is free to mate with any or all of the males of her choosing. This shows that very different types of behavior can have very similar fitness consequences.

Courtship

Courtship is behavior that occurs prior to mating. It can be as simple as finding a partner (although 'finding' may not always be simple) or it can involve elaborate displays. The behavior of the male elephant seal is an example of the former: the females in his harem are available and there need be no preliminaries to mating. The male merely mounts the very much smaller female and copulates. Male black grouse, on the other hand, have evolved striking plumage and ornaments as well as behavioral displays that all play an important role in their courtship behavior on the lek.

During breeding season, males on the lek spend most of their time moving about their territories in the 'horizontal display posture', in which the tail is raised and spread, exposing the white tail feathers; the wingbows are taken out of the supporting feathers, the primaries of both wings are extended downward slightly, and the body and neck are held horizontally (see Figure 9.3). While in this posture, males continuously 'rookoo', a far-reaching vocalization that sounds similar to the

Figure 9.3 Male black grouse on the lek. Male in the center is in the 'squat'. The other males are in the normal threat posture. Each male is in his own territory.
Courtesy of Jaap Kruijt.

cooing of a dove. The combination of posture and sound makes the male very conspicuous. From our observations, it seems likely that this display serves both as a threat to other males and as an advertisement signal to females.

When a conspecific is spotted flying toward the lek, all the males on the lek fly up about a meter from the ground with a display called 'flutter-jumping' that exposes the white underside of their wings and includes a crowing sort of call. If the arrival is a female that lands near a male's territory, he will usually 'squat' (a variant of the horizontal display posture with bent legs) facing the female and remain stationary while continuing to rookoo. If the female crosses into his territory, he usually begins 'circling' her. This is also a variant of the horizontal display posture in which the primary feathers of both wings are lowered to the ground (cf. waltzing in roosters, p. 93). If the female is ready to mate, she will crouch to a circling male, and he will approach her from the rear, mount, and copulate. Females chose the most 'charming' males to mate with in accordance with Darwin's theory of sexual selection (discussed in Chapter 10).

Tinbergen (1959) has categorized the function of displays as either threat (a display that tends to space out individuals) or appeasement (a display that has a distance-reducing effect). In black grouse courtship, 'squat' would be an appeasement display, while 'circling' is a display with an ambivalent function: if the female is not ready to mate, circling is a threat display, but if she is ready to mate, it is an appeasement display.

Many male displays serve not only to threaten other males, but also to attract females, such as the 'horizontal display posture' in black grouse, the 'long-call' in gulls, and song in many bird species. All these functions are generally assumed, but it is possible to test these assumptions experimentally, though it is seldom done. An example, in which the outcome was unexpected, is a study by Daly (1977) on the functional significance of scent marking by gerbils (*Meriones unguiculatus*). As mentioned above, many mammals mark their territories with scent. Scent marks are usually considered to function as threats to other males. In Daly's experiments, male gerbils were exposed to male and female odors and were found to mark and groom more to the female odors; in other experiments, females were selectively less aggressive toward familiar-smelling males in comparison with unfamiliar-smelling males. He concluded (p. 1082): "These results, in conjunction with field observations of related species, call into question the hypothesis that gerbil scent-marks function territorially and instead suggest that the primary targets are adult females."

Finally, I should point out that, in many species, courtship behaviors are necessary for the females' eggs to develop or for successful fertilization. The experiments of Lehrman on doves, discussed in Chapter 3, provide a well-analyzed example.

Care of Offspring

To care for one's young is obviously functional, but even in this situation, trade-offs exist. Consider, for example, a parent faced with several mouths to feed, but insufficient food to provide adequate nourishment to all of them. One solution, adopted by many species of rodents, is simply to kill or eat the excess young. A less drastic method is to reduce the number of young that are produced. For example, in many species of birds, such as the kestrel (see above), clutch size depends on food availability. In other species, siblings kill each other when the food supply is insufficient. Another problem is related to life history strategies. In species with more than one reproductive event in a typical lifetime, parents are faced with the problem of when to stop caring for current offspring in order to invest in future offspring. A related issue is that the fitness interests of the offspring are different from the interests of the parents (Trivers, 1972, 1974). These considerations have led to many theoretical and experimental studies that I will not discuss here, but these issues are reviewed and discussed in an excellent chapter by Davies *et al.* (2012, Ch. 8).

WHEN FITNESSES COLLIDE

The studies of the survival value of behavior discussed above have focused almost exclusively on the fitness of a single trait or groups of related traits. It is actually good scientific practice to isolate a single variable for study while keeping all other possible variables constant or while controlling in various ways for those other variables. Many traits, however, have more than one function (cf. Beer, 1975). For example, eating provides nutrients and energy, and many advertisement displays attract females and repel other males. Johnson *et al.* (2000) considered eight possible functions for the postcopulatory displays of dabbling ducks discussed in Chapter 8. They found pair bond maintenance, individual identification, and signaling successful copulation to be the three most likely functions, with pair bond maintenance probably the most important. Further, the value of different traits may conflict with each other. In the words of Mayr (1982, p. 589): "Since the phenotype as a whole is the target of selection, it is impossible to improve simultaneously all aspects of the phenotype to the same degree." An obvious case is the possible conflict between behaviors that promote individual survival and behaviors that promote reproduction. Some kind of compromise must be reached that benefits both functions as much as possible. Since most extant species do survive *and* reproduce, such a compromise has generally been reached. But it is necessary to keep this problem in mind when doing and interpreting experiments on survival value.

Behavioral ecology is awash with theories of function that come and go quite regularly. I have not considered most of them, but I will give one example of the difficulty of providing evidence for a specific theory of function. The rock ptarmigan, *Lagopus muta*, is a species of grouse that inhabits arctic regions globally. These birds are often cited as the classic example of seasonal camouflage: they are pure white in the winter (when the ground is covered in snow) and they molt and become mottled brown/grey in the summer (after the snow melts on the tundra). In both seasons, they blend into the environment and are very inconspicuous, at least to humans. The problem is that while the females molt into summer plumage as the snow retreats, the males remain conspicuously white until about three weeks later. What is the function of the male's white plumage in the spring? At least seven theories have been proposed (see Montgomerie et al., 2001). Plumage, of course, is not a behavior, but all the more probable theories of its function do involve behavior.

Males occupy territories in the spring that are used for food, shelter, mating, and care of the young (this species is mostly monogamous). One theory of its function, favored by Montgomerie, is that the white spring plumage of the male attracts females and repels males. Another theory is that the conspicuous plumage attracts predators, like the broken-wing display of the killdeer, so that the male can lead the predators away from the female and chicks (Bergerud & Mossop, 1984). Predators, especially gyrfalcons (Falco rusticolus), are a real danger for the males, so why do the males 'choose' conspicuous plumage rather than camouflaged plumage? The rock ptarmigan story becomes even more confused because, in Scotland, it is the males that molt into cryptic summer plumage before the females (Watson, 1973). Further, in the Nunavut, Canada population studied by Montgomerie, the males soil themselves with substrate when their female begins laying eggs and is no longer fertile. This behavior makes the males more cryptic many days before their summer molt begins. Presumably, all these variations in plumage molt and behavior are adaptive, but untangling all the functions and their selection pressures will not be easy.

Untangling functions becomes even more complicated when the benefits of an individual's behavior are conditional on the behavior of another individual. To attack this problem, Maynard Smith & Price (1973) introduced the concept of Evolutionarily Stable Strategy, or ESS. The ESS is a strategy that, if all members of the group adopt it, no mutant strategy can invade; that is, no other strategy would be more fit. They also used the mathematical tool of game theory to analyze the problem. Davies et al. (2012, Ch. 5) and Giraldeau (2005, in press) describe and explain these concepts and review many examples of how they are used. This approach has been used to analyze agonistic encounters, alternative resource-harvesting strategies, occupation of space, and many other situations. I will use the producer-scrounger game and the sequential assessment game in fighting as examples.

We saw above, in Bijleveld's experiment on foraging in knots, that some individuals (the producers) searched for the hidden mussels while others (the scroungers) waited and joined the group after the mussels were found. Barnard & Sibly (1981) recognized that such mixed strategies are quite common and proposed a general game they called the 'producer-scrounger game' to analyze them. This game assumes that resources (mates, food, nest sites, etc.) can be gained in two mutually exclusive ways: producer invests effort in making the resource available, and scrounger exploits the resources that producer has made available. This game is characterized by strong negative frequency-dependent payoffs: scroungers do well when they are rare, because they have the opportunity of exploiting a large number of producers while suffering little competition from each other. However, when they become common, the numerous scroungers must compete with each other for the dwindling resources due to fewer producers. If the population would consist only of scroungers, there would be no resources; if it consists only of producers, scroungers can invade. So the ESS solution is a mixture of strategies that occur at a frequency for which the payoffs to each strategy are equal.

Luc-Alain Giraldeau and his colleagues have used this game for many years to study alternative foraging tactics within flocks of seed-eating birds (Giraldeau & Dubois, 2008). As with the optimality models discussed above, models based on the producer-scrounger game have been successful at predicting the change in frequency of the scrounger tactic within bird flocks qualitatively, but usually not quantitatively. Nonetheless, Giraldeau points out that the producer-scrounger game helps us understand that scroungers exist despite the cost they impose upon other group members: evolutionary stability is not always synonymous with maximization of benefits. Further, ESS theory assumes that different alternative strategies reflect genetic differences, but the foraging studies show that adjustment to changing payoffs can occur within a few hours (Morand-Ferron & Giraldeau, 2010). Giraldeau suggests that 'behavioral stable strategy' may be a more appropriate expression for solutions to evolutionary games solved by behavioral assessment. A recent paper (Hills et al., 2015) has pointed out that the problem of exploration versus exploitation is common to fields as diverse as visual attention, foraging, memory, and artificial intelligence. Strategies for handling exploration-exploitation trade-offs in one field can inform strategies in the other fields.

My final example is an experiment on fighting in the cichlid fish Nannacara anomala (Enquist et al., 1990). Many species have a repertoire of behavior patterns that function in the assessment of asymmetries between contestants such as size or strength. Enquist & Leimar (1983) have used a version of the sequential assessment game to develop a model of how an animal should behave when a series of behavior patterns changes as an agonistic interaction escalates. At each stage of the encounter, the animal is assessing its likelihood of winning or losing, and the model specifies when the animal should move on to the next phase or give up. In *N. anomala*, fighting normally begins with the two contestants erecting their fins and changing color; they assume a lateral orientation to each other and begin 'tail beating'; fighting continues with 'biting' and then 'mouth wrestling'; finally, if a decision is still not reached, the fish 'circle' with interspersed bites. The fight can end at any point when one fish 'gives up' (folds its fins and changes color) and swims away.

In the experiment, 102 fights were arranged between fish that differed in relative weights (asymmetry) from no difference to the heavier fish weighing about five times more than the lighter fish. In this species, relative weight is known to be the primary factor in predicting the winner, and, in fact, a lighter fish won the encounter only three times. The duration of the fights ranged from a few seconds to more than two hours, with a very strong negative correlation between duration and relative weight (evenly matched fish fought the longest). With respect to the model, the most interesting result, as predicted, was that changes from one phase to the next occurred at the same time in the fight irrespective of the fight duration: fights with greater asymmetry ended in an earlier phase with less dangerous fighting and less injury. As usual, however, not all aspects of the model were supported quantitatively by the results. Even relatively complex models need to become more complex if they are going to reflect reality.

10

Phylogenetic Consequences
Adaptations and Historical Origins

Charles Darwin (1809–1882). Photo by The Print Collector/Getty Images.

In Chapter 1, I outlined the four ways that Wouters (2003) character-ized biologists' use of the word function. Chapter 9 approached func-tion from the perspective of his third meaning: the value for the organism (fitness) of an item having a certain character rather than another. This chapter will consider the way a trait acquired and has maintained its current share in the population. The perspicacious reader of Chapter 9 may have noticed, however, that I very seldom explicitly compared the values of alternative characters. That is par-tially due to few studies actually testing alternative characters. The clutch size studies and some of the foraging and defense studies are exceptions. Another, and more frequent, reason alternative char-acters were not mentioned is that the fitness of possible alternative characters often seems obviously less than the fitness of the character being studied. Once the fitness of the target character has been experimentally verified, it then seems unnecessary to explicitly consider alternatives. That is bad practice. A contemporary example is instructive. Feeding is necessary for survival and there is no doubt that eating is adaptive. Further, we saw in Chapter 9 that the rate of energy intake is often considered the best measure of the fitness of foraging behavior. In humans, however, maximizing the rate of energy intake is maladaptive: it leads to obesity. In order to understand how eating can be both adaptive and maladaptive, we need to consider the evolutionary history of the behavior. That is what I do in this chapter. I first discuss the concepts of adaptation and adaptive radiation, and then look at the role played in adaptive thinking by sexual selection and kin selection. I then consider exaptation, a supplementary explanation for evolutionary change, and finally discuss some issues related to animal personality and human behavior.

THE CONCEPT OF ADAPTATION

The idea of adaptation is ancient, and until fairly recently, it was usually thought to implicate the work of some deity. It was especially Darwin in the mid-nineteenth century who proposed that adaptation is the outcome of natural selection. In modern biology, the word refers to either the process (a mechanism) or the outcome of the process (a trait) that fits the animal to its environment. Like fitness, adaptation is defined by its consequences.

As a mechanism, adaptation is actually a synonym for the process of evolution; as a trait, it contributes to fitness, but it also has additional specific characteristics. One characteristic of an adaptation is that, at some time in the past, with respect to the environment at that time (selection pressures), it contributed more than other traits to reproductive success. (I apologize for the cumbersome formulation, but simpler formulations can be misleading). A second characteristic is that it has been maintained in the population because it continued to be superior to other traits. Whether its current adaptiveness should be part of the definition is moot. I am using this formulation to emphasize that natural selection is the *outcome* of the process and not its cause (cf. Chapter 1). Nonetheless, however one wishes to define an adaptation, the essence of any definition of adaptation for behavior is the idea of history (Gould & Vrba, 1982). This is Wouter's fourth meaning of the word function.

Using this definition of adaptation, are the examples of behaviors with survival value I discussed in Chapter 9 also examples of adaptations? Most of them probably are adaptations, but the evidence necessary to support that conclusion was not presented or was not available. As Tinbergen said (p. 273), "the selection pressures which must be assumed to have moulded a species' past evolution can never be subjected to experimental proof, and must be traced indirectly". Here, I will discuss some putative examples of adaptation, both process and trait.

Adaptations of the Kittiwake

The work on gulls by Tinbergen's group, discussed in Chapter 8, included a now classic study on adaptations in the kittiwake to cliff nesting. E. Cullen (1957) studied the breeding behavior of these gulls on a small island off the coast of England for five seasons. Unlike the great majority of gull species that mate and nest on the ground, kittiwakes mate and nest on tiny cliff ledges. Cullen saw that the kittiwakes' behavior had many peculiarities compared to the behavior of ground-nesting gull species, and her study showed that these could all be related to their cliff-nesting habit. The peculiarities included aspects of nest construction, sexual behavior, chick behavior, and parent-chick relations. All these 'unusual' behaviors could be understood as adaptations to cliff nesting. I will discuss one behavioral difference, aggression to intruders, in some detail to point out the evidence that Cullen used to conclude that the peculiar behaviors were adaptations. But first, here is the evidence that cliff nesting itself may be an adaptation.

> Since the great majority of gulls nest on the ground one may presume that this was also the ancestral breeding habitat of the Kittiwakes. This view is supported by two facts: (1) Kittiwakes' eggs retain to some extent the cryptic pattern of blotching although this can be of little value, as every nest is marked conspicuously by a flag of white droppings; (2) the young Kittiwakes are able to run under suitable conditions, though not quite as well as young ground-nesting gulls. This is an unusual feature for a species nesting on such precarious sites.
>
> The advantage of cliff-nesting is certainly that it reduces predation. The nests seem fairly safe not only from ground-predators but also from such aerial ones as large gulls. [There follows three paragraphs summarizing evidence of predation reduction.]

> Throughout this paper I have contrasted the behaviour of the Kittiwake with that of the ground-nesting gulls as a group. Only two species, the Black-headed Gull *Larus ridibundus* and Herring Gull have been at all fully studied, but quite a lot is known about the other gulls in a more fragmentary way. Although more studies are desirable, on the existing information there seems a good indication that in many respects the ground-nesters behave alike (E. Cullen, 1957, pp. 275–276).

So, we can conclude that kittiwakes are derived from a precursor gull species that was ground nesting, based on two vestigial features of the precursor. We can also conclude that cliff nesting is a trait that reduces predation compared to ground nesting. However, we do not know why kittiwakes evolved a cliff-nesting habit and other gulls, which also have high predation rates, did not. Maybe nesting spaces were scarce, and the kittiwakes were being crowded out by other species of gulls (kittiwakes are smaller than many compatriot species); maybe kittiwakes were more agile flyers than compatriot species and were therefore able to exploit small cliff ledges; maybe … etc. These, and other hypotheses may be correct, but the bottom line is: we don't know, we can only speculate. This is a general problem with many adaptationist hypotheses, especially about behavior: collecting relevant evidence is very difficult, and sometimes impossible. Current fitness and survival value can be studied experimentally, but adaptation can be studied experimentally only in a limited way, and cannot be proved by experiments. Returning to the kittiwakes, I should point out that Cullen never claimed that cliff nesting was an adaptation. Her argument began with the fact that kittiwakes nest on cliffs; she then asked whether differences in behavior between the kittiwakes and ground-nesting gulls could be understood as adaptations to this environmental difference.

Fighting in most gulls generally comprises pecking down at the opponent, wing beating, and grasping the opponent, usually by the wing but occasionally by the beak, and pulling. Kittiwakes grasp the opponent by the beak and twist. The structure of the cliffs usually does not allow a bird to get above the other to peck downwards and wing beat, and the size of the ledges would often result in the resident bird falling into the water if it pulled backwards. The beak-twisting movement often results in the intruder being thrown into the sea, and beak twisting is very rarely seen in other gull species. Kittiwakes have also lost the aggressive upright threat posture. These observations all give strong support to considering fighting by beak twisting an adaptation: the selection pressures have changed and the aggression behavior system has become reorganized, and this change has clear survival value.

A secondary adaptation to fighting has also evolved. In the process of adaptation to fighting on cliff ledges, the beak became an important releasing stimulus in aggressive encounters. This led to the evolution of a new appeasement display: 'beak hiding', in which the head is turned and the beak is hidden in the breast feathers. This display has presumably been derived from the display 'facing away' (or 'head flagging') seen in other gull species, which would be often impossible to perform on the cliff ledges due to space limitations. Beak hiding is performed by a female when she lands on the ledge of an aggressive male. If her beak is showing, she is often attacked and twisted away. A kittiwake chick also shows beak hiding when being attacked by its older sibling (kittiwakes normally lay only two eggs), which inhibits the attack. Cullen did an experiment in which she added a herring gull chick or a black-headed gull chick to a kittiwake nest. These chicks showed no trace of beak hiding and were vigorously attacked by their foster siblings. These results all support the adaptation hypothesis for 'beak hiding': the reorganized aggression system provided a selection pressure for a change in the escape system of the birds, and this change has clear survival value.

Adaptations of Cuckoos and Their Hosts

Adaptation is a continuing dynamic process. It is a mechanism of change. The environment of organisms is continuously changing: climate changes, food sources fluctuate, competitors and predators invade, competitors and predators go extinct, etc. Animals either adapt to these changing conditions or go extinct themselves. A well-studied and interesting example is the coevolution of adaptations in cuckoos and their hosts. About 1% of all bird species are brood parasites, as are 40% of cuckoos (Cuculidae): they lay their eggs in the nests of other species and leave incubation and rearing of young to the host species. In the case of the Old World common cuckoo (*Cuculus canorus*), the female removes one egg from the host's nest and replaces it with one of her eggs. When the cuckoo chick hatches, it ejects the unhatched eggs or young of the host from the nest, and remains by itself in the nest, where it obtains food and protection from its hosts. The behavior of the female cuckoo and the young chick varies from species to species, but the general pattern is always the same.

Nicholas Davies and his colleagues have studied adaptations in cuckoos for many years, and he has analyzed his and related work in an excellent review article (Davies, 2011). He suggests that the cuckoos' adaptations are of two kinds: 'trickery' and tuning. Trickery involves the coevolution of adaptations that counter successive improvements of host discrimination: changes in the behavior mechanisms or physiology of the sender and in the perceptual mechanisms of the receiver. As an example of coevolution, cuckoo egg mimicry (adapting the appearance of the egg to match that of the host) evolves in response to egg rejection by the host: in those species that are best able to discriminate the cuckoos' eggs, the cuckoos' eggs most closely resemble the host's eggs.

Tuning involves adaptations that exploit the resources of the host. Tuning would require adapting to a host's life history, including adapting to the host's nest, its length of incubation, its willingness to accept the cuckoo chick when it hatches, and the type of food it provides (insects or grains) for the young hatchling. Davies (2011, p. 1) concludes: "The two hurdles of effective trickery in the face of evolving host defense and difficulties of tuning into another species' life history may together explain why obligate brood parasitism is relatively rare." I should add that there is still no convincing explanation of how or why any species has evolved brood parasitism in the first place.

Another question is why hosts of the common cuckoo that are so discriminating against eggs unlike their own readily accept a cuckoo chick, which is larger and has a different gape color from their own chicks. An interesting explanation was proposed by Lotem (1993). He suggested that parents of the host species imprint on their young, just as the young imprint on their parents. If the host parents would imprint on the young cuckoo, they might reject their own young in subsequent breeding events. Such an explanation implies differences between the incubation behavior system (egg recognition) and the filial behavior system (chick recognition) of the host species.

Several cases are known of differences between closely related species in chick recognition that are correlated with ecological differences. Noddy terns (*Anous stolidus*) nest in low tree branches and their chicks do not leave the nest until about two weeks posthatch; sooty terns (*Sterna fuscata* – now *Onychoprion fuscatus*) nest on the ground and their chicks leave the nest temporarily from about four days posthatch. Chick-replacement experiments have shown that noddy parents do not recognize their own chicks before two weeks posthatch, whereas sooty parents recognize them after four days (Watson, 1908). Similarly, kitti-wake parents accept foreign chicks until fledging at about three weeks, while the ground-nesting gulls expel foreign chicks after three or four days. Interestingly, kittiwake chicks are excellent at discriminating non-sibling chicks even though their parents are not (Cullen, 1957). These results give strong support to accepting these differences as adaptations.

Displays (Signals) and Coevolution

In Chapter 8, I defined a display as a movement specifically adapted to serve as a signal and gave several examples. Most displays are thought to be derived from movements already present in the species' reper-toire. However, they are often different in form from the original movements from which they are derived, and can also be motivated by different causal factors from the ones that caused the original move-ments. The evolutionary process by which changes in the form of a display occur as an adaptation to a signal function is usually called *ritualization* (Huxley, 1923). The process by which changes in the causal factors controlling the occurrence of a display occur as an adaptation to a signal function is usually called *emancipation* (Tinbergen, 1952).

The upright posture of the herring gull and the oblique and forward postures of the black-headed gull (p. 245) would be considered ritualized, but not necessarily emancipated. The waltzing display of male junglefowl (see Figure 3.10) is an example in which both pro-cesses play a role. Following Kruijt's (1964) analysis, this display is derived primarily from attack and escape elements. Nonetheless, some elements, such as the scratching movements through the feath-ers of the extended wing, are clearly embellishments to the move-ments from which the display evolved; the movement as a whole is ritualized. In addition, waltzing is also at least partially emancipated because sexual causal factors, as well as attack and escape causal factors, play an important role in its occurrence. It is a general rule that most displays comprise some combination of ritualized and non-ritualized elements and of emancipated and non-emancipated causal factors.

The evolution of displays usually requires coevolution of the motor mechanisms of the sender and the perceptual mechanisms of the receiver. In the older ethological literature, most authors assumed that the current function of a display reflected the selection pressures acting during its evolution, though specific trajectories were seldom considered. Displays have evolved because they influence the behavior of the receiver in ways that benefit the sender. However, the benefits of a display can be different for a sender and a receiver, and R. Dawkins & Krebs (1978) pointed out that there is often a conflict of interest between the two parties. They used the metaphor of an arms race between senders trying to manipulate receivers, and receivers mind-reading senders. There followed a flurry of theoretical and experimental papers examining the question of how 'honest' signals could evolve if senders were always trying to 'deceive' receivers. The scenario is somewhat different depending on the function of the signal being considered.

In a courtship situation, both parties are looking for a high quality mate of the correct species (see sexual selection below). In theory, high quality means high potential reproductive success, but in practice, it usually means what potential mates find attractive. With respect to species recognition, any signal that attracted an individual of the wrong species would quickly disappear due to the process of natural selection: any such union would provide at most infertile offspring. With respect to high quality, a male can advertise his quality with a signal that cannot be faked, such as the loudness or frequency of a frog's croak that indicates his size, or with a signal that would be too costly to fake, such as the peacock's tail (the 'handicap' principle, Zahavi, 1975). Other possibilities include occupying a location that involves male-male competition to achieve, such as a central territory on a black grouse lek. Further, there are often successive steps in pair formation, and female assessment can occur at each step. Females would coevolve recognition mechanisms that recognize high quality. As we have seen, however, in many cases, the perceptual mechanisms of the female do not need to evolve because the males' signals are evolving to conform to pre-existing biases or preferences of the females (p. 34 and p. 250). In mated pairs, the selection pressure for evolving displays would be coordinating the behavior of the pair to ensure copulation and fertilization. In all these cases, the benefits of the displays are basically equal for the two sexes.

In agonistic situations, displays generally evolve that reduce the damage that would be done by actual fighting. Rivals need to assess each other and estimate their chances of winning before engaging in combat: actual fighting can be costly, and running away too soon can also be costly in terms of missed opportunities. The assessment signals seen in the fights of the cichlid fish discussed in Chapter 9 are an example (Enquist et al., 1990). In those experiments, encounters that

ended early showed little or no physical damage to either of the contestants. Most agonistic displays are derived from intention movements or components of attack and escape behaviors, and
evolutionary changes can usually be categorized as ritualization and
emancipation. Here too, the benefits of the displays are basically equal
for the two rivals. In general, the ESS solution to coevolutionary games
is a mixed strategy in which the fitness benefits to each participant are
equal. Studies such as those discussed in Chapter 9 can be done to
estimate the fitness benefits of the various strategies. Whether anthropomorphic terms such as arms races, honesty, deception, trickery,
manipulation, and mind-reading are useful for understanding the process of coevolution is moot.

Alarm Signals

Alarm signals are a subset of signals that have been studied extensively
because they present problems for evolutionary theory and cognitive
psychology. A paragraph in an early paper by Marler (1955, p. 6) illustrates many of the issues:

> There is a conflict in many animals between the need to appear
> conspicuous to their own species and inconspicuous to predators or prey,
> which is manifest in their coloration and also in their vocalizations. Small
> Passerines have two distinct responses to a hawk or owl. If the bird of prey
> is perched, they make themselves conspicuous by 'mobbing' behaviour,
> which attracts other birds' attention to the predator; if it is in flight, they
> dash for cover, and hide. Chaffinches use quite different calls in each case.
> In the first they give a 'chink' note, which provides abundant location
> clues; in the second, males give a high, thin 'seeet' note. The former is
> easily located by man, the latter only with difficulty, and the same is
> probably true for predators.

To begin, why do the birds make any call at all when they see
a predator? Why not just quietly hide? Perhaps calling is an 'emotional'
response to seeing a predator: the birds simply cannot help themselves
(cf. Marler & Evans, 1996). What would the historical evolutionary
antecedents (selection pressures) be for such a response? And why are
there two distinctly different calls in the two contexts? Perhaps calling
does have a current function. Mobbing does have a function for the
gulls (p. 284), but gulls and most passerine species live and nest in very
different environments: gulls on the open shore and the passerines in
the forest. Did mobbing evolve independently (convergently) in all
these species? The 'seeet' call has properties that suggest it may have

evolved as a sound that is difficult to locate, but, as Marler noted, at least half a dozen other unrelated passerines also have a very similar 'hawk' call. Has the 'seeet-type' call found in current species evolved from a similar call in an *ur*-passerine species (descent with modification)? Or has it evolved independently (convergently)? Or some combination? Many other species from insects to anurans to primates also have calls that are difficult to locate. Such calls must certainly have evolved independently. A phylogenetic analysis would be a good step toward answering some of these questions.

At least one phylogenetic analysis of alarm calling has actually been carried out. Daniel Blumstein and his students have been studying alarm communication in eight of the 14 (now 15) species of marmots (*Marmota* spp. - large squirrel-like rodents) in the field and laboratory for more than 20 years. In a very thoughtful review of his work, Blumstein (2007) noted that in thousands of hours watching marmots, he has only seen one complete predation event, and in many more thousands of hours, his assistants have seen only a few more. On the basis of his observations and many experiments, he concluded that "alarm calling seems to have initially evolved as a means of detection signaling to predators. Conspecific warning functions are thus an exaptation [see below]" (p. 381). Further, Blumstein was only able to find good evidence for a warning function in female yellow-bellied marmots (*M. flaviventris*) when they had young. Other females and males seldom gave an alarm call.

In an extensive review of animal alarm calling, Zuberbühler (2009) considers three main groups of evolutionary hypotheses: kin selection (alarm calling improves the survival of individuals that share copies of a certain proportion of the caller's own genes - see below), sexual selection (females are attracted to males that risk calling - the handicap principle again), and individual selection (detection signaling: 'I'm here and I know where you are, so don't even bother trying to catch me'). He finds some evidence for each of these functions in some cases, and concludes: "alarm calling is found in a large number of species, demonstrating its adaptive value in predation avoidance" (p. 313). Unfortunately, the prevalence of alarm calling, by itself, is not evidence of its adaptive value in predator avoidance, as Blumstein's results indicate. The major problem is that alarm calling is a functional category. It is unreasonable to expect a uniform mechanism for all exemplars. This means that whether an alarm call is adaptive in predator avoidance and an adaptation needs to be investigated in each case. I will have more to say about this in the section on exaptation.

Alarm calls often contain information about predators. For example, we saw above that chaffinches give distinct calls for perched predators and for predators in flight, and chickens give distinct calls for ground and aerial predators. In the 1960s, working in the Amboseli reserve in Kenya, Struhsaker (1967) noted that vervet monkeys (*Cercopithecus aethiops*, now *Chlorocebus aethiops*) gave acoustically different alarm calls in response to different types of predators, and that each call evoked a different and seemingly adaptive response. His suggestions were followed up in a 14-month field study by Seyfarth *et al.* (1980) who observed the responses of vervet monkeys to predators in natural encounters and to playbacks of recorded alarm calls. Their results confirmed the observations of Struhsaker: "animals on the ground respond to leopard alarms by running into trees, to eagle alarms by looking up, and to snake alarms by looking down" (p. 1070).

These findings started a conversation about the problem of meaning in alarm calls (Macedonia & Evans, 1993). Do non-human animal vocalizations have 'functional reference': do the callers only give the leopard alarm when they spot a leopard, or do they give the leopard alarm when they spot any ground predator or non-predator; do the receivers always behave appropriately when they hear the alarm for a predator they cannot see? Do the monkeys (or birds) have representations of the predator analogous to human representations? Subsequent studies of other primate species (Diana monkeys, *C. diana*, Zuberbühler, 2000; titi monkeys, *Callicebus nigrifrons*, Cäsar *et al.*, 2013) and several avian species (review in Gill & Bierema, 2013) support the hypothesis of functional reference, but the meaning of meaning is still a controversial issue. It is logically an aspect of the search for the adequate (sign or key) stimulus.

Seyfarth & Cheney (1997) also studied the development of alarm calling and responses to alarm calls in young vervet monkeys from three to seven months of age. Their results showed that both the motor mechanism of the sender and the perceptual mechanism of the receiver require functional experience (i.e. learning) in order to attain adult form. These findings raise issues of social learning and cultural transmission, as well as a host of developmental and evolutionary issues. Should alarm calling be considered a learned adaptation? There is still much to be learned about alarm calls.

ADAPTIVE RADIATION

In his book recounting his experiences during the voyage of the *Beagle*, Darwin commented on the birds he saw in the Galapagos Islands (1845, p. 380): "Seeing this gradation and diversity of structure in one small, intimately related group of birds, one might really fancy that from an original paucity of birds in this archipelago, one species had been taken and modified for different ends." He called the birds 'finches' (though they actually belong to the tanager family and are not closely related to the true finches). They have been known as 'Darwin's finches' since a paper published by Lowe (1936) and popularized by Lack (1947). They are the classic example of adaptive radiation and their study continues to provide insights into evolutionary mechanisms (e.g. Grant & Grant, 2006).

"Adaptive radiation refers to diversification from an ancestral species that produces descendants adapted to use a great variety of distinct ecological niches" (Losos, 2010, p. 623). It can be contrasted to straight-line adaptation (anagenesis) in which a species produces descendants that gradually adapt to a single new environment – one-to-many versus one-to-one. There are many reasons that adaptive radiation is of interest to evolutionary biologists (Schluter, 1996), but for my present purpose, it is important for understanding the process of adaptation. Adaptive radiation tends to be seen in environments with underexploited resources such as are found on remote archipelagos or newly formed lakes – one aspect of the 'ecological opportunity' hypothesis. A few million years ago (not long in evolutionary time), one species of 'finch' arrived in the Galapagos. Most of the other birds on the islands were sea and shore birds, so there was little competition from them for the seeds, insects, and plants endemic to the islands. But there was competition for food among the original 'finches'. In a series of steps, the original species 'radiated' into 14 separate species, each specializing on different aspects of the available ecological resources: trees and ground; seeds, insects, fruits, cactus. In the process, each species evolved physiological and morphological adaptations, most noticeable of which was the structure of beak.

Tinbergen's (1959) comparative studies of the behavior of gulls, discussed in Chapter 8, provide several examples of adaptive radiation of displays. As we saw, in the 14 species he studied, he had evidence that all the various agonistic and courtship displays were derived from the displays of the ancestral species. However, except for the difference between the cliff-nesting kittiwake and the ground-nesting species, he did not speculate about possible ecological selection pressures responsible for differences between the displays of the different species. He did propose conspicuousness and distinctness as selection pressures within species, but that would not account for the radiation of species.

Studies of ducks by Lorenz (1941) and continued by McKinney (1975) provide better evidence for the ecological selection pressures leading to adaptive radiation of their agonistic and courtship displays. McKinney asked three questions: Why do ducks need displays (i.e. what were the selection pressures)? Why do displays have the characteristics they have (current fitness)? What factors have been most important in producing specific differences in display repertoires (comparative selection pressures)? To find answers, he surveyed the behavior of 36 species of the genus *Anas* (dabbling ducks).

McKinney first noted that all these species must have descended from ancestral stock that had pre-adaptive commitments to exploiting ecological niches associated with marshes, ponds, rivers, and lagoons. They had also evolved specific differences in habitat preferences and feeding habits, as well as social systems in which only the female incubates the eggs and cares for the young. Given these constraints, males need signals for courtship and establishment of bonds with females, for bond maintenance and defense of the mate, for territory defense, and for pre- and post-copulatory rituals. Females need signals for pairing and bond maintenance. These needs are common to all the dabbling duck species, which should result in many similar display characteristics (as in the gulls), but because these species have also exploited different habitats, courtship and bonding displays are expected to be different and to reflect the requirements of each species' niche.

An additional feature of the ducks' social systems is that males usually have 'time on their hands' during the incubation period; so, as well as forming pair bonds, males are always looking for extra-pair copulations both with single and mated females. Even mated females are available because they have to leave the nest and fly to the water to feed, drink, and bathe. Thus a male must have displays to defend his mate while, at the same time, he may be courting other females. ('Rape' is a regular feature of dabbling duck society.) Given all these putative functions of duck displays, tracing their evolutionary trajectory is an enormous task. After several years' work, McKinney was able to make a simplified attempt to classify green-winged teal (*Anas crecca carolinensis*) displays according to their probable functions, but the teal is only one of the 36 species. Answering his questions for all the duck species

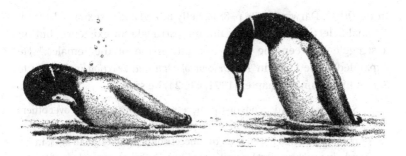

Figure 10.1 Two motor phases of the 'grunt-whistle' of the mallard drake.
Note the arc of upward-flying water droplets
From Lorenz, 1941.

will take a long time. Still, the work of Johnson (2000) was able to give some answers. Using the phylogenetic approach discussed in Chapter 8, he found that one common courtship display, the 'grunt-whistle' (see Figure 10.1), occurred only in those species whose courtship took place exclusively in the water. His analysis also showed that it was the less derived species that courted both on land and in water, while the more derived species courted only in water. He concluded that both ecological and social factors were important in driving the evolution of displays in the dabbling ducks. Another conclusion from these studies is that speciation and social system evolution occurred prior to the evolution of most of the displays.

The radiations of the gulls and ducks took place over millions of years and over much of the earth's surface, whereas the 'finch' radiations happened much more quickly in a much more restricted area. A radiation that occurred in even less time is that of the East African cichlid fishes: more than 500 cichlid species have evolved in Lake Victoria from a single precursor species in less than 15,000 years (Meyer, 2015). Similar cichlid radiations also occurred earlier in the other two major African lakes. As pointed out above (p. 252), these fish have evolved adaptations to eating every conceivable food source in their environment. The other remarkable feature of these fish is that some highly specialized traits have evolved independently in all three lakes (parallel evolution), including species that feed almost exclusively on the scales of other fishes, and enlarged lips in species that target prey found in rock crevices. Such a feature suggests that the phenomenon of 'deep homology' (p. 269) is actually widespread.

SEXUAL SELECTION

In *The Origin*, Darwin (1859, p. 88) briefly talked about sexual selection: "Sexual Selection … depends, not on a struggle for existence, but on a struggle between the males for possession of the females." He expanded these ideas in *The Descent of Man and Selection in Relation to Sex*. In his summary chapter (1871, Ch. 21) he says:

> Sexual selection depends on the success of certain individuals over others of the same sex, in relation to the propagation of the species; whilst natural selection depends on the success of both sexes, at all ages, in relation to the general conditions of life. The sexual struggle is of two kinds; in the one it is between individuals of the same sex, generally the males, in order to drive away or kill their rivals, the females remaining passive; whilst in the other, the struggle is likewise between the individuals of the same sex, in order to excite or charm those of the opposite sex, generally the females, which no longer remain passive, but select the more agreeable partners.

The behavior of the elephant seals and the black grouse discussed in Chapter 9 are classic examples of each kind of sexual selection.

Sexual selection is the outcome of nonrandom variance in the reproductive success of one or both sexes. Darwin developed his theory to account for the evolution and maintenance of exaggerated traits that do not benefit the individual's survival. An example is an experiment by Møller & de Lope (1994) on male barn swallows (*Hirundo rustica*) carried out in Denmark and Spain. Females of this species prefer males with longer tails (Møller, 1988). In the experiment, some males had their tails shortened, and others lengthened; control males kept their original length. Survival was measured by the number of males returning to their nest boxes the following year. The relative survival rates were 49% and 29% for the shorter- and longer-tailed males, respectively, and 39% for the controls. One probable source of the differences is that shorter-tailed birds are able to capture larger insects than the longer-tailed birds. This case is instructive because it shows how sexual selection (females preference for long tails) is opposed by natural selection (survival rate of males with long tails) so that the ESS is a compromise, presumably represented by the tail length of the control birds.

Nonrandom mating is probably the rule rather than the exception in most vertebrate and many invertebrate species, so there is a very large literature on studies dealing with sexual selection. Most of these studies investigate behavior that has evolved to assess the quality of a potential mate and to facilitate or hamper mating. These studies are usually more concerned with current fitness (cf. Chapter 9) than with historical origins; reviews and discussions of these studies can be found in any modern textbook of behavioral ecology. I will consider one example in which ideas about sexual selection have been used in an attempt to understand the evolution of an exceptional mating system: the behavior of the ruff.

Sexual Selection in the Ruff

The ruff (*Philomachus pugnax*) is a medium-sized wader (shorebird) that breeds in grassy marshes and moist meadows across northern Eurasia. It is a migratory, gregarious sandpiper that winters mostly in Africa and southern Asia. Males are larger than females, but during most of the year, there is nothing especially remarkable about the species. Things change dramatically in the spring when the birds return to their breeding grounds. The males develop a striking nuptial plumage comprising a large pectoral ruff, two tufts on top of the head, and long ornamental feathers and wattles on the head between bill and eyes (see cover of this book). There is great individual diversity with respect to color and pattern of the nuptial feathers and color of the wattles, such that humans (and presumably the birds themselves) can easily recognize males individually. Females also develop a nuptial plumage, but it is only slightly different from winter plumage. At this time, males are often seen fighting with each other, hence the species' name, *pugnax*. The striking appearance and behavior of these birds has long drawn the attention of naturalists, and Darwin (1871, Ch. 13) cites the species, with a drawing, as an example of sexual selection. The English naturalist Selous (1906), studying sexual selection, observed ruffs in Holland and published the earliest accounts of their behavior. Later observations were published by Danish investigators, but the most comprehensive account was a four-year study published in 1966 by Lidy Hogan-Warburg (my late wife). This work was followed up by studies of van Rhijn (1991) and Verkuil (2010).

Courtship and mating in the ruff occur on a lek (communal display ground); no sounds are uttered during these activities. A lek site consists of a number of bare spots about 30 cm in diameter called residences, which are about 1 to 1½ m apart; the inter-residential areas are generally covered with grass. The seven leks observed during the course of Hogan-Warburg's study ranged in size from three to nineteen residences. Within the male community, she distinguished two groups: independent males and satellite males. Independent males were further subdivided into resident and marginal males. Resident males possess a residence (territory) that they occupy most of the time during the breeding season and that they defend from other independent males. Inter-residential areas are not defended. Marginal males visit the lek sporadically and also visit other leks. When on the lek, they usually remain at the edge, but very occasionally can be seen fighting with, and at least once usurping, a resident male.

Satellite males (about 40% of males in Hogan-Warburg's study) may visit several leks, and are generally seen when females are also present. They never possess a residence: they occupy a residence jointly with its owner. Satellite males also never show overt aggressive behavior on the lek; their behavior toward resident males is in many ways similar to the behavior shown by females toward resident males. Satellite plumage also differs from independent plumage: satellite males usually have white or very light-colored ruffs and head tufts, while independent males are typically much darker. Hogan-Warburg gave arguments for considering the behavior and plumage differences between independent and satellite males a balanced, genetically-based, behavioral polymorphism in which the fitness of each morph is frequency-dependent in a way that prevents either from going extinct. This suggestion was highly controversial at the time. Later, however, Lank et al. (1995) showed that these differences are indeed due to a genetic polymorphism. A third morph, called a 'faeder', has also been discovered recently that is a female mimic (Jukema & Piersma, 2006); it constitutes 1% or less of males. A genomic study has now shown that all three male morphs are controlled by a single 'supergene' (Küpper et al., 2016).

Resident males, alone on the lek, may show aggressive display toward their neighbors, or even fighting, but mostly they remain relaxed on their residences. When satellite males are present, they approach and enter one or more residences and may be either chased away or tolerated. These interactions involve a series of specific displays. As many as three satellite males may be tolerated on a single residence, but some resident males never tolerate even one satellite male. Females visit the lek for short periods and normally approach the lek by air. On these occasions, all the males on the lek react with a reception ceremony that includes wing fluttering and up and down movements, and ends with the males freezing in a squat posture. To the inexperienced observer, the reception ceremony gives the impression of (silent) pandemonium breaking out, followed by

Figure 10.2 Ruff behavior on the lek. Resident male (left) and satellite male (right) together on a residence, with female nearby.
From Hogan-Warburg, 1966.

a weird sudden cessation of movement. After landing, females place themselves near a residence where they usually remain during their stay on the lek (see Figure 10.2). They may also move around in the inter-residential areas and visit other residences. They sometimes step on a residence and crouch and copulate. Only males on a residence copulate, and only if the female is crouching. Both resident and satellite males copulate, but copulation frequency is low (only about once per hour in Hogan-Warburg's study) so relative frequency is highly variable. Relative copulation frequency also varies with size of the lek. It seems likely that resident males copulate more than satellite males, and resident males with satellites often copulate more than resident males without satellites. The choice of the female appears to depend mainly on individual characteristics in the behavior and plumage of the males, especially rising from the squat.

The breeding behavior of the ruff is highly unusual in many respects, and there has been considerable speculation about its evolutionary origin. Plausible hypotheses have been hampered by the fact that there are no closely related species showing similar behavior. Van Rhijn (1991) cast the net wide by first considering the conditions under which the entire order to which ruffs belong (Charadriiformes) evolved. He then considered the evolution of the ruff's various behavioral characteristics in highly differing taxonomic groups. These included promiscuity, communal display, satellite strategies, terrestrial display, and silence. Unfortunately, no clear, obvious hypothesis emerged. One conclusion was that the satellite strategy of the ruff appeared after or concurrently with the development of the lek. This means that the lek, especially its high density, must provide the necessary conditions for the evolution of the satellite strategy. Sexual selection has clearly been important in the evolution of the ruff's mating system, but specifying the selection pressures for the ruff's current behavior is difficult because we do not even know in what order the various aspects of its behavior evolved. Hogan-Warburg (1992, p. 402), in a review and critique of van Rhijn's (1991) book, concluded: "All things taken together, however, I suspect that there are as many hypotheses about the evolutionary development of the satellite strategy as there are experienced students of Ruff behavior."

INCLUSIVE FITNESS, KIN SELECTION, AND COOPERATION

In 1964, Hamilton published a set of papers on the genetical evolution of social behavior in which he introduced the concept of *inclusive fitness*. He pointed out that individuals share copies of their genes not only with their offspring, but also with siblings and other relatives, and showed, mathematically, that altruistic behavior (e.g. helping the offspring of relatives rather than having one's own offspring) could evolve by natural selection if certain conditions were met. These conditions are summarized in 'Hamilton's rule': altruistic cooperation can be favored if the benefits to the recipient, weighted by the genetic relatedness of the recipient to the actor, outweigh the costs to the actor. Maynard Smith (1964) coined the term *kin selection* to describe this process. These concepts are important because they have allowed biologists to understand social evolution without invoking concepts of group selection. They are also essential for understanding many aspects of social behavior as adaptations, including the complex social communities of many species of ants, bees, wasps, and termites. In vertebrates, kin selection has been invoked to explain the evolution and maintenance of alarm signaling in many species (see above) and aspects of cooperation in mammals. Davies *et al.* (2012, Ch. 11) provide a good introduction to these concepts and give many examples.

EXAPTATIONS AND CONSTRAINTS ON EVOLUTION

Adaptation has been defined and recognized by two different criteria: historical genesis (features built by natural selection for their present role) and current utility (features now enhancing fitness no matter how they arose). Biologists have often failed to recognize the potential confusion between these different definitions ... [But there are many features of organisms] that now enhance fitness but were not built by natural selection for their current role. We propose that such features be called *exaptations* and that adaptation be restricted, as Darwin suggested, to features built by selection for their current role (Gould & Vrba, 1982, p. 4).

This apparently straightforward proposal has been and still is a source of acrimony and controversy. Why? Nuances of language are one reason, but the main reason is sociological. At a very public meeting of eminent behavioral biologists at the Royal Society in London in 1979, Stephen Gould and Richard Lewontin presented a paper that was a very pointed critique of what they called the "adaptationist programme", a research strategy based on "faith in the power of natural selection as an optimizing agent". Their argument was that there were many agents of evolutionary change in addition to natural selection that the adaptationists were not considering, and that many of their conclusions were mere storytelling ("just-so stories") unsupported by evidence. Many of those present at the meeting were 'adaptationists', and nobody likes being criticized, especially not publically. And so the war began. Of course, there were also many other aspects of the controversy, but I will not discuss them here. Important review articles subsequently appeared (Buss *et al.*, 1998; Andrews *et al.*, 2002) and Gould himself died in 2002; things have quieted down since then. Nonetheless, the controversy still continues (e.g. Pievani & Serrelli, 2011). Most behavioral biologists today accept the concept of exaptation, though many still disagree with or have reservations about some of Gould's views. Subtleties of definition also still play a role.

The standard example of exaptation is the evolution of the feather: it evolved as a means of temperature regulation, but was later co-opted for flying. Behavioral examples that I have discussed include the warning function of the marmot's alarm call (p. 307) and the chuck component of the túngara frog's advertisement call (p. 250). Many of the other adaptive traits discussed in Chapter 9 are also probably exaptations, but the evidence necessary to come to such a conclusion has still not been gathered. An exceptional example could be human language. If the central behavior mechanism 'merge' evolved under selection for a non-linguistic function, as suggested by Bolhuis *et al.* (see p. 253), then human language itself is an exaptation.

Related to the concept of exaptation is the idea of pre-adaptation. Zeder (p. 255) used it to explain why some species can be domesticated easily and other species cannot. McKinney (above) suggested that the dabbling duck species must have descended from ancestral stock that had pre-adaptive commitments to exploiting ecological niches associated with marshes, ponds, rivers, and lagoons. The tuning characteristics of female frogs' perceptual mechanisms also exist prior to the evolution of the males' advertisement call, as do preferences of many female birds for long tails in the male. Traits that evolved subsequently would be exaptations. It is likely that many of the ruff's behavior mechanisms are also exaptations.

These examples all suggest that there are constraints under which natural selection must work. This has been increasingly recognized in the behavioral ecology literature as papers such as McNamara & Houston (2009) and Fawcett *et al.* (2013) illustrate (see p. 22). Such recognition is less obvious in the evolutionary psychology literature. In many ways, the factors constraining evolution are analogous to the factors constraining development: both must start from a point that already has much history behind it, and that often means that the best solution is simply not attainable. An example should make this clear.

Daly (1979) asked the question: Why don't male mammals lactate? They could then take a more active role in raising the young. Daly discussed the question from both physiological and reproductive strategy viewpoints. He concluded:

> Functional male lactation would require changes in sexually differentiated ontogenetic processes at both prepubertal and circumpubertal stages, as well as some male analogue of lactogenic events in late pregnancy. None of these modifications seems impossible, but together they constitute a formidable barrier to the evolution of male lactation [cf. Waddington's example of changing the car into two motorcycles (p. 188)]. Moreover, it is by no means clear that such an evolutionary development could enhance its bearer's fitness. Other factors than maternal lactational capacity evidently limit the reproductive potential of monogamous mammals (p. 325).

For somewhat different reasons, Davies (2011) concluded that obligate brood parasitism (as found in cuckoos) is relatively rare. Both of these cases, and many others, illustrate the fact that natural selection cannot always, and maybe can never, provide an optimal solution.

ANIMAL PERSONALITY AND INDIVIDUAL VARIATION

Differences in behavior among members of a species have been known since ancient times. Such differences in humans have been studied scientifically by psychologists since the nineteenth century (see Ashton, 2013), but it is only since the last half of the twentieth century that individual variation in behavior within non-human species has become a matter of interest for evolutionary biologists. Differences in the behavior of dog breeds have been one focus of attention (e.g. Scott & Fuller, 1965; Hart, 1995), but there has also been much interest more generally in the fitness consequences of individual variation within species (Slater, 1981). An early example of this sort is an experimental field study of individual differences in social behavior and reproductive success in yellow-bellied marmots by Armitage (1986). His results suggested to him that marmots may have a strategy of phenotypic plasticity: by producing young that are variable in 'sociality', a female increases the probability that some of her descendants will survive in unpredictable social and ecological environments.

Studies of this kind were rather sporadic until the turn of the century when a psychologist published an extensive review of non-human animal studies and asked the question "What Can We Learn about Personality from Animal Research?" (Gosling, 2001). Shortly thereafter, Sih et al. (2004) published an important review paper in which they urged their behavioral ecology colleagues to take a broader view of adaptation. They noted that behavioral ecologists typically treat all individual animals as being potentially able to exhibit optimal behavior; they also tend to study ecological situations in isolation from each other, and they ignore individual variation. These considerations led many behavioral ecologists to investigate what many now call personality or temperament or plasticity (e.g. Dingemanse et al., 2004).

Results of optimality studies, as we have seen, usually do not support the models quantitatively. Sih et al. suggested that consideration of 'behavioral syndromes' could account for many of the anomalies in the results. They pointed out:

> Populations and species often exhibit behavioral syndromes; that is suites of correlated behaviors across situations. An example is an aggression syndrome where some individuals are more aggressive, whereas others are less aggressive across a range of situations and contexts. The existence of behavioral syndromes focuses the attention of behavioral ecologists on limited (less than optimal) behavioral plasticity and behavioral carryovers across situations, rather than on optimal plasticity in each isolated situation (p. 372).

In other words, an aggressive individual may be at a disadvantage in a mating situation, but the same individual may have an advantage in a feeding situation. We have already seen that a particular behavior (trait) can have multiple functions and that using a game theory approach can lead to an ESS in which the balance between functions is as optimal as possible. But such a solution with respect to aggression would imply that all individuals in the species should stabilize at the same level, which is almost never the case. A frequency dependent behavioral polymorphism, in which variation itself is being selected, may be a better solution: when food is in short supply, more aggressive individuals will have an advantage; when food is plentiful, mating opportunities may be available and more aggressive individuals will be at a disadvantage. This suggestion is similar to Armitage's suggestion above.

As studies multiplied, it became clear that mutual understanding of the concepts and issues among workers in the field was lacking. Several major papers have appeared trying to define terms and clarify issues (Réale et al., 2007; Stamps & Groothuis, 2010; Carter et al., 2013; Snell-Rood, 2013), evidently to no avail: Stamps (2016) has just published a paper that presents a major reorganization of "individual differences in behavioural plasticities". Behavioral plasticities, for Stamps, apparently means any change in behavior, and since, by definition, behavior is always active and changing unless the animal is dead, we return to the topic of 'individual differences in behavior'. I can only wish her success, but time will tell whether this new reorganization leads to new insights into evolution and behavior.

EVOLUTIONARY PERSPECTIVES ON HUMAN BEHAVIOR

Humans are animals, and human behavior can be understood in exactly the same ways as the behavior of other animals, and as described throughout this book (cf. Daly & Wilson, 1999). Humans are also special, but so are all other animal species. The reason there are so many studies of humans is that we are more interested in ourselves than we are in other species. That's just natural! I have not discussed humans separately in the book, however, because for the topics I have discussed, human behavior is just another example of the behavior relevant to the topic. I am making this one exception because interest in evolutionary perspectives on human behavior has spread to many disparate fields, so a few words on the topic seem relevant to the purpose of this book.

During the last half of the nineteenth century, several intellectuals promoted and extended Darwin's theory of natural selection, especially with respect to human activities. Prominent among them were Herbert Spencer (1820–1903) and Francis Galton (1822–1911). Spencer promoted Darwin's idea of natural selection, but reintroduced Lamarck's ideas about evolution being progressive and leading to an ever more perfect state. Later in his life and after his death, his ideas led to what became known as Social Darwinism, in which 'survival of the fittest' was applied to social and economic aspects of human behavior. Such ideas were used, for example, to support the idea that capitalism is a better system than socialism. Similar types of ideas were also promoted by Galton. Among other things, he studied the inheritance of intelligence and espoused eugenics as the way to improve the human race. These ideas were often popular at the time, but fell out of favor for various reasons by mid-twentieth century. This state of affairs suited most social scientists who tended to believe that biology had little or no place in understanding human behavior, and most behavioral biologists who had never been especially concerned with human behavior in the first place. Boakes (1984) and Laland & Brown (2011, Ch. 2) provide a more detailed history of this period.

The Social Darwinists were using evolutionary arguments about the current fitness of human behavior, but in the latter half of the twentieth century, discussions about adaptations became more prominent. Lorenz (1966), in his book *On Aggression*, suggested that many aspects of current human aggressive behavior were carryovers from behavior that was adaptive earlier in human history. In my terminology, a central coordinating behavior mechanism (aggression) had evolved as an adaptation to the selection pressures current many thousands (or millions) of years ago, and there had not been enough time for that mechanism to evolve as an adaptation to current environmental conditions. Many of Lorenz' ideas were denounced for mostly non-biological reasons, but Lorenz' logic actually underlies the current theoretical framework of the field of evolutionary psychology ('a modern man with a Stone Age brain').

It was not Lorenz' influence, however, that spurred the current proliferation of evolutionary explanations of human behavior, but rather the publication of Wilson's book *Sociobiology* in 1975. Its last chapter, on human behavior, was also denounced for many of the same reasons that Lorenz' ideas had been denounced, but Wilson's last chapter was preceded by a masterful summary of the theories and evidence for the evolution of social behavior throughout the animal kingdom. It was this presentation that was embraced by many biologists, especially ecologists, as well as by a number of anthropologists and psychologists. This has led to an influence of evolutionary thinking in many fields that study human behavior. There is no space here to discuss this material, but the book by Laland & Brown (2011) presents a very readable discussion of many of these various fields and is highly recommended for both students and professionals. I will, however, briefly discuss the question of what is special about an evolutionary perspective on human behavior, and also discuss one recent example.

What Is an Evolutionary Perspective?

The two basic evolutionary questions one can ask about any behavior are: (1) what is its survival value (current fitness)? and (2) how did it evolve (is it an adaptation)? In Chapter 9 and this chapter, I discussed these questions and gave many examples including some from humans. So, one could say that an evolutionary perspective on human behavior merely means asking similar questions and using similar methods to the ones that are asked about the evolution of non-human behavior, which is basically the approach I have taken in this book. There are, however, some special problems with respect to human behavior.

For non-human behavior, we have seen that it is necessary to choose a currency of fitness, but whatever currency is chosen, the ultimate theoretical currency is reproductive success. For humans, in their current environment, reproductive success is often not a useful currency for understanding the evolution of human behavior; wealth, intellectual achievement, or social status might be better currencies. We do not know how many children Shakespeare sired, but we do know that his behavior changed the English language forever; similar logic applies to Darwin, Einstein and many others. A related problem is that, whatever currency is being considered, selection pressures on humans are changing much more rapidly than in the past, and humans also have more means to deal with those changes (cognitive abilities and technology) than other animals. This means that standard evolutionary explanations may need to be modified. A further problem is that many researchers who study evolutionary aspects of human behavior often have only a superficial understanding of evolutionary biology and tend to make unwarranted assumptions about putative selection pressures or fail to acknowledge the role developmental factors may have played in the evolution of particular traits or modules. These problems are not peculiar to students of human behavior, but seem to affect them more than students of non-human behavior. One more problem, especially rampant in evolutionary studies of human behavior, is that all too often causal and functional explanations are confused (cf. Bolhuis & Wynne, 2009; Bolhuis, 2015). All these, and other, problems notwithstanding, an evolutionary perspective can lead to progress in understanding some aspects of human behavior. I will discuss one example.

Why Do People Kill Each Other?

Martin Daly and his late wife Margo Wilson have been studying homicide for many years. In their research, homicide refers to an individual murdering another individual, usually one whom they know. It does not refer to killing that occurs during war or other occasions of social unrest. One possible conclusion from their work is that homicide is an evolved adaptation in human males. In a recent publication, Daly (2016, p. 2) states:

> in developed countries with stable governments and judicial systems, both those who kill and those whom they kill are primarily young, disadvantaged men, [and] the contexts in which these men kill one another are *competitive* contexts: sexual rivalries, turf wars, robberies, and, above all else, contests over the limited social resources of dominance, respect, and "face."

Using small, 'undeveloped' societies as proxies for our ancestral environment, Daly marshals evidence that selectively killing your competitors did bring you wives and other bounties. He argues that these days, the local homicide rate reflects the severity of competition for limited resources and that economic inequality is a major cause of violence and homicide. This argument is straightforward and compelling, but here, too, there has been much vociferous dispute. Most of Daly's book, in fact, is devoted to discrediting other explanations for violent behavior. In this example, an evolutionary perspective led to a testable hypothesis that was not widely held previously.

POSTSCRIPT: CAUSE AND FUNCTION AGAIN

It seems fitting to end this book by going back to the beginning. In Chapter 1, I discussed the concepts of cause and function (consequences), and reiterated at the end of that chapter my opinion that cause and function are separate questions. Both are legitimate questions to ask about behavior, and answers to both increase our understanding of behavior. But causal and functional explanations should not be confused. I will end with one example in which, if not confused, the argument is at least ambiguous. In the penultimate section of their book, *An Introduction to Behavioural Ecology*, Davies *et al.* (2012) consider causal and functional explanations of behavior and make the laudable statement: "causal and functional explanations complement each other." Their example is the burrow of the black-tailed prairie dog, *Cynomys ludovicianus*. "A typical burrow has two entrances, one with a low, round 'dome' at its entrance and the other with a taller, steep-sided 'crater'. The different heights and shapes of the burrow entrances cause air to be sucked out of the crater end and, therefore, in through the dome" (p. 437). One function of the burrow is to provide fresh air, which it does because of its design; but whereas Davies *et al.* allude to how the burrow's structure 'causes' a certain pattern of air flow, they say nothing about the causes of the burrow's structure itself, which reside in the behavior mechanisms of the prairie dog's burrowing behavior.

The analysis of behavior is a complicated matter, but I hope this book has made some of those complications easier to understand.

Afterword

Looking over what I have written, I was reminded of a course I took as an undergraduate at the University of Chicago: OMP. It was a rather special kind of philosophy course called Organization, Methods, and Principles of Knowledge that I found myself taking as a naïve, 18-year-old, second-year student. OMP was meant to be a sort of summing up of the undergraduate curriculum, and I was not really prepared. I found the course very intimidating, and I admit that I didn't understand most of the readings. But somehow I passed, and it seems that the ideas behind the course must have stayed with me, because I realize that, in many ways, the title of this book could have been *Organization, Methods, and Principles of Behavior*. That was not what I had in mind when I started writing, but that's the way it turned out. I am not displeased with the result. And, it fits very well as a subtitle.

What I had planned to do was discuss the various fields of behavioral biology and experimental psychology in a modified ethological framework. That is what I have done, although I know that my coverage has been spotty. Two topics that deserved more space than they received are the mathematical models being developed in learning, cultural evolution, and behavioral ecology, as well as the whole area of insect social behavior. I have never been persuaded by most of the mathematical models, because they are usually based on simplistic or unrealistic assumptions, and I find analog explanations more satisfying. Omitting the insect societies was largely a matter of space, but also I don't have anything new to say about them. Wilson's (1971) book remains a classic. Behavior genetics (e.g. Sokolowski, 2001, 2010) is another area that should have found its way into the book, but didn't. On the topics I did cover, I perhaps emphasized my own work excessively, and I generally cited examples from work that I was familiar with or that I found especially interesting; I undoubtedly shortchanged other work. I am aware of that, but I don't apologize; overall, I think I have been fairly evenhanded. I hope that I have explained my framework sufficiently so that material I did not cover can easily be discussed using the behavior system concepts I have developed.

Among people who know me, I have a reputation for being opinionated. I'm sure that many of my opinions are fairly obvious throughout the book. I like to think that they are based on evidence, but I know many of my colleagues do not agree. I have tried to keep fact and opinion separate, though I also know I have not always succeeded. I have learned many things while writing this book, one of which is that many 'facts' change with time. What was true yesterday is not necessarily true today. And opinions have an even greater tendency to change with time. One of the few positive aspects of getting old is gaining perspective. I think I have done that and hope I have been able to express that perspective in this book.

References

Adams, C. D. & Dickinson, A. (1981). Instrumental Responding Following Reinforcer Devaluation. *Quarterly Journal of Experimental Psychology*, **33B**, 109–122.

Adler, N. T & Hogan, J. A. (1963). Classical Conditioning and Punishment of an Instinctive Response in *Betta splendens*. *Animal Behaviour*, **11**, 351–354.

Adrian, E. D., Cattell, M. & Hoagland, H. (1931). Sensory Discharges in Single Cutaneous Nerve Fibres. *Journal of Physiology*, **72**, 377–391.

Agnvall, B., Katajamaa, R., Altimiras, J. & Jensen, P. (2015). *Biology Letters*, **11**, 20150509.

Akers, K. G. *et al.* (2014). Hippocampal Neurogenesis Regulates Forgetting during Adulthood and Infancy. *Science*, **344**, 598–602.

Alcock, J. S. (1989). *Animal Behavior: An Evolutionary Approach*. Sunderland, MA: Sinauer.

Alem, S., Perry, C. J., Zhu, X., Loukola, O. J., Ingraham, T., Søvika, E. & Chittka, L. (2016). Associative Mechanisms Allow for Social Learning and Cultural Transmission of String Pulling in an Insect. *PLoS Biology*, **14**(10).

Allport, G. (1937). *Personality: A Psychological Interpretation*. New York: Holt, Rinehart & Winston.

Andrew, R. J. (1966). Precocious Adult Behaviour in the Young Chick. *Animal Behaviour*, **14**, 485–500.

Andrews, P. W., Gangestad, S. W. & Matthews, D. (2002). Adaptationism – How to Carry out an Exaptationist Program. *Behavioral and Brain Sciences*, **25**, 489–553.

Armitage, K. B. (1986). Individuality, Social Behavior, and Reproductive Success in Yellow-Bellied Marmots. *Ecology*, **67**, 1186–1193.

Armstrong, E. A. (1954). *Bird Display and Behaviour*. Cambridge, UK: Cambridge University Press.

Aronson, L. R., Tobach, E., Lehrman, D. S. & Rosenblatt, J. S. (Eds.). (1970). *Development and Evolution of Behavior*. San Francisco: Freeman.

Aschoff, J. (1960). Exogenous and Endogenous Components in Circadian Rhythms. *Cold Spring Harbor Symposia on Quantitative Biology*, **25**, 11–28.

Aschoff, J. (1989). Temporal Organization: Circadian Clocks in Animals and Humans. *Animal Behaviour*, **37**, 881–896.

Ashton, M. C. (2013). *Individual Differences and Personality*, 2nd ed. San Diego, CA: Academic Press.

Ashton, M. C. (2015). Hogan's Framework for the Study of Behavior as Applied to Personality Psychology. *Behavioural Processes*, **117**, 48–51.

Aslin, R. N. (2014). Infant Learning: Historical, Conceptual, and Methodological Challenges. *Infancy*, **19**, 2–27.

Avarguès-Weber, A., Dyer, A. G., Combe, M. & Giurfa, M. (2012). Simultaneous Mastering of Two Abstract Concepts by the Miniature Brain of Bees. *Proceedings of the National Academy of Sciences*, **109**, 7481–7486.

Baerends, G. P. (1941). *Fortpflanzungsverhalten und Orientierung der Grabwespe Ammophila campestris*. *Tijdschrift voor Entomologie*, **84**, 81–275.

Baerends, G. P. (1958). Comparative Methods and the Concept of Homology in the Study of Behaviour. *Archives Néerlandaises de Zoologie*, **13**, Suppl 1, pp. 401–417.

Baerends, G. P. (1970). A Model of the Functional Organization of Incubation Behaviour. *Behaviour*, Suppl. **17**, pp. 263–312.

Baerends, G. P. (1975). An Evaluation of the Conflict Hypothesis as an Explanatory Principle for the Evolution of Displays. In Baerends *et al.*, pp. 187–227.

Baerends, G. P. (1976). The Functional Organization of Behaviour. *Animal Behaviour*, **24**, 726–738.

Baerends, G. P. (1982). The Relative Effectiveness of Different Egg-features in Responses Other Than Egg-retrieval. *Behaviour*, **82**, 225–246.

Baerends, G. P. & Kruijt, J. P. (1973). Stimulus Selection. In Hinde & Stevenson-Hinde, pp. 23–49.

Baerends, G. P. & Drent, R. H. (Eds.). (1982). The Herring Gull and Its Eggs. *Behaviour*, **82**, 1–416.

Baerends, G. P., Beer, C. & Manning, A. (Eds.). (1975). *Function and Evolution in Behaviour*. London: Oxford University Press.

Baerends, G. P., Brouwer, R. & Waterbolk, H. T. (1955). Ethological Studies on *Lebistes reticulatus*: 1. An Analysis of the Male Courtship Pattern. *Behaviour*, **8**, 249–334.

Baerends-van Roon, J. M. & Baerends, G. P. (1979). The Morphogenesis of the Behaviour of the Domestic Cat, with a Special Emphasis on the Development of Prey-catching. *Verhandelingen der Koninklijke Nederlandse Akademie van Wetenschappen, Afd. Natuurkunde, Tweede Reeks* (Proceedings of the Royal Netherlands Academy of Sciences, Section Physics, Second Series), Part 72.

Baker, J. R. (1938). The Evolution of Breeding Seasons. In G. R. de Beer (Ed.), *Evolution: Essays on Aspects of Evolutionary Theory*. Oxford: Oxford University Press, pp. 161–171.

Balleine, B. W. & O'Doherty, J. P. (2010). Human and Rodent Homologies in Action Control: Corticostriatal Determinants of Goal-directed and Habitual Action. *Neuropsychopharmacology*, **35**, 48–69.

Balsam, P. D & Silver, R. (1994). Behavioral Change as a Result of Experience: Toward Principles of Learning and Development. In Hogan & Bolhuis, pp. 327–357.

Balsam, P. D., Graf, J. S. & Silver, R. (1992). Operant and Pavlovian Contributions to the Ontogeny of Pecking in Ring Doves. *Developmental Psychobiology*, **25**, 389–410.

Baptista, L. F. & Gaunt, S. L. L. (1997). Social Interaction and Vocal Development in Birds. In Snowdon & Hausberger, pp. 23–40.

Barbour, M. A. & Clark, R. W. (2012). Ground Squirrel Tail-flag Displays Alter Both Predatory Strike and Ambush Site Selection Behaviours of Rattlesnakes. *Proceedings of the Royal Society B*, **279**, 3827–3833.

Barlow, G. W. (1968). Ethological Units of Behavior. In D. Ingle (Ed.), *Central Nervous System and Fish Behavior.* Chicago: University of Chicago Press, pp. 217–232.

Barnard, C. J. (2004). *Animal Behaviour: Mechanism, Development, Function and Evolution.* London: Pearson Education Ltd./Prentice-Hall.

Barnard, C. J. & Sibly, R. M. (1981). Producers and Scroungers: A General Model and Its Application to Captive Flocks of House Sparrows. *Animal Behaviour*, **29**, 543–550.

Barrett, H. C. & Kurzban, R. (2006). Modularity in Cognition: Framing the Debate. *Psychological Review*, **113**, 628–647.

Barrett, L. F. (2013). Psychological Construction: The Darwinian Approach to the Science of Emotion. *Emotion Review*, **5**, 379–389.

Barrett, L. F. & Russell, J. A. (Eds.). (2015). *The Psychological Construction of Emotion.* New York: Guilford Press.

Bateson, P. P. G. (1979). How Do Sensitive Periods Arise and What Are They For? *Animal Behaviour*, **27**, 470–486.

Bateson, P. P. G. (1987). Imprinting as a Process of Competitive Exclusion. In J. P. Rauschecker & P. Marler (Eds.), *Imprinting and Cortical Plasticity.* New York: Wiley, pp. 151–168.

Bateson, P. P. G. (1999). Foreword. In Bolhuis & Hogan, pp. ix–xi.

Bateson, P. P. G. & Hinde, R. A. (1987). Developmental Changes in Sensitivity to Experience. In M. H. Bornstein (Ed.), *Sensitive Periods in Development.* Hillsdale, NJ: Erlbaum, pp. 19–34. (Reprinted in Bolhuis & Hogan, 1999.)

Bateson, P. P. G. & Laland, K. N. (2013). Tinbergen's Four Questions: An Appreciation and an Update. *Trends in Ecology and Evolution*, **28**, 712–718.

Battley, P. F. et al. (2012). Contrasting Extreme Long-distance Migration Patterns in Bar-tailed Godwits *Limosa Lapponica*. *Journal of Avian Biology*, **43**, 1–12.

Baylis, J. R. (1982). Avian Vocal Mimicry: Its Function and Evolution. In D. E. Kroodsma, E. H. Miller & H. Ouellet (Eds.), *Acoustic Communication in Birds*, vol. 2. New York: Academic Press, pp. 51–83.

Beach, F. A. (1942). Analysis of Factors Involved in the Arousal, Maintenance, and Manifestation of Sexual Excitement in Male Animals. *Psychosomatic Medicine*, **4**, 173–198.

Beach, F. A. (1948). *Hormones and Behavior.* New York: Hoeber.

Beach, F. A. (1955). The Descent of Instinct. *Psychological Review*, **62**, 401–410.

Beach, F. A. (1956). Characteristics of the Masculine "Sex Drive". In M. R. Jones (Ed.), *The Nebraska Symposium on Motivation.* Lincoln, NE: University of Nebraska Press, pp. 1–32.

Beer, C. G. (1975). Multiple Functions and Gull Displays. In Baerends et al., pp. 16–54.

Bergerud, A. T. & Mossop, D. H. (1984). The Pair Bond in Ptarmigan. *Canadian Journal of Zoology*, **62**, 2129–2141.

Berlyne, D. E. (1960). *Conflict, Arousal, and Curiosity.* New York: McGraw-Hill.

Berridge, K. C. (1990). Comparative Fine Structure of Action: Rules of Form and Sequence in the Grooming Patterns of Six Rodent Species. *Behaviour*, **113**, 21–56.

Berridge, K. C. (1994). The Development of Action Patterns. In Hogan & Bolhuis, pp. 147–180.

Berridge, K. C. (2004). Motivation Concepts in Behavioral Neuroscience. *Physiology and Behavior*, **81**, 179–209.

Berridge, K. C. (2012). From Prediction Error to Incentive Salience: Mesolimbic Computation of Reward Motivation. *European Journal of Neuroscience*, **35**, pp. 1124–1143.

Berridge, K. C., Fentress, J. C. & Parr, H. (1987). Natural Syntax Rules Control Action Sequence of Rats. *Behavioural Brain Research*, **23**, 59–68.

Berridge, K. C. & Robinson, T. E. (2003). Parsing Reward. *Trends in Neurosciences*, **26**, 507–513.

Berry, C. S. (1908). An Experimental Study of Imitation in Cats. *Journal of Comparative Neurology and Physiology*, **18**, 1–12.

Berthold, P. (1973). Relationships between Migratory Restlessness and Migration Distance in Six *Sylvia* Species. *Ibis*, **115**, 594–599.

Berwick, R. C., Friederici, A. D., Chomsky, N. & Bolhuis, J. J. (2013). Evolution, Brain, and the Nature of Language. *Trends in Cognitive Sciences*, **17**, 89–98.

Bijleveld, A. I., Folmer, E. O. & Piersma, T. (2012). Experimental Evidence for Cryptic Interference among Socially Foraging Shorebirds. *Behavioral Ecology*, **23**, 806–814.

Bijleveld, A. I., van Gils, J. A., Jouta, J. & Piersma, T. (2015). Benefits of Foraging in Small Groups: An Experimental Study on Public Information Use in Red Knots *Calidris canutus*. *Behavioural Processes*, **117**, 74–81.

Bischof, H.-J. (1994). Sexual Imprinting as a Two-stage Process. In Hogan & Bolhuis, pp. 82–97.

Bissonette, G. B., Bryden, D. W. & Roesch, M. R. (2014). You Won't Regret Reading This. *Nature Neuroscience*, **17**, 892–893.

Blaisdell, A., Sawa, K., Leising, K. J. & Waldmann, M. R. (2006). Causal Reasoning in Rats. *Science*, **311**, 1020–1022.

Blakemore, C. (1973). Environmental Constraints on Development in the Visual System. In Hinde & Stevenson-Hinde, pp. 51–73.

Blankenship, A. G. & Feller, M. B. (2010). Mechanisms Underlying Spontaneous Patterned Activity in Developing Neural Circuits. *Nature Reviews Neuroscience*, **11**, 18–29.

Blass, E. M. (1999). The Ontogeny of Human Infant Face Recognition: Orogustatory, Visual, and Social Influences. In P. Rochat (Ed.), *Early Social Cognition*. Mahway, NJ: Erlbaum, pp. 35–65.

Blass, E. M. (2015). Energy Conservation in Infants. *Behavioural Processes*, **117**, 35–41.

Blass, E. M., Ganchrow, J. R. & Steiner, J. E. (1984). Classical Conditioning in Newborn Humans 2–48 Hours of Age. *Infant Behavior and Development*, **7**, 125–134.

Blass, E. M., Hall, W. G. & Teicher, M. H. (1979). The Ontogeny of Suckling and Ingestive Behaviors. *Progress in Psychobiology and Physiological Psychology*, **8**, 243–299.

Blest, A. D. (1957). The Function of Eyespot Patterns in the Lepidoptera. *Behaviour*, **11**, 209–256.

Bloch, G., Barnes, B. M., Gerkema, M. P. & Helm, B. (2013). Animal Activity around the Clock with No Overt Circadian Rhythms: Patterns, Mechanisms and Adaptive Value. *Proceedings of the Royal Society B*, **280**, 30130019.

Blüm, V. & Fiedler, K. (1965). Hormonal Control of Reproductive Behavior in Some Cichlid Fish. *General and Comparative Endocrinology*, **5**, 186–196.

Blumstein, D. T. (2007). The Evolution, Function, and Meaning of Marmot Alarm Communication. *Advances in the Study of Behavior*, **37**, 371–401.

Boakes, R. (1984). *From Darwin to Behaviourism*. Cambridge, UK: Cambridge University Press.

Bock, W. J. & von Wahlert, G. (1965). Adaptation and the Form-Function Complex. *Evolution*, **19**, 269–299.

Bohacek, J. & Mansuy, I. M. (2015). Molecular Insights into Transgenerational Non-genetic Inheritance of Acquired Behaviours. *Nature Reviews Genetics*, **16**, 641–652.

Bolhuis, J. J. (1991). Mechanisms of Avian Imprinting: A Review. *Biological Reviews*, **66**, 303–345.

Bolhuis, J. J. (1994). Neurobiological Analyses of Behavioural Mechanisms in Development. In Hogan & Bolhuis, pp. 16–46.

Bolhuis, J. J. (1996). Development of Perceptual Mechanisms in Birds: Predispositions and Imprinting. In C. F. Moss & S. J. Shettleworth (Eds.), *Neuroethological Studies of Cognitive and Perceptual Processes*. Boulder, CO: Westview Press, pp. 158–184. (Reprinted in Bolhuis & Hogan, 1999.)

Bolhuis, J. J. (2009). Function and Mechanism in Neuroecology: Looking for Clues. In J. J. Bolhuis & S. Verhulst (Eds.), *Tinbergen's Legacy*. Cambridge, UK: Cambridge University Press, pp. 163–196.

Bolhuis, J. J. (2015). Evolution Cannot Explain How Minds Work. *Behavioural Processes*, **117**, 82–91.

Bolhuis, J. J. & Eda-Fujiwara, H. (2003). Bird Brains and Songs: Neural Mechanisms of Birdsong Perception and Memory. *Animal Biology*, **53**, 129–145.

Bolhuis, J. J. & Everaert, M. (Eds.). (2013). *Birdsong, Speech, and Language*. Cambridge, MA: MIT Press.

Bolhuis, J. J. & Giraldeau, L.-A. (2005). *The Behavior of Animals*. New York: Wiley.

Bolhuis, J. J. & Giraldeau, L.-A. (in press). *The Behavior of Animals*, 2nd edn. New York: Wiley.

Bolhuis, J. J. & Hogan, J. A. (1999). *The Development of Animal Behaviour: A Reader*. Oxford: Blackwell.

Bolhuis, J. J. & Wynne, C. D. L. (2009). Can Evolution Explain How Minds Work? *Nature*, **458**, 832–833.

Bolhuis, J. J., Tattersall, I., Chomsky, N. & Berwick, R. C. (2014). How Could Language Have Evolved? *PLoS Biology*, **12**(8).

Bolhuis, J. J., Zijlstra, G. G. O., den Boer-Visser, A. M. & van der Zee, E. A. (2000). Localised Neuronal Activation in the Zebra Finch Brain Is Related to the Strength of Song Learning. *Proceedings of the National Academy of Sciences*, **97**, 2282–2285.

Bolhuis, J.J., Brown, G.R., Richardson, R.C. & Laland, K.N. (2011). Darwin in Mind: New Opportunities for Evolutionary Psychology. *PLoS Biology*, **9**(7).

Bolles, R. C. (1967). *Theory of Motivation*. New York: Harper & Row.

Bols, R. J. (1977). Display Reinforcement in the Siamese Fighting Fish, *Betta splendens*: Aggression Motivation or Curiosity? *Journal of Comparative and Physiological Psychology*, **91**, 233–244.

Bonner, J. T. (1980). *The Evolution of Culture in Animals*. Princeton: Princeton University Press.

Borbély, A. A. (1982). A Two-process Model of Sleep. *Human Neurobiology*, **1**, 195–204.

Borbély, A. A., Dijk, D.-J., Achermann, P. & Tobler, I. (2001). Processes Underlying the Regulation of the Sleep-wake Cycle. In J. S. Takahashi, F. W. Turek & R. Y. Moore (Eds.), *Handbook of Behavioral Neurobiology*, vol. **12**: *Circadian Clocks*. New York: Kluwer Academic/Plenum, pp. 458–479.

Borchelt, P. L. (1975). The Organization of Dustbathing Components in Bobwhite Quail (*Colinus virginianus*). *Behaviour*, **53**, 217–237.

Bowlby, J. (1991). Ethological Light on Psychoanalytical Problems. In P. Bateson (Ed.), *The Development and Integration of Behaviour*. Cambridge, UK: Cambridge University Press, pp. 301–313.

Boyd, R. & Richerson, P. J. (1985). *Culture and the Evolutionary Process*. Chicago: Chicago University Press.

Brainard, M. S. & Doupe, A. J. (2000). Auditory Feedback in Learning and Maintenance of Vocal Behaviour. *Nature Reviews Neuroscience*, **1**, 31–40.

Brembs, B. (2009). The Importance of Being Active. *Journal of Neurogenetics*, **23**, 120–126.

Bronstein, P. M. (1981). Commitments to Aggression and Nest Sites in Male *Betta splendens*. *Journal of Comparative and Physiological Psychology*, **95**, 436–449.

Brown, J. S. (1948). Gradients of Approach and Avoidance Responses and Their Relation to Level of Motivation. *Journal of Comparative and Physiological Psychology*, **41**, 450–485.

Brown, M. C., Hopkins, W. G. & Keynes, R. J. (1991). *Essentials of Neural Development*. Cambridge, UK: Cambridge University Press.

Bruce, R. H. (1935). A Further Study of the Effect of Reward and Drive upon the Maze Performance of Rats. *Journal of Comparative Psychology*, **20**, 157–182.

Buhusi, C. V. & Meck, W. H. (2005). What Makes Us Tick? Functional and Neural Mechanisms of Interval Timing. *Nature Reviews Neuroscience*, **6**, 755–765.

Buntin, J. D. (1996). Neural and Hormonal Control of Parental Behavior in Birds. *Advances in the Study of Behavior*, **25**, 161–213.

Burghagen, H. & Ewert, J.-P. (in press). Stimulus Perception. In Bolhuis & Giraldeau, 2nd edn.

Burghardt, G. M. (2001). Play: Attributes and Neural Substrates. In E. M. Blass (Ed.), *Handbook of Behavioral Neurobiology*, Vol. **13**:7 *Developmental Psychobiology*, New York: Kluwer Academic/Plenum, pp. 317–356.

Buss, D. M. *et al.* (1998). Adaptations, Exaptations, and Spandrels. *American Psychologist*, **53**, 533–548.

Buss, R. R., Sun, W. & Oppenheim R. W. (2006). Adaptive Roles of Programmed Cell Death during Nervous System Development. *Annual Review of Neuroscience*, **29**, 1–35.

Butler, R. A. & Harlow, H. F. (1954). Persistence of Visual Exploration in Monkeys. *Journal of Comparative and Physiological Psychology*, **47**, 257–263.

Caldwell. C. A. & Millen, A. E. (2008). Studying Cumulative Cultural Evolution in the Laboratory. *Philosophical Transactions of the Royal Society B*, **363**, 3529–3539.

Cannon, W. B. (1927). The James-Lange Theory of Emotions: A Critical Examination and an Alternative Theory. *American Journal of Psychology*, **39**, 106–124.

Cannon, W. B. (1929). Organization for Physiological Homeostasis. *Physiological Reviews*, **9**, 399–431.

Cannon, W. B. (1932). *The Wisdom of the Body*. New York: Norton.

Capranica, R. R. (1965). *The Evoked Vocal Response of the Bullfrog: A Study of Communication by Sound*. Cambridge, MA: MIT Press.

Caro, T. (2005). *Antipredator Defences in Birds and Mammals*. Chicago: University of Chicago Press.

Carroll, S. B. (2008). Evo-Devo and an Expanding Evolutionary Synthesis: A Genetic Theory of Morphological Evolution. *Cell*, **134**, 25–36.

Carter, A. J., Feeney, W. E., Marshall, H. H., Cowlishaw, G. & Heinsohn, R. (2013). Animal Personality: What Are Behavioural Ecologists Measuring? *Biological Reviews*, **88**, 465–475.

Cartwright, B. A. & Collett, T. S. (1982). How Honey Bees Use Landmarks to Guide Their Return to a Food Source. *Nature*, **295**, 560–564.

Cäsar, C., Zuberbühler, K., Young, R. J. & Byrne, R. W. (2013). Titi Monkey Call Sequences Vary with Predator Location and Type. *Biology Letters*, **9**, 20130535.

Catchpole, C. K. & Slater, P. J. B. (2008). *Bird Song: Biological Themes and Variations.* Cambridge, UK: Cambridge University Press.

Cavalli-Sforza, L. L. & Feldman, M. W. (1981). *Cultural Transmission and Evolution: A Quantitative Approach.* Princeton: Princeton University Press.

Chase, I. D. (1982). Dynamics of Hierarchy Formation: The Sequential Development of Dominance Relationships. *Behaviour*, **80**, 218–240.

Chittka, L. & Niven, J. (2009). Are Bigger Brains Better? *Current Biology*, **19**, R995–R1008.

Chomsky, N. (1965). *Aspects of the Theory of Syntax.* Cambridge, MA: MIT Press.

Chomsky, N. (2012). *The Science of Language.* Cambridge, UK: Cambridge University Press.

Chow, A. & Hogan, J. A. (2005). The Development of Feather Pecking in Burmese Red Junglefowl: The Influence of Early Experience with Exploratory-rich Environments. *Applied Animal Behaviour Science*, **93**, 283–294.

Clayton, N. S. (2015). Ways of Thinking: From Crows to Children and Back Again. *Quarterly Journal of Experimental Psychology*, **68**, 209–241.

Clutton-Brock, J. (1995). Origins of the Dog: Domestication and Early History. In J. Serpell (Ed.), *The Domestic Dog.* Cambridge, UK: Cambridge University Press, pp. 7–20.

Colgan, P. (1989). *Animal Motivation.* London: Chapman & Hall.

Collett, M. & Collett, T. S. (2000). How Do Insects Use Path Integration for Their Navigation? *Biological Cybernetics*, **83**, 245–259.

Collett, M., Chittka, L. & Collett, T. S. (2013). Spatial Memory in Insect Navigation. *Current Biology*, **23**, R789–R800.

Collett, T. S. & Collett, M. (2002). Memory Use in Insect Visual Navigation. *Nature Reviews Neuroscience*, **3**, 542–552.

Colonnese, M. T., Stallman, E. L. & Berridge, K. C. (1996). Ontogeny of Action Syntax in Altricial and Precocial Rodents: Grooming Sequences of Rat and Guinea Pig Pups. *Behaviour*, **133**, 1165–1195.

Colwill, R. C. & Rescorla, R. A. (1985). Postconditioning Devaluation of a Reinforcer Affects Instrumental Responding. *Journal of Experimental Psychology: Animal Behavior Processes*, **11**, 120–132.

Coppinger, R. & Schneider, R. (1995). Evolution of Working Dogs. In J. Serpell (Ed.), *The Domestic Dog.* Cambridge, UK: Cambridge University Press, pp. 21–47.

Coss, R. G. & Biardi, J. E. (1997). Individual Variation in the Antisnake Behavior of California Ground Squirrels (*Spermophilus beecheyi*). *Journal of Mammalogy*, **73**, 294–310.

Craig, W. (1918). Appetites and Aversions as Constituents of Instincts. *Biological Bulletin*, **34**, 91–107.

Cramer, C. P. & Blass, E. M. (1983). Mechanisms of Control of Milk Intake in Suckling Rats. *American Journal of Physiology*, **245**, R154–R159.

Crews, D. & Groothuis, T. (2009). Tinbergen's Fourth Question, Ontogeny: Sexual and Individual Differentiation. In J. J. Bolhuis & S. Verhulst (Eds.), *Tinbergen's Legacy.* Cambridge, UK: Cambridge University Press, pp. 54–81.

Cullen, E. (1957). Adaptations in the Kittiwake to Cliff-nesting. *Ibis*, **99**, 275–302.

Curio, E., Ernst, U. & Vieth, W. (1978). Cultural Transmission of Enemy Recognition: One Function of Mobbing. *Science*, **202**, 899–901.

Cuthill, I. (2009). The Study of Function in Behavioral Ecology. In J J. Bolhuis & S. Verhulst (Eds.), *Tinbergen's Legacy*. Cambridge, UK: Cambridge University Press, pp. 107–126.

Czeisler, C. A. & Dijk, D.-J. (2001). Human Circadian Physiology and Sleep-Wake Regulation. In J. S. Takahashi, F. W. Turek & R. Y. Moore (Eds.), *Handbook of Behavioral Neurobiology*, vol. 4: Circadian Clocks. New York: Plenum, pp. 531–569.

Daan, S. (2000). Learning and Circadian Behavior. *Journal of Biological Rhythms*, **15**, 296–299.

Daan, S., Dijkstra, C. & Tinbergen, J. M. (1990). Family Planning in the Kestrel (*Falco tinnunculus*): The Ultimate Control of Covariance of Laying Date and Clutch Size. *Behaviour*, **114**, 83–116.

Daly, M. (1977). Some Experimental Tests of the Functional Significance of Scent-marking by Gerbils (*Meriones unguiculatus*). *Journal of Compartive and Physiological Psychology*, **91**, 1082–1094.

Daly, M. (1979). Why Don't Male Mammals Lactate? *Journal of Theoretical Biology*, **78**, 325–345.

Daly, M. (2015). On Function, Cause, and Being Jerry Hogan's Student. *Behavioural Processes*, **117**, 70–73.

Daly, M. (2016). *Killing the Competition*. New Brunswick, NJ: Transaction

Daly, M. & Daly, S. (1974). Spatial Distribution of a Leaf-eating Saharan Gerbil (*Psammomys obesus*) in Relation to Its Food. *Mammalia*, **38**, 591–603.

Daly, M. & Wilson, M. (1978). *Sex, Evolution, and Behavior*. Scituate, MA: Duxbury Press.

Daly, M. & Wilson, M. (1983). *Sex, Evolution, and Behavior*, 2nd edn. Boston, MA: Willard Grant Press.

Daly, M. & Wilson, M. (1999). Human Evolutionary Psychology and Animal Behaviour. *Animal Behaviour*, **57**, 509–519.

Darwin, C. (1845). *Journal of Researches into the Natural History and Geology of the Countries Visited during the Voyage of H. M. S. Beagle Round the World*, 2nd edn. London: Murray.

Darwin, C. (1859). *On the Origin of Species*. London: Murray.

Darwin, C. (1868). *The Variation of Animals and Plants under Domestication*. London: Murray.

Darwin, C. (1872). *The Expression of the Emotions in Man and Animals*. London: Murray.

Darwin, C. (1873). Origin of Certain Instincts. *Nature*, **7**, 417–418.

Darwin, C. (1896). *The Descent of Man and Selection in Relation to Sex*, 2nd ed. New York: D. Appleton and Company.

Davies, N. B. (2011). Cuckoo Adaptations: Trickery and Tuning. *Journal of Zoology* (London), **284**, 1–14.

Davies, N. B. Krebs, J. R. & West, S. A. (2012). *An Introduction to Behavioural Ecology*. Chichester: Wiley-Blackwell.

Davis, F. C. & Reppert, S. M. (2001). Development of Mammalian Rhythms. In J. S. Takahashi, F. W. Turek & R. Y Moore (Eds.). (2001). Handbook of Behavioral Neurobiology. Vol. 12: Circadian Clocks. New York: Kluwer Academic/Plenum, pp. 247–290.

Dawkins, M. S. (2008). The Science of Animal Suffering. *Ethology*, **114**, 937–945.

Dawkins, M. S. (2015). Animal Welfare and the Paradox of Animal Consciousness. *Advances in the Study of Behavior*, **47**, 5–37.

Dawkins, M. S. & Guilford, T. (1995). An Exaggerated Preference for Simple Neural Network Models of Signal Evolution? *Proceedings of the Royal Society of London* B, **261**, 357–360.

Dawkins, R. (1976a). Hierarchical Organization: A Candidate Principle for Ethology. In P. P. G. Bateson & R. A. Hinde (Eds.), *Growing Points in Ethology*. Cambridge, UK: Cambridge University Press, pp. 7–54.

Dawkins, R. (1976b). *The Selfish Gene*. Oxford: Oxford University Press.

Dawkins, R. & Krebs, J. R. (1978). Animal Signals: Information or Manipulation? In J. R. Krebs & N. B. Davies (Eds.), *Behavioural Ecology*. Oxford: Blackwell, pp. 282–309.

De Bruyn, P. J. N., Tosh, C. A., Bester, M. N. *et al.* (2011). Sex at Sea: Alternative Mating System in an Extremely Polygynous Mammal. *Animal Behaviour*, **82**, 445–451.

Dean, L. G., Kendal, R. L., Schapiro, S. J., Thierry, B. & Laland, K. N. (2012). Identification of the Social Processes underlying Human Cumulative Culture. *Science*, **335**, 1114–1118.

Dewsbury, D. A. (1992). On the Problems Studied in Ethology, Comparative Psychology, and Animal Behavior. *Ethology*, **92**, 89–107.

Dewsbury, D. A. (1994). On the Utility of the Proximate-Ultimate Distinction in the Study of Animal Behavior. *Ethology*, **96**, 63–68.

Dewsbury, D. A. (1999). The Proximate and the Ultimate: Past, Present, and Future. *Behavioural Processes*, **46**, 189–199.

Dezfouli, A. & Balleine, B. W. (2012). Habits, Action Sequences and Reinforcement Learning. *European Journal of Neuroscience*, **35**, 1036–1051.

Dickinson, A. (1980). *Contemporary Animal Learning Theory*. Cambridge, UK: Cambridge University Press.

Dickinson, A. (1994). Instrumental Conditioning. In Mackintosh, pp. 45–79.

Dickinson, A. & Balleine, B. W. (1994). Motivational Control of Goal-directed Action. *Animal Learning and Behavior*, **22**, 1–18.

Diedrichsen J. & Kornysheva, K. (2015). Motor Skill Learning between Selection and Execution. *Trends in Cognitive Sciences*, **19**, 227–233.

Diekelmann, S. & Born, J. (2010). The Memory Function of Sleep. *Nature Reviews Neuroscience*, **11**, 115–126.

Dijk, D.-J. & von Schantz, M. (2005). Timing and Consolidation of Human Sleep, Wakefulness, and Performance by a Symphony of Oscillators. *Journal of Biological Rhythms*, **20**, 279–290.

Dijk, D.-J. & Archer, S. N. (2009). Light, Sleep, and Circadian Rhythms: Together Again. *PLoS Biology*, **7**(6).

Dijk, D.-J., Duffy, J. F. & Czeisler, C. A. (1992). Circadian and Sleep/Wake Dependent Aspects of Subjective Alertness and Cognitive Performance. *Journal of Sleep Research*, **1**, 112–117.

Dijkstra, C., Bult, A. Bijlsma, S., Daan, S., Meijer, T & Zijlstra, M. (1990). Brood Size Manipulations in the Kestrel (*Falco tinnunculus*): Effects on Offspring and Parent Survival. *Journal of Animal Ecology*, **59**, 269–285.

Dingemanse, N, J., Both, C., Drent, P. J. & Tinbergen, J. M. (2004). Fitness Consequences of Avian Personalities in a Fluctuating Environment. *Proceedings of the Royal Society of London* B, **271**, 847–852.

Dingle, H. & Drake, V. A. (2007). What Is Migration? *BioScience*, **57**, 113–121.

Dollard, J. & Miller, N. E. (1950). *Personality and Psychotherapy*. New York: McGraw-Hill.

Domjan, M. & Holloway, K. S. (1998). Sexual Learning. In G. Greenberg & M. M. Haraway (Eds.), *Comparative Psychology: A Handbook*. New York: Garland, pp. 602–613.

Domjan, M., Cusato, B. & Krause, M. (2004). Learning with Arbitrary versus Ecological Conditioned Stimuli: Evidence from Sexual Conditioning. *Psychological Bulletin and Review*, **11**, 232–246.

Doty, R. W. (1976). The Concept of Neural Centers. In J. C. Fentress (Ed.), *Simpler Networks and Behavior*. Sunderland, MA: Sinauer, pp. 251–265.

Doupe, A. J. & Kuhl, P. K. (1999). Birdsong and Human Speech: Common Themes and Mechanisms. *Annual Review of Neuroscience*, **22**, 567–631.

Drachman, D. B. & Sokoloff, L. (1966). The role of Movement in Embryonic Joint Development. *Developmental Biology*, **14**, 401–420.

Duncan, I. J. H., Widowski, T. M., Malleau, A. E., Lindberg, A. C. & Petherick, J. C. (1988). External Factors and Causation of Dustbathing in Domestic Hens. *Behavioural Processes*, **43**, 219–228.

Durkin, J. M. & Aton, S. J. (2016). Sleep-dependent Potentiation in the Visual System Is at Odds with the Synaptic Homeostasis Hypothesis. *Sleep*, **39**, 155–159.

Eibl-Eibesfeldt, I. (1972). Similarities and Differences between Cultures in Expressive Movements. In R. A. Hinde (Ed.), *Non-verbal Communication*. Cambridge, UK: Cambridge University Press, pp. 297–311.

Eimas, P. D., Miller, J. L. & Jusczyk, P. W. (1987). On Infant Speech Perception and the Acquisition of Language. In S. Harnad (Ed.), *Categorical Perception*. Cambridge, UK: Cambridge University Press, pp. 161–195.

Eimas, P. D., Siqueland, P., Jusczyk, P. & Vigorito, J. (1971). Speech Perception in Infants. *Science*, **171**, 303–306.

Ekman, P. (1982). *Emotion in the Human Face*, 2nd edn. Cambridge, UK: Cambridge University Press.

Ekman, P. & Davidson, R. J. (Eds.). (1994). *The Nature of Emotion: Fundamental Questions*. New York: Oxford University Press.

Elliot, A. J., Eder, A. B. & Harmon-Jones, E. (2013). Avoidance Motivation and Emotion: Convergency and Divergence. *Emotion Review*, **5**, 308–311.

Emlen, S. T. (1967). Migratory Orientation in the Indigo Bunting, Passerina cyanea. *Auk*, **84**, 309–342, 463–489.

Emlen, S. T. (1969). Bird Migration: Influence of Physiological State upon Celestial Orientation. *Science*, **165**, 716–718.

Enquist, M. & Arak, A. (1993). Selection of Exaggerated Male Traits by Female Aesthetic Senses. *Nature*, **361**, 446–448.

Enquist, M. & Ghirlanda, S. (2005). *Neural Networks and Animal Behavior*. Princeton, NJ: Princeton University Press.

Enquist, M. & Ghirlanda, S. (2007). Evolution of Social Learning Does Not Explain the Origin of Human Cumulative Culture. *Journal of Theoretical Biology*, **246**, 129–135,

Enquist, M. & Leimar, O. (1983). Evolution of Fighting Behaviour: Decision Rules and Assessment of Relative Strength. *Journal of Theoretical Biology*, **102**, 387–410.

Enquist, M. Leimar, O., Ljungberg, T., Mallner, Y. & Segerdahl, N. (1990). A Test of the Sequential Game: Fighting in the Cichlid Fish Nannacara anomala. *Animal Behaviour*, **40**, 1–14.

Etienne, A. S. & Jeffery, K. J. (2004). Path Integration in Mammals. *Hippocampus*, **14**, 180–192.

Etienne, A. S., Maurer, R. & Séguinot, V. (1996). Path Integration in Mammals and Its Interaction with Visual Landmarks. *Journal of Experimental Biology*, **199**, 201–209.

Evans, R. M. (1968). Early Aggressive Responses in Domestic Chicks. *Animal Behaviour*, **16**, 24–28.

Ewert, J.-P. (1997). Neural Correlates of Key Stimulus and Releasing Mechanism. *Trends in Neurosciences*, **20**, 332–339.

Ewing, A. W. & Miyan, J. A. (1986). Sexual Selection, Sexual Isolation and the Evolution of Song in the *Drosophila repleta* Group of Species. *Animal Behaviour*, **34**, 421–429.

Falk, J. L. (1961). Production of Polydipsia in Normal Rats by an Intermittent Food Schedule. *Science*, **133**, 195–196.

Falk, J. L. (1966). Schedule-induced Polydipsia as a Function of Fixed Interval Length. *Journal of the Experimental Analysis of Behavior*, **9**, 37–39.

Falk, J. L. (1971). The Nature and Determinants of Adjunctive Behavior. *Physiology and Behavior*, **6**, 577–588.

Falótico, T. & Ottoni, E. B. (2014). Sexual Bias in Probe Tool Manufacture and Use by Wild Bearded Capuchin Monkeys. *Behavioural Processes*, **108**, 117–122.

Falótico, T. & Ottoni, E. B. (2016). The Manifold Use of Pounding Stone Tools by Wild Capuchin Monkeys of Serra da Capivara National Park, Brazil. *Behaviour*, **153**, 421–442.

Fanselow, M. S. (1994). Neural Organization of the Defensive Behavior System Responsible for Fear. *Psychonomic Bulletin and Review*, **1**, 429–438.

Fanselow, M. S. & De Oca, B. M. (1988). Defensive Behaviors. In G. Greenberg & M. M. Haraway (Eds.), *Comparative Psychology: A Handbook*. New York: Garland, pp. 653–665.

Fantz, R. L. (1957). Form Preferences in Newly Hatched Chicks. *Journal of Comparative and Physiological Psychology*, **50**, 422–430.

Farris, H. E. (1967). Classical Conditioning of Courting Behavior in the Japanese Quail (*Coturnix c. japonica*). *Journal of the Experimental Analysis of Behavior*, **10**, 213–217.

Fawcett, T. W., Hamblin, S. & Giraldeau, L.-A. (2013). Exposing the Behavioral Gambit: The Evolution of Learning and Decision Rules. *Behavioral Ecology*, **24**, 2–11.

Feekes, F. (1972). "Irrelevant" Ground Pecking in Agonistic Situations in Burmese Red Junglefowl (*Gallus gallus spadiceus*). *Behaviour*, **43**, 186–326.

Fehér, O. & Tchernichovski, O. (2013). Vocal Culture in Songbirds: An Experimental Approach to Cultural Evolution. In Bolhuis & Everaert, pp. 143–156.

Fentress, J. C. & Gadbois, S. (2001). The Development of Action Sequences. In E. M. Blass (Ed.), *Handbook of Behavioral Neurobiology*, Vol. 13: *Developmental Psychobiology*. New York: Kluwer Academic/Plenum, pp. 393–431.

Fentress, J. C. & Stilwell, F. P. (1973). Grammar of a Movement Sequence in Inbred Mice. *Nature*, **224**, 52–53.

Field, T. M., Woodson, R., Greenberg, R. & Cohen, D. (1982). Discrimination and Imitation of Facial Expressions by Neonates. *Science*, **218**, 179–181.

Finn, A. S. *et al.* (2016). Developmental Dissociation between the Maturation of Procedural Memory and Declarative Memory. *Journal of Experimental Child Psychology*, **142**, 212–220.

Fleming, A. S. & Blass, E. M. (1994). Psychobiology of the Early Mother-Young Relationship. In Hogan & Bolhuis, pp. 212–241.

Fleming, A. S., Ruble, D., Krieger, H. & Wong, P. Y. (1997). Hormonal and Experiential Correlates of Maternal Responsiveness during Pregnancy and the Puerperium in Human Mothers. *Hormones and Behavior*, **31**, 145–158.

Fraenkel, G. S. & Gunn, D. L. (1940/1961). *The Orientation of Animals*. New York: Dover.

Francis, R. C. (1990). Causes, Proximate and Ultimate. *Biology and Philosophy*, **5**, 401–415.

Frankland, P. W., Köhler, S. & Josselyn, S. A. (2013). Hippocampal Neurogenesis and Forgetting. *Trends in Neuroscience*, **36**, 497–503.

Freedman, A. H. et al. (2014). Genome Sequencing Highlights the Dynamic Early History of Dogs. PLoS Genetics, **10**(1).

Freud, S. (1905). Three Contributions to the Theory of Sexuality. London: Hogarth Press.

Freud, S. (1915). Instincts and Their Vicissitudes. London: Hogarth Press.

Freud, S. (1940/1949). An Outline of Psycho-analysis. New York: Norton.

Frishkopf, L. S. & Goldstein, Jr., M. H. (1963). Responses to Acoustic Stimuli from Single Units in the Eighth Nerve of the Bullfrog. Journal of the Acoustical Society of America, **35**, 1219–1228.

Fugazza, C, Pogány, Á. & Miklósi, Á. (2016). Recall of Others' Actions after Incidental Encoding Reveals Episodic-like Memory in Dogs. Current Biology, **26**, 1–5.

Gadbois, S., Sievert, O., Reeve, C., Harrington, F. H. & Fentress, J. C. (2015). Revisiting the Concept of Behavior Patterns in Animal Behavior with an Example from Food-caching Sequences in Wolves (Canis lupus), Coyotes (Canis latrans), and Red Foxes (Vulpes vulpes). Behavioural Processes, **110**, 3–14.

Galef Jr., B. G. (1988). Imitation in Animals: History, Definition and Interpretation of Data from the Psychological Laboratory. In T. R. Zentall & B. G. Galef Jr. (Eds.), Social Learning: Psychological and Biological Perspectives. Hillsdale, NJ: Erlbaum, pp. 3–28.

Galef Jr., B. G. (1996). Social Enhancement of Food Preferences in Norway Rats. In C. M. Heyes & B. G. Galef Jr. (Eds.), Social Learning in Animals: The Roots of Culture. New York: Academic Press, pp. 49–64.

Galef Jr., B. G. (2013). Imitation and Local Enhancement: Detrimental Effects of Consensus Definitions on Analyses of Social Learning in Animals. Behavioural Processes, **100**, 123–130.

Gallistel, C. R. (1973). Self-stimulation: The Neurophysiology of Reward and Motivation. In J. A. Deutsch (Ed.), The Physiological Basis of Memory. New York: Academic Press, pp. 175–267.

Gallistel, C. R. (1980). The Organization of Action. Hillsdale, NJ: Erlbaum.

Gallistel, C. R. (1981). Précis of Gallistel's The Organization of Action: A New Synthesis. Behavioral and Brain Sciences, **4**, 609–650.

Gallistel, C. R. (1990a). Representations in Animal Cognition: An Introduction. Cognition, **37**, 1–22.

Gallistel, C. R. (1990b). The Organization of Learning. Cambridge, MA: MIT Press.

Gallistel, C. R. & Matzel, L. D. (2013). The Neuroscience of Learning: Beyond the Hebbian Synapse. Annual Review of Psychology, **63**, 169–200.

Gendron, M. & Barrett, L. F. (2009). Reconstructing the Past: A Century of Ideas about Emotion in Psychology. Emotion Review, **1**, 316–339.

Gerhardt, H. C. (2001). Acoustic Communication in Two Groups of Closely Related Treefrogs. Advances in the Study of Behavior, **30**, 99–167.

Ghirlanda, S. & Enquist, M. (2003). A Century of Generalization. Animal Behaviour, **66**, 15–36.

Ghirlanda, S., Jansson, L. & Enquist, M. (2002). Chickens Prefer Beautiful Humans. Human Nature, **13**, 383–389.

Gil, S. A. & Bierema, A. M. K. (2013). On the Meaning of Alarm Calls: A Review of Functional Reference in Avian Alarm Calling. Ethology, **119**, 449–461.

Gilbert, S. F. (2012). Ecological Developmental Biology: Environmental Signals for Normal Animal Development. Evolution and Development, **14**, 20–28.

Giraldeau, L.-A. (2005). The Function of Behavior. In Bolhuis & Giraldeau, pp. 199–225.

Giraldeau, L.-A. & Dubois,F. (2008). Social Foraging and the Study of Exploitative Behaviour. *Advances in the Study of Behavior*, **38**, 59-104.

Giraldeau, L.-A. & Kramer, D. L. (1982). The Marginal Value Theorem: A Quantitative Test Using Load Size Variation in a Central Place Forager the Eastern Chipmunk, *Tamias* striatus. *Animal Behaviour*, **30**, 1036-1042.

Glickman, S. E. (1973). Responses and Reinforcement. In Hinde & Stevenson-Hinde, pp. 207-241.

Glickman, S. E. & Schiff, B. B. (1967). A Biological Theory of Reinforcement. *Psychological Review*, **74**, 81-109.

Glimcher, P. W. (2011). Understanding Dopamine and Reinforcement Learning: The Dopamine Reward Prediction Error Hypothesis. *Proceedings of the National Academy of Sciences*, **108**, 15647-15654.

Gobes, M. H., Fritz, J. B. & Bolhuis, J. J. (2013). Neural Mechanisms of Auditory Learning in Songbirds and Mammals. In Bolhuis & Everaert, pp. 295-315.

Goldin-Meadow, S. (1997). The Resilience of Language in Humans. In Snowdon & Hausberger, pp. 293-311.

Gosling, S. D. (2001). From Mice to Men: What Can We Learn about Personality from Animal Research? *Psychological Bulletin*, 127, 45-86.

Gottlieb, G. (1978). Development of Species Identification in Ducklings: IV. Change in Species-specific Perception Caused by Auditory Deprivation. *Journal of Comparative and Physiological Psychology*, **92**, 375-387.

Gottlieb, G. (1980). Development of Species Identification in Ducklings: VI. Specific Embryonic Experience Required to Maintain Species-typical Perception in Peking Ducklings. *Journal of Comparative and Physiological Psychology*, **94**, 579-587. (Reprinted in Bolhuis & Hogan, 1999.)

Gottlieb, G. (1992). *Individual Development and Evolution: The Genesis of Novel Behavior*. New York: Oxford University Press.

Gottlieb, G. (1997). *Synthesizing Nature-Nurture: Prenatal Roots of Instinctive Behavior*. Mahwah, NJ: Erlbaum.

Gould, S. J. & Lewontin, R. C. (1979). The Spandrels of San Marco and the Panglossian Paradigm: A Critique of the Adaptationist Programme. *Proceedings of the Royal Society of London* B, **205**, 581-598.

Gould, S. J. & Vrba, E. S. (1988). Exaptation-A Missing Term in the Science of Form. *Paleobiology*, **8**, 4-15.

Graf, J. S., Balsam, P. D. & Silver, R. (1985). Associative Factors and the Development of Pecking in the Ring Dove. *Developmental Psychobiology*, **18**, 447-460.

Grant, P. R. & Grant, B. R. (2006). Evolution of Character Displacement in Darwin's Finches. *Science*, **313**, 224-226.

Gray, R. D., Atkinson, Q. D. & Greenhill, S. J. (2011). Language Evolution and Human History: What a Difference a Day Makes. *Philosophical Transactions of the Royal Society* B, **366**, 1090-1100.

Graybiel, A. M. & Smith, K. S. (2014). Good Habits, Bad Habits. *Scientific American*, **310**, 38-43.

Greenough, W. T., Black, J. E. & Wallace, C. S. (1987). Experience and Brain Development. *Child Development*, **58**, 539-559.

Griesemer, J. (2000). Development, Culture, and the Units of Inheritance. *Philosophy of Science*, 67 (Suppl.), S348-S368.

Groothuis, T. G. G. (1992). The Influence of Social Experience on the Development and Fixation of the Form of Displays in the Black-headed Gull. *Animal Behaviour*, **43**, 1-14.

Groothuis, T. G. G. (1994). The Ontogeny of Social Displays: Interplay between Motor Development, Development of Motivational Systems and Social Experience. In Hogan & Bolhuis, pp. 183-211.

Groothuis, T. G. G. & Meeuwissen, G. (1992). The Influence of Testosterone on the Development and Fixation of the Form of Displays in Two Age Classes of Young Black-headed Gulls. *Animal Behaviour*, **43**, 189–208.

Gross, C. T. & Canteras, N S. (2012). The Many Paths to Fear. *Nature Reviews Neuroscience*, **13**, 651–658.

Grossman, S. P. (1967). *A Textbook of Physiological Psychology*. New York: Wiley.

Gwinner, E. (1986). Circannual Rhythms in the Control of Avian Migrations. *Advances in the Study of Behavior*, **16**, 191–228.

Gwinner, E. (1996). Circadian and Circannual Programmes in Avian Migration. *Journal of Experimental Biology*, **199**, 39–48.

Gwinner, E. & Brandstätter, R. (2001). Complex Bird Clocks. *Philosophical Transactions of the Royal Society London B*, **356**, 1801–1810.

Hafting, T., Fyhn, M., Molden, S. Moser, M.-B. & Moser, E. I. (2005). Microstructure of a Spatial Map in the Entorhinal Cortex. *Nature*, **436**, 801–806.

Hale C. & Green, L. (1979). Effect of Initial Pecking Consequences on Subsequent Pecking in Young Chicks. *Journal of Comparative and Physiological Psychology*, **93**, 730–735.

Hall, C. S. (1934). Emotional Behavior in the Rat: 1. Defecation and Urination as Measures of Individual Differences in Emotionality. *Journal of Comparative Psychology*, **18**, 385–403.

Hall, W. G. & Browde, J. A. (1986). The Ontogeny of Independent Ingestion in Mice: Or, Why Won't Infant Mice Feed? *Developmental Psychobiology*, **19**, 211–222.

Hall, W. G. & Williams, C. L. (1983). Suckling Isn't Feeding, or Is It? A Search for Developmental Continuities. *Advances in the Study of Behavior*, **13**, 219–254.

Hamburger, V. (1973). Anatomical and Physiological Basis of Embryonic Motility in Birds and Mammals. In G. Gottlieb (Ed.), *Behavioral Embryology*. New York: Academic Press, pp. 52–76.

Hamilton, W. D. (1964). The Genetical Evolution of Social Behaviour. I & II. *Journal of Theoretical Biology*, **7**, 1–52.

Hamilton, W. D. (1971). Geometry for the Selfish Herd. *Journal of Theoretical Biology*, **31**, 295–311.

Harlow, H. F. (1950). Learning and Satiation of Response in Intrinsically Motivated Complex Puzzle Performance by Monkeys. *Journal of Comparative and Physiological Psychology*, **43**, 289–294.

Harlow, H. F. & Harlow, M. K. (1962). Social Deprivation in Monkeys. *Scientific American*, Nov. (Reprinted in Bolhuis & Hogan, 1999.)

Harlow, H. F. & Harlow, M. K. (1965). The Affectional Systems. In A. M. Schrier, H. F. Harlow & F. Stollnitz (Eds.), *Behavior of Nonhuman Primates*, Vol. 2. New York: Academic Press.

Harlow, H. F. & Suomi, S. J. (1971). Social Recovery by Isolation-reared Monkeys. *Proceedings of the National Academy of Sciences*, **68**, 1534–1538.

Hart, B. L. (1995). Analysing Breed and Gender Differences in Behaviour. In J. Serpell (Ed.), *The Domestic Dog*. Cambridge, UK: Cambridge University Press, pp. 65–77.

Haslam, M., Luncz, L. V. & Falótico, T. (2016). Pre-Columbian Monkey Tools. *Current Biology*, **26**, R515–R522.

Hauser, M. D. (1996). *The Evolution of Communication*. Cambridge, MA: MIT Press.

Hauser, M.D., Yang, C., Berwick, R. C., Tattersall, I., Ryan, M. J., Watumull, J., Chomsky, N. & Lewontin, R. C. (2014). The Mystery of Language Evolution. *Frontiers in Psychology*, **5**, May, article 401.

Hebb, D. O. (1946). On the Nature of Fear. *Psychological Review*, **53**, 259–276.

Hebb, D. O. (1949). *The Organization of Behavior.* New York: Wiley.

Heiligenberg, W. (1974). Processes Governing Behavioral States of Readiness. *Advances in the Study of Behavior,* **5**, 173–200.

Heinroth, O. (1911). *Beiträge zur Biologie: namentlich Ethologie und Psychologie der Anatiden.* Verhandlungen des 5. Internationalen Ornitologen-Kongresses in Berlin, 1910, pp. 389–702.

Henning, F. & Meyer, A. (2014). The Evolutionary Genomics of Cichlid Fishes: Explosive Speciation and Adaptation in the Postgenomic Era. *Annual Review of Genomics and Human Genetics,* **15**, 417–441.

Herrnstein, R. J. (1977). The Evolution of Behaviorism. *American Psychologist,* **32**, 593–603.

Hess, E. H. (1956). Natural Preferences of Chicks and Ducklings for Objects of Different Colors. *Psychological Reports,* **2**, 477–483.

Heyes, C. M. (2012). New Thinking: The Evolution of Human Cognition. *Philosophical Transactions of the Royal Society B,* **367**, 2091–2096.

Heyes, C. M. (1994). Social Learning in Animals: Categories and Mechanisms. *Biological Reviews,* **69**, 207–231.

Heyes, C. M. (2012). What's Social about Social Learning? *Journal of Comparative Psychology,* **126**, 109–113.

Hills, T. T., Todd, P. M., Lazer, D. *et al.* (2015). Exploration versus Exploitation in Space, Mind, and Society. *Trends in Cognitive Sciences,* **19**, 46–54.

Hinde, R. A. (1959). Unitary Drives. *Animal Behaviour,* **7**, 130–141.

Hinde, R. A. (1960). Energy Models of Motivation. *Symposia of the Society for Experimental Biology,* **14**, 199–213.

Hinde, R. A. (1970). *Animal Behavior,* 2nd edn. New York: McGraw-Hill.

Hinde, R. A. (1975). The Concept of Function. In Baerends *et al.,* pp. 2–15.

Hinde, R. A. (1977). Mother-Infant Separation and the Nature of Inter-individual Relationships: Experiments with Rhesus Monkeys. *Proceedings of the Royal Society of London B,* **196**, 29–50. (Reprinted in Bolhuis & Hogan, 1999.)

Hinde, R. A. & Stevenson-Hinde, J. G. (Eds.). (1973). *Constraints on Learning.* London: Academic Press.

Hodos, W. & Campbell, C. B. G. (1969). Scala Naturae: Why There Is No Theory in Comparative Psychology. *Psychological Review,* **76**, 337–350.

Hofer, M. A. (1987). Early Social Relationships: A Psychobiologist's View. *Child Development,* **58**, 633–647.

Hofer, M. A. (1996). On the Nature and Consequences of Early Loss. *Psychosomatic Medicine,* **58**, 570–581.

Hogan, J. A. (1961). *Motivational Aspects of Instinctive Behavior in* Betta splendens. Ph.D. thesis, Harvard University.

Hogan, J. A. (1964). Operant Control of Preening in Pigeons. *Journal of the Experimental Analysis of Behavior,* **7**, 351–354.

Hogan, J. A. (1965). An Experimental Study of Conflict and Fear: An Analysis of Behavior of Young Chicks to a Mealworm. Part I: The Behavior of Chicks Which Do Not Eat the Mealworm. *Behaviour,* **25**, 45–97.

Hogan, J. A. (1966). An Experimental Study of Conflict and Fear: An Analysis of Behavior of Young Chicks to a Mealworm. Part II: The Behavior of Chicks Which Eat the Mealworm. *Behaviour,* **27**, 273–289.

Hogan, J. A. (1971). The Development of a Hunger System in Young Chicks. *Behaviour,* **39**, 128–201.

Hogan, J. A. (1973). How Young Chicks Learn to Recognize Food. In Hinde & Stevenson-Hinde, pp. 119–139.

Hogan, J. A. (1974). Responses in Pavlovian Conditioning Studies. *Science*, **186**, 156–157.

Hogan, J. A. (1977). The Ontogeny of Food Preferences in Chicks and Other Animals. In L. M. Barker, M. Best & M. Domjan (Eds.), *Learning Mechanisms in Food Selection*. Waco, TX: Baylor University Press.

Hogan, J. A. (1978). An Eccentric View of Development: A Review of *The Dynamics of Behavior Development* by Zing-Yang Kuo. *Contemporary Psychology*, **23**, 690–691.

Hogan, J. A. (1980). Homeostasis and Behaviour. In F. M. Toates & T. R. Halliday (Eds.), *Analysis of Motivational Processes*. London: Academic Press, pp. 3–21.

Hogan, J. A. (1981). Hierarchy and Behavior. *Behavioral and Brain Sciences*, **4**, 625.

Hogan, J. A. (1984a). Cause, Function, and the Analysis of Behavior. *Mexican Journal of Behavior Analysis*, **10**, 65–71.

Hogan, J. A. (1984b). Pecking and Feeding in Chicks. *Learning and Motivation*, **15**, 360–376.

Hogan, J. A. (1988). Cause and Function in the Development of Behavior Systems. In E. M. Blass (Ed.), *Handbook of Behavioral Neurobiology*, vol. 9. New York: Plenum, pp. 63–105.

Hogan, J. A. (1989). The Interaction of Incubation and Feeding in Broody Junglefowl Hens. *Animal Behaviour*, **38**, 121–128.

Hogan, J. A. (1994a). The Concept of Cause in the Study of Behavior. In Hogan & Bolhuis, pp. 3–15.

Hogan, J. A. (1994b). Development of Behavior Systems. In Hogan & Bolhuis, pp. 242–264.

Hogan, J. A. (1994c). Structure and Development of Behavior Systems. *Psychonomic Bulletin and Review*, **1**, 439–450.

Hogan, J. A. (1997). Energy Models of Motivation: A Reconsideration. *Applied Animal Behaviour Science*, **53**, 89–105.

Hogan, J. A. (1998). Motivation. In G. Greenberg & M. M. Haraway (Eds.), *Comparative Psychology: A Handbook*, pp. 164–175. New York: Garland.

Hogan, J. A. (2001). Development of Behavior Systems. In E. M. Blass (Ed.), *Handbook of Behavioral Neurobiology*, Vol. **13**: *Developmental Psychobiology*, New York: Kluwer Academic/Plenum, pp. 229–279.

Hogan, J. A. (2005). Motivation. In Bolhuis & Giraldeau, pp. 41–70.

Hogan, J. A. (2015). A Framework for the Study of Behavior. *Behavioural Processes*, **117**, 105–113.

Hogan, J. A. & Bolhuis, J. J. (2009). Tinbergen's Four Questions and Contemporary Behavioral Biology. In J. J. Bolhuis & S. Verhulst (Eds.), *Tinbergen's Legacy*. Cambridge, UK: Cambridge University Press, pp. 25–34.

Hogan, J. A. & Bolhuis, J. J. (Eds.). (1994). *Causal Mechanisms of Behavioural Development*. Cambridge, UK: Cambridge University Press.

Hogan, J. A. & Bols, R. J. (1980). Priming of Aggressive Motivation in *Betta splendens*. *Animal Behaviour*, **28**, 135–142.

Hogan, J. A. & Roper, T. J. (1978). A Comparison of the Properties of Different Reinforcers. *Advances in the Study of Behavior*, **8**, 155–255.

Hogan, J. A. & van Boxel, F. (1993). Causal Factors Controlling Dustbathing in Burmese Red Junglefowl: Some Results and a Model. *Animal Behaviour*, **46**, 627–635.

Hogan, J. A., Hogan-Warburg, A. J., Panning, L. & Moffatt, C. A. (1998). Causal Factors Controlling the Brooding Cycle of Broody Junglefowl Hens with Chicks. *Behaviour*, **135**, 957–980.

Hogan, J. A., Honrado, G. I. & Vestergaard, K. S. (1991). Development of a Behavior System: Dustbathing in the Burmese Red Junglefowl (*Gallus gallus spadiceus*): II. Internal Factors. *Journal of Comparative Psychology*, **195**, 269–273.

References   343

Hogan-Warburg, A. J. (1966). Social Behavior of the Ruff, *Philomachus pugnax* (L.). *Ardea*, **54**, 109–229.

Hogan-Warburg, A. J. (1992). Female Choice and the Evolution of Mating Strategies in the Ruff *Philomachus pugnax* (L.). *Ardea*, **80**, 395–403.

Hogan-Warburg, A. J. & Hogan, J. A. (1981). Feeding Strategies in the Development of Food Recognition in Young Chicks. *Animal Behaviour*, **29**, 143–154.

Hogan-Warburg, A. J., Hogan, J. A. & Ashton, M. C. (1995). Locomotion and Grooming in Crickets: Competition or Time Sharing? *Animal Behaviour*, **49**, 531–533.

Hogan-Warburg, A. J., Panning, L. & Hogan, J. A. (1993). Analysis of the Brooding Cycle of Broody Junglefowl Hens with Chicks. *Behaviour*, **125**, 21–37.

Hollis, K. L. (1984). The Biological Function of Pavlovian Conditioning: The Best Defence Is a Good Offence. *Journal of Experimental Psychology: Animal Behavior Processes*, **10**, 413–425.

Hollis, K. L., ten Cate, C. & Bateson, P. (1991). Stimulus Representation: A Subprocess of Imprinting and Conditioning. *Journal of Comparative Psychology*, **105**, 307–317.

Hopkins, B. & Butterworth, G. (1990). Concepts of Causality in Explanations of Development. In G. Butterworth & P. Bryant (Eds.), *Causes of Development*, London: Harvester Wheatsheaf, pp. 3–33.

Horn, G. (1985). *Memory, Imprinting, and the Brain*. Oxford: Oxford University Press.

Hughes, G. M. (Ed.). (1964). *Symposia of the Society for Experimental Biology: Homeostasis and Feedback Mechanisms*, vol.18. Cambridge, UK: Cambridge University Press.

Hull, C. L. (1933). Differential Habituation to Internal Stimuli in the Albino Rat. *Journal of Comparative Psychology*, **16**, 255–273.

Hull, C. L. (1943). *Principles of Behavior*. New York: Appleton-Century-Crofts.

Hultsch, H. (1993). Tracing the Memory Mechanisms in the Song Acquisition of Nightingales. *Netherlands Journal of Zoology*, **43**, 155–171.

Huxley, J. S. (1923). Courtship Activities in the Red-throated Diver *(Colymbus stellatus* Pontopp.); Together with a Discussion of the Evolution of Courtship in Birds. *Zoological Journal of the Linnean Society of London*, **35**, 253–292.

Huxley, J. S. (1942). *Evolution: The Modern Synthesis*. London: Allen & Unwin.

Huxley, T. H. (1893). *Collected Essays*, Vol. **1**, *Method and Results*. London: Macmillan.

Ibbotson, P. & Tomasello, M. (2016). Language in a New Key. *Scientific American*, **315** (5), 70–75.

Immelmann, K. (1972). The Influence of Early Experience upon the Development of Social Behaviour in Estrildine Finches. In K. H. Voous (Ed.), *Proceedings of the XV International Ornithological Congress, Den Haag, 1970*. Leiden: Brill, pp. 316–338.

Immelmann, K., Lassek, R., Pröve, R. & Bischof, H.-J. (1991). Influence of Adult Courtship Experience on the Development of Sexual Preferences in Zebra Finch Males. *Animal Behaviour*, **42**, 83–89.

Ince, S. A., Slater, P. J. B. & Weismann, C. (1980). Changes with Time in the Songs of a Population of Chaffinches. *Condor*, **82**, 285–290.

Ingold, T. (2004). Beyond Biology and Culture. The Meaning of Evolution in a Relational World. *Social Anthropology*, **12**, 209–221.

Insel, N. & Frankland, P. W. (2015). Mechanism, Function, and Computation in Neural Systems. *Behavioural Processes*, **117**, 4–11.

Jablonka, E. & Lamb, M. J. (2014). *Evolution in Four Dimensions*, revised edn. Cambridge, MA: MIT Press.

Jackson, R. R. & Cross, F. R. (2011). Spider Cognition. *Advances in Insect Physiology*, **41**, 115–174.

Jackson, R. R. & Wilcox, R. S. (1998). Spider-eating Spiders. *American Scientist*, **86**, 350–357.

James, W. (1890). *Principles of Psychology*, vol. **2**. New York: Holt.

Japyassú, H. F. & Malange, J. (2014). Plasticity, Stereotypy, Intro-individual Variability and Personality: Handle with Care. *Behavioural Processes*, **109**, 40–47.

Jarvis E. D. & Nottebohm, F. (1997). Motor-driven Gene Expression. *Proceedings of the National Academy of Sciences*, **94**, 4097–4102.

Johanson, I. B. & Terry, L. M. (1988). Learning in Infancy: A Mechanism of Behavioral Change in Development. In E. M. Blass (Ed.), *Handbook of Behavioral Neurobiology*, vol. **9**. New York: Plenum, pp. 245–281.

Johnson, A. & Redish, A. D. (2007). Neural Ensembles in CA3 Transiently Encode Paths Forward of the Animal at a Decision Point. *Journal of Neuroscience*, **27**, 12176–12189.

Johnson, K. P. (2000). The Evolution of Courtship Display Repertoire Size in the Dabbling Ducks (Anatini). *Journal of Evolutionary Biology*, **13**, 634–644.

Johnson, M. H. & Bolhuis, J. J. (2000). Predispositions in Perceptual and Cognitive Development. In J. J. Bolhuis (Ed.), *Brain, Perception, Memory: Advances in Cognitive Neuroscience*. Oxford: Oxford University Press, pp. 69–84.

Johnson, M. H., Bolhuis, J. J. & Horn, G. (1985). Interaction between Acquired Preferences and Developing Predispositions during Imprinting. *Animal Behaviour*, **33**, 1000–1006.

Johnson, M. H., Davies, D. C. & Horn, G. (1989). A Critical Period for the Development of a Predisposition in the Chick. *Animal Behaviour*, **37**, 1044–1046.

Johnson, R. N. & Johnson, L. D. (1973). Intra- and Interspecific Social and Aggressive Behaviour in the Siamese Fighting Fish, *Betta* splendens. *Animal Behaviour*, **21**, 665–672.

Josselyn, S. A., Köhler, S. & Frankland, P. W. (2015). Finding the Engram. *Nature Reviews Neuroscience*, **16**, 521–534.

Josselyn, S. A., Köhler, S. & Frankland, P. W. (2017). Heroes of the Engram. *Journal of Neuroscience*, 17, 4647–4657.

Jukema, J. & Piersma, T. (2006). Permanent Female Mimics in a Lekking Shorebird. *Biology Letters*, **2**, 161–164.

Jusczyk, P. W. (1997). Finding and Remembering Words: Some Beginnings by English-learning Infants. *Current Directions in Psychological Science*, **6**, 170–174.

Kacelnik, A. (1984). Central-place Foraging in Starlings (*Sturnus vulgaris*). I. Patch Residence Time. *Journal of Animal Ecology*, **53**, 283–299.

Kahneman, D. (2011). *Thinking Fast and Slow*. Canada: Doubleday.

Karmiloff-Smith, A. (1992). *Beyond Modularity: A Developmental Perspective on Cognitive Science*. Cambridge, MA: MIT Press.

Karmiloff-Smith, A. (1998). Development Itself Is the Key to Understanding Developmental Disorders. *Trends in Cognitive Sciences*, **2**, 389–398.

Kellman, P. J. (2002). Perceptual Learning. In C. R. Gallistel (Ed.), *Stevens' Handbook of Experimental Psychology*, vol. **3**: *Learning, Motivation, and Emotion*. New York: Wiley, pp. 259–299.

Kenny, J. T., Stoloff, M. L., Bruno, J. P. & Blass, E. M. (1979). The Ontogeny of Preferences for Nutritive over Nonnutritive Suckling in the Albino Rat. *Journal of Comparative and Physiological Psychology*, **93**,752–759.

King, A. P. & West, M. J. (1977). Species Identification in the North American Cowbird: Appropriate Responses to Abnormal Song. *Science*, **195**, 1002-1004.

Kirkpatrick, K. & Hall, G. (2005). Learning and Memory. In Bolhuis & Giraldeau, pp. 146-169.

Kish, G. B. (1966). Studies of Sensory Reinforcement. In W. K. Honig (Ed.), *Operant Behavior: Areas of Research and Application*. New York: Appleton, pp. 109-159.

Kissileff, H. R. (1969). Food-associated drinking in the rat. *Journal of Comparative and Physiological Psychology*, **67**, 284-300.

Konishi, M. (1965). The Role of Auditory Feedback in the Control of Vocalizations in the White-crowned Sparrow. *Zeitschrift für Tierpsychologie*, **22**, 770-783.

Kortlandt, A. (1940). Wechselwirkung zwischen Instinkten. *Archives Néerlandaises de Zoologie*, **4**, 443-520.

Kraemer, G. W. (1992). A Psychobiological Theory of Attachment. *Behavioral and Brain Sciences*, **15**, 493-511.

Kramer, D. L. & Bonenfant, M. (1997). Direction of Predator Approach and the Decision to Flee to a Refuge. *Animal Behaviour*, **54**, 289-295.

Kramer, G. (1952). Experiments on bird orientation. *Ibis*, **94**, 265-285.

Krebs, J. R., Erichsen, J. T., Webber, M. I. & Charnov, E. L. (1977). Optimal Prey-selection by the Great Tit (*Parus major*). *Animal Behaviour*, **25**, 30-38.

Kristiansen. M. & Ham, J. (2014). Programmed Cell Death during Neuronal Development: The Sympathetic Neuron Model. *Cell Death and Differentiation*, **21**, 1025-1035.

Kruijt, J. P. (1962a). On the Evolutionary Derivation of Wing Display in Burmese Red Junglefowl and Other Gallinaceous Birds. *Symposia of the Zoological Society, London*, **8**, 25-35.

Kruijt, J. P. (1962b). Imprinting in Relation to Drive Interactions in Burmese Red Junglefowl. *Symposia of the Zoological Society, London*, **8**, 219-226.

Kruijt, J. P. (1964). Ontogeny of Social Behaviour in Burmese Red Junglefowl (*Gallus gallus spadiceus*). *Behaviour* (Suppl. 9).

Kruijt, J. P. (1985). On the Development of Social Attachments in Birds. *Netherlands Journal of Zoology*, **35**, 45-62.

Kruijt, J. P. & Hogan, J. A. (1967). Social Behavior on the Lek in Black Grouse, *Lyrurus tetrix tetrix* (L.). *Ardea*, **55**, 203-240.

Kruijt, J. P. & Meeuwissen, G. B. (1991). Sexual Preferences of Male Zebra Finches: Effects of Early Adult Experience. *Animal Behaviour*, **42**, 91-102.

Kruijt, J. P. & Meeuwissen, G. B. (1993). Consolidation and Modification of Sexual Preferences in Adult Male Zebra Finches. *Netherlands Journal of Zoology*, **43**, 68-79.

Kruuk, H. (1964). Predators and Anti-predator Behaviour of the Black-headed Gull, *Larus ridibundus*. *Behaviour* (Suppl. **11**), 1-129.

Kuhl, P. K. (1994). Learning and Representation in Speech and Language. *Current Opinion in Neurobiology*, **4**, 812-822.

Kuhl, P. K. (2004). Early Language Acquisition: Cracking the Speech Code. *Nature Reviews Neuroscience*, **5**, 831-843.

Kuhl, P. K. (2015). How Babies Learn Language. *Scientific American*, **313**(5).

Kuhl, P. K., Williams, K. A., Lacerda, R., Stevens, K. N. & Lindblom, B. (1992). Linguistic Experience Alters Phonetic Perception in Infants by 6 Months of Age. *Science*, **255**, 606-608.

Kühn, A. (1919). *Die Orientierung der Tiere im Raum*. Jena.

Kuo, Z. Y. (1921). Giving up Instincts in Psychology. *Journal of Philosophy*, **18**, 645–664.

Kuo, Z. Y. (1967). *The Dynamics of Behavioral Development*. New York: Random House. (Excerpt reprinted in Bolhuis & Hogan, 1999.)

Küpper, C., Stocks, M., Risse, J. E. *et al.* (2016). A Supergene Determines Highly Divergent Male Reproductive Morphs in the Ruff. *Nature Genetics*, **48**, 79–83.

Lack, D. (1954). *The Natural Regulation of Animal Numbers*. Oxford: Oxford University Press.

Lack. D. (1947). *Darwin's Finches*. Cambridge, UK: Cambridge University Press.

Laland, K. N. (2015). On Evolutionary Causes and Evolutionary Processes. *Behavioural Processes*, **117**, 97–104.

Laland, K. N. & Brown, G. R. (2011). *Sense and Nonsense: Evolutionary Perspectives on Human Behaviour*, 2nd edn. Oxford: Oxford University Press.

Laland, K. N. & Galef Jr, B. G. (Eds.). (2009). *The Question of Animal Culture*. Cambridge, MA: Harvard University Press.

Laland, K. N. *et al.* (2014). Does Evolutionary Theory Need a Rethink? Yes, Urgently. *Nature*, **514**, 161–164.

Laland, K. N., Sterelny, K., Odling-Smee, J., Hoppitt, W. & Uller, T. (2011). Cause and Effect in Biology Revisited: Is Mayr's Proximate-Ultimate Dichotomy Still Useful? *Science*, **334**, 1512–1516.

Land, M. F. (2014). Do We Have an Internal Model of the Outside World? *Philosophical Transactions of the Royal Society B*, **369** (doi: 20130045).

Langtimm, C. A. & Dewsbury, D. A. (1991). Phylogeny and Evolution of Rodent Copulatory Behaviour. *Animal Behaviour*, **41**, 217–225.

Lank, D. B., Smith, C. M., Hanotte, O., Burke, T. & Cooke, F. (1995). Genetic Polymorphism for Alternative Mating Behaviour in Lekking Male Ruff *Philomachus pugnax*. *Nature*, **378**, 59–62.

Larsen, B. H., Hogan, J. A. & Vestergaard, K. S. (2000). Development of Dustbathing Behavior Sequences in the Domestic Fowl: The Significance of Functional Experience. *Developmental Psychobiology*, **37**, 5–12.

Larson, G. & Fuller, D. Q. (2014). The Evolution of Animal Domestication. *Annual Review of Ecology, Evolution, and Systematics*, **45**, 115–136.

Lashley, K. S. (1938). Experimental Analysis of Instinctive Behavior. *Psychological Review*, **45**, 445–471.

Lashley, K. S. (1950). In Search of the Engram. *Symposia of the Society for Experimental Biology*, **4**, 454–482.

Lashley, K. S. (1951). The Problem of Serial Order in Behavior. In L. A. Jeffress (Ed.), *Cerebral Mechanisms in Behavior*. New York: Wiley, pp. 112–136.

Leadbeater, E. (2015). What Evolves in the Evolution of Social Learning. *Journal of Zoology*, **295**, 4–11.

Leadbeater, E. & Chittka, L. (2007). Social Learning in Insects – From Miniature Brains to Consensus Building. *Current Biology*, **17**, R703–R713.

LeBeouf, B. J. (1974). Male-male Competition and Reproductive Success in Elephant Seals. *American Zoologist*, **14**, 163–176.

LeDoux, J. (2012). Rethinking the Emotional Brain. *Neuron*, **73**, 653–676.

Lefebvre, L. (2015). Should Neuroecologists Separate Tinbergen's Four Questions? *Behavioural Processes*, **117**, 92–96.

Lehrman, D. S. (1953). A Critique of Konrad Lorenz's Theory of Instinctive Behaviour. *Quarterly Review of Biology*, **28**, 337–363.

Lehrman, D. S. (1955). The Physiological Basis of Parental Feeding Behaviour in the Ring Dove (*Streptopelia risoria*). *Behaviour*, **7**, 241–286.

Lehrman, D. S. (1965). Interaction between Internal and External Environments in the Regulation of the Reproductive Cycle of the Ring Dove. In F. A. Beach (Ed.), *Sex and Behavior*. New York: Wiley, pp. 355–380.

Lehrman, D. S. (1970). Semantic and Conceptual Issues in the Nature-Nurture Problem. In Aronson *et al.*, pp. 17–52. (Reprinted in Bolhuis & Hogan, 1999.)

Lenneberg, E. H. (1967). *Biological Foundations of Language*. New York: Wiley.

Lettvin, J. Y., Maturana, H. R., McCulloch, W. S. & Pitts, W. H. (1959). What the Frog's Eye Tells the Frog's Brain. *Proceedings of the Institute of Radio Engineers*, **47**, 1940–1951.

Levinson, S. C. & Gray, R. D. (2012). Tools from Evolutionary Biology Shed New Light on the Diversification of Languages. *Trends in Cognitive Sciences*, **16**, 167–173.

Lewontin, R. C. (1970). The Units of Selection. *Annual Review of Ecology and Systematics*, **1**, 1–18.

Lieberman, P. (2016). The Evolution of Language and Thought. *Journal of Anthropological Sciences*, **94**, 127–146.

Locke, J. L. (1993). *The Child's Path to Spoken Language*. Cambridge, MA: Harvard University Press.

Locke, J. L. (1994). The Biological Building Blocks of Spoken Language. In Hogan & Bolhuis, pp. 300–324.

Locke, J. L. & Snow, C. (1997). Social Influences on Vocal Learning in Human and Nonhuman Primates. In Snowdon & Hausberger, pp. 274–292.

Loeb, J. (1918). *Forced Movements, Tropisms and Animal Conduct*. Philadelphia: Lippincott.

Logan, C. A. (1983). Biological Diversity in Avian Vocal Learning. In M. D. Zeiler & P. Harzem (Eds.), *Advances in Analysis of Behavior*, vol. **3**: *Biological Factors in Learning*. Chichester, UK: Wiley.

Lorenz, K. (1935). Der Kumpan in der Umwelt des Vogels. *Journal für Ornithologie*, **83**, 137–213, 289–413. [Tr. as: Companions as Factors in the Bird's Environment. In Lorenz, 1970, pp. 101–258 - excerpt reprinted in Bolhuis & Hogan, 1999.]

Lorenz, K. (1937). Über die Bildung des Instinktbegriffes. *Naturwissenschaften*, **25**, 289–300, 307–318, 324–331. [Tr. as: The Establishment of the Instinct Concept. In Lorenz, 1970, pp. 259–315.]

Lorenz, K. (1941). *Vergleichende Bewegungsstudien an Anatinen. Journal für Ornithologie*, **89**, Sonderheft. (Tr. by R. D. Martin as: Comparative Studies of the Motor Patterns of Anatinae. In Lorenz (1971). *Studies in Animal and Human Behaviour*, vol. 2. London: Methuen, pp. 14–114.)

Lorenz, K. (1950). The Comparative Method in Studying Innate Behaviour Patterns. *Symposia of the Society for Experimental Biology*, **4**, 221–268.

Lorenz, K. (1956). Plays and Vacuum Activities. In *L'instinct dans le comportement des animaux et de l'homme*. Paris: Masson, pp. 633–637.

Lorenz, K. (1961). Phylogenetische Anpassung und adaptive Modifikation des Verhaltens. *Zeitschrift für Tierpsychologie*, **18**, 139–187.

Lorenz, K. (1965). *Evolution and Modification of Behavior*. Chicago: University of Chicago Press. (Excerpt reprinted in Bolhuis & Hogan, 1999.)

Lorenz, K. (1966). *On Aggression*. London: Methuen.

Lorenz, K. (1970). *Studies in Animal and Human Behaviour*, vol. **1**. (Tr. by R. Martin). London: Methuen.

Lorenz, K. (1981). *The Foundations of Ethology*. New York: Springer.

Lorenz, K. & Tinbergen, N. (1939). *Taxis und Instinkthandlung in der Eirollbewegung der Graugans. Zeitschrift für Tierpsychologie*, **2**, 1–29. [Tr. as: Taxis and Instinctive Behavior Pattern in Egg-rolling by the Greylag Goose. In Lorenz, 1970, pp. 316–350.]

Losos, J. B. (2010). Adaptive Radiation, Ecological Opportunity, and Evolutionary Determinism. *American Naturalist*, **175**, 623-639.

Lotem, A. (1993). Learning to Recognize Nestlings Is Maladaptive for Cuckoo *Cuculus canorus* Hosts. *Nature*, **362**, 743-745.

Lovic, V. & Fleming, A. S. (2015). Propagation of Maternal Behavior across Generations Is Associated with Changes in Non-maternal Cognitive and Behavioral Processes. *Behavioural Processes*, **117**, 42-47.

Lowe, P. R. (1936). The Finches of the Galapagos in Relation to Darwin's Conception of Species. *Ibis*, **78**, 310-321.

Macedonia, J. M. & Evans, C. S. (1993). Variation among Mammalian Alarm Call Systems and the Problem of Meaning in Animal Signals. *Ethology*, **93**, 177-197.

Macintosh, N. J. (1994). *Animal Learning and Cognition*. San Diego, CA: Academic Press.

Marler, P. (1955). Characteristics of Some Animal Calls. *Nature*, **176**, 6-8.

Marler, P. (1970a). A Comparative Approach to Vocal Learning: Song Development in White-crowned Sparrows. *Journal of Comparative and Physiological Psychology* (monograph suppl.), **71**, 1-25.

Marler, P. (1970b). Birdsong and Speech Development: Could There Be Parallels? *American Scientist*, **58**, 669-673.

Marler, P. (1976). Sensory Templates in Species-specific Behavior. In J. C. Fentress (Ed.), *Simpler Networks and Behavior*. Sunderland, MA: Sinauer, pp. 314-329. (Reprinted in Bolhuis & Hogan, 1999.)

Marler, P. (1984). Song Learning: Innate Species Differences in the Learning Process. In P. Marler & H. S. Terrace (Eds.), *The Biology of Learning*. Dahlem Workshop Reports-Life Sciences Research Report 29. Berlin: Springer.

Marler, P. (1991). Song Learning Behavior: The Interface with Neuroethology. *Trends in Neurosciences*, **14**, 199-206.

Marler, P. & Evans, C. (1996). Bird Calls: Just Emotional Displays or Something More? *Ibis*, **138**, 26-33.

Marler, P. & Peters, S. (1982). Subsong and Plastic Song: Their Role in the Vocal Learning Process. In D. E. Kroodsma & E. H. Miller (Eds.), *Acoustic Communication in Birds*, vol. 2. New York: Academic Press, pp. 25-50.

Marler, P. & Tamura, M. (1964). Culturally Transmitted Patterns of Vocal Behavior in Sparrows. *Science*, **146**, 1483-1486.

Marr, D. (1982). *Vision*. New York: Freeman

Martin, P. & Caro, T. M. (1985). On the Functions of Play and Its Role in Behavioral Development. *Advances in the Study of Behavior*, **15**, 59-103.

Mason, G. J. (2010). Species Differences in Responses to Captivity: Stress, Welfare and the Comparative Method. *Trends in Ecology and Evolution*, **25**, 713-721.

Maye, J., Werker, J. F. & Gerken, L. (2002). Infant Sensitivity to Distributional Information Can Affect Phonetic Discrimination. *Cognition*, **82**, B101-B111.

Maynard Smith, J. (1964). Group Selection and Kin Selection. *Nature*, **201**, 1145-1147.

Maynard Smith, J. & Price, G. R. (1973). The logic of Animal Conflict. *Nature*, **246**, 15-18.

Mayr, E. (1961). Cause and Effect in Biology. *Science*, **134**, 1501-1506.

Mayr, E. (1982). *The Growth of Biological Thought: Diversity, Evolution, and Inheritance*. Cambridge, MA: Harvard University Press.

Mayr, E. (1988). The Multiple Meanings of Teleological. *Toward a New Philosophy of Biology*. Cambridge, MA: Harvard University Press, pp. 38-66.

Mayr, E. (1993). Proximate and Ultimate Causations. *Biology and Philosophy*, **8**, 93–94.

McAleer, K. & Giraldeau, L.-A. (2006). Testing Central Place Foraging in Eastern Chipmunks, *Tamias Striatus*, by Altering Loading Functions. *Animal Behaviour*, **71**, 1447–1453.

McDougall, W. (1923). *An Outline of Psychology*. London: Methuen.

McFarland, D. J. (1970). Behavioural Aspects of Homeostasis. *Advances in the Study of Behavior*, **3**, 1–26.

McFarland, D. J. (1974). Time-sharing as a Behavioral Phenomenon. *Advances in the Study of Behavior*, **5**, 201–225.

McFarland, D. J. & Houston, A. (1981). *Quantitative Ethology: The State Space Approach*. London: Pitman.

McKeon, R. (Ed.). (1947). *Introduction to Aristotle*. New York: Random House.

McKinney, F. (1975). The Evolution of Duck Displays. In Baerends *et al.*, pp. 331–357.

McLaren, I. P. L. (1994). Representation Development in Associative Systems. In Hogan & Bolhuis, pp. 377–402.

McNamara, J. M. & Houston, A. I. (2009). Integrating Function and Mechanism. *Trends in Ecology and Evolution*, **24**, 670–675.

McNamara, J. M., Houston, A. I. & Collins, E. J. (2001). Optimality Models in Behavioral Biology. *SIAM Review*, **42**, 413–466.

Meijer, T., Daan, S. & Hall, M. (1990). Family Planning in the Kestrel (*Falco tinnunculus*): The Proximate Control of Covariation of Laying Date and Clutch Size. *Behaviour*, **114**, 117–160.

Mendl, M. T., Burman, O. H. P. & Paul, E. S. (2010). An Integrative and Functional Framework for the Study of Animal Emotion and Mood. *Proceedings of the Royal Society B*, **277**, 2895–2904.

Mendl, M. T. & Paul, E. S. (2016). Bee Happy. *Science*, **353**, 1499–1500.

Menzel, R. & Fischer, J. (2011). *Animal Thinking: Contemporary Issues in Comparative Cognition*. Cambridge, MA: MIT Press.

Meyer, A. (1993). Phylogenetic Relationships and Evolutionary Processes in East African Cichlid Fishes. *Trends in Ecology and Evolution*, **8**, 279–284.

Meyer, A. (2015). Extreme Evolution. *Scientific American*, **313**(4), 70–75.

Miller, N. E. (1957). Experiments on Motivation. *Science*, **126**, 1271–1278.

Miller, N. E. (1959). Liberalization of Basic S-R Concepts: Extensions to Conflict Behavior, Motivation, and Social Learning. In S. Koch (Ed.), *Psychology: A Study of a Science*, vol. **2**, pp. 196–292. New York: McGraw-Hill.

Ming, G. & Song, H. (2011). Adult Neurogenesis in the Mammalian Brain: Significant Answers and Significant Questions. *Neuron*, **70**, 687–702.

Mirmiran, M., Maas, Y. G. H. & Ariagno, R. L. (2003). Development of Fetal and Neonatal Sleep and Circadian Rhythms. *Sleep Medicine Reviews*, **7**, 321–334.

Mistleberger, R. E. (2011). Neurobiology of Food Anticipatory Circadian Rhythms. *Physiology and Behavior*, **104**, 535–545.

Mistleberger, R. E. & Rusak, B. (2005). Biological Rhythms and Behavior. In Bolhuis & Giraldeau, pp. 71–96.

Mittelstaedt, M.-L. & Mittelstaedt, H. (1980). Homing by Path Integration in a Mammal. *Naturwissenschaften*, **67**, 566–567.

Moczek, A. P. *et al.* (2015). The Significance and Scope of Evolutionary Developmental Biology: A Vision for the 21st Century. *Evolution and Development*, **17**, 198–219.

Moffatt, C. A. & Hogan, J. A. (1992). Ontogeny of Chick Responses to Maternal Food Calls in the Burmese Red Junglefowl (*Gallus gallus spadiceus*). *Journal of Comparative Psychology*, **106**, 92–96.

Møller, A. P. (1988). Female Choice Selects for Male Sexual Tail Ornaments in a Swallow. *Nature*, **332**, 640–642.

Møller, A. P. & de Lope, F. (1994). Differential Costs of a Secondary Sexual Character: An Experimental Test of the Handicap Principle. *Evolution*, **48**, 1676–1683.

Montgomerie, R., Lyon, B. & Holder, K. (2001). Dirty Ptarmigan: Behavioral Modification of Conspicuous Male Plumage. *Behavioral Ecology*, **12**, 429–438.

Mook, D. G. (1987). *Motivation: The Organization of Action*. New York: Norton.

Moore, B. R. (1973). The Role of Directed Pavlovian Reactions in Simple Instrumental Learning in the Pigeon. In Hinde & Stevenson-Hinde, pp. 159–188.

Moran, T. H. (1986). Environmental and Neural Determinants of Behavior in Development. In M. S. Gazzaniga (Ed.), Handbook of Behavioral Neurobiology, *vol. 2*: Neuropsychology. New York: Plenum, pp. 99–128.

Morand-Ferron, J. & Giraldeau, L.-A. (2010). *Behavioral Ecology*, **21**, 343–348.

Morgan, C. L. (1896). *Habit and Instinct*. London: Arnold.

Morgan, T. J. H., Rendell, L. E., Ehn, M., Hoppitt, W. & Laland, K. N. (2012). The Evolutionary Basis of Human Social Learning. *Proceedings of the Royal Society B*, **279**, 653–662.

Moro, A. (2014). On the Similarity between Syntax and Actions. *Trends in Cognitive Sciences*, **18**, 109–110.

Moro, A. (2014). Response to Pulvermüller: The Syntax of Actions and Other Metaphors. *Trends in Cognitive Sciences*, **18**, 221.

Morris, D. (1956). [Discussion following Lorenz] In *L'instinct dans le comportement des animaux et de l'homme*. Paris: Masson, pp. 642–643.

Morris, W. (Ed.). (1969). *The American Heritage Dictionary of the English Language*. New York: Houghton Mifflin.

Moser, M-B. & Moser, E. I. (2016). Where Am I? Where Am I Going? *Scientific American*, **314**(1), pp. 26–33.

Mulder, C. K., Gerkema, M. P. & Van der Zee, E. A. (2013). Circadian Clocks and Memory: Time-Place Learning. *Frontiers in Molecular Neuroscience*, **6**, article 8.

Neill, S. R. & Cullen, J. M. (1974). Experiments on Whether Schooling by Their Prey Affects the Hunting Behaviour of Cephalopods and Fish Predators. *Journal of Zoology*, **172**, 549–569.

Nelson, D. A. (1997). Social Interaction and Sensitive Phases for Song Learning: A Critical Review. In Snowdon & Hausberger, pp. 7–22.

Nelson, K. (1964). The Temporal Patterning of Courtship Behaviour in the Glandulocaudine Fishes. *Behaviour*, **24**, 90–146.

Nelson, R. J. (2016). *An Introduction to Behavioral Endocrinology*, 4th edn. Sunderland, MA: Sinauer.

Newman, A. J., Supalla, T., Fernandez, N., Newport, E. L. & Bavelier, D. (2015). Neural Systems Supporting Linguistic Structure, Linguistic Experience, and Symbolic Communication in Sign Language and Gesture. *Proceedings of the National Academy of Sciences*, **112**, 11684–11689.

Novak, M. & Harlow, H. F. (1975). Social Recovery of Monkeys Isolated for the First Year of Life: I. Rehabilitation and Therapy. *Developmental Psychology*, **11**, 453–465.

O'Keefe, J. & Nadel, L. (1978). *The Hippocampus as a Cognitive Map*. Oxford: Clarendon.

Odling-Smee, F. J., Laland, K. N. & Feldman, M. W. (1996). Niche Construction. *American Naturalist*, **147**, 641–648.

Olkowicz, S, Kocourek, M., Lučan, R. K. *et al.* (2016). Birds have Primate-like Numbers of Neurons in the Forebrain. *Proceedings of the National Academy of Sciences*, **113**, 7255–7260.

Olds, J. (1958). Self-stimulation of the Brain. *Science*, **127**, 315–324.

Oppenheim, R. W. (1974). The Ontogeny of Behavior in the Chick Embryo. *Advances in the Study of Behavior*, **5**, 133–172.

Oppenheim, R. W. (1981). Ontogenetic Adaptations and Retrogressive Processes in the Development of the Nervous System and Behaviour: A Neuroembryological Perspective. In K. J. Connolly & H. F. R. Prechtl (Eds.), *Maturation and Development: Biological and Psychological Perspectives*. Philadelphia: Lippincott, pp. 73–109. (Reprinted in Bolhuis & Hogan, 1999.)

Oppenheim, R. W. (1991). Cell Death during Development of the Nervous System. *Annual Review of Neuroscience*, **14**, 453–501.

Orr, H. A. (2009). Fitness and Its Role in Evolutionary Genetics. *Nature Reviews Genetics*, **10**, 531–539.

Oster, H. (1978). Facial Expression and Affect Development. In M. Lewis & L. A. Rosenblum (Eds.), *The Development of Affect*. New York: Plenum, pp. 43–76.

Ottoni, E. B. & Izar, P. (2008). Capuchin Monkey Tool Use: Overview and Implications. *Evolutionary Anthropology*, **17**, 171–178.

Oyama, S. (1985). *The Ontogeny of Information: Developmental Systems and Evolution*. Cambridge, UK: Cambridge University Press.

Oyama, S., Griffiths, P. E. & Gray, R. D. (2001). *Cycles of Contingency: Developmental Systems and Evolution*. Cambridge, MA: MIT Press.

Pagel, M., Atkinson, Q. D. & Meade, A. (2007). Frequency of Word-use Predicts Rates of Lexical Evolution throughout European History. *Nature*, **449**, 717–720.

Pagnotta, M. (2014). On the Controversy over Non-human Culture: The Reasons for Disagreement and Possible Directions toward Consensus. *Behavioural Processes* **109**, 95–100.

Palameta, B. & Lefebvre, L. (1985). The social Transmission of a Food-finding Technique in Pigeons: What Is Learned? *Animal Behaviour*, **33**, 892–896.

Panksepp, J. (2005). Affective Consciousness: Core Emotional Feelings in Animals and Humans. *Consciousness and Cognition*, **14**, 30–80.

Panksepp, J. (2010). Affective Consciousness in Animals: Perspectives on Dimensional and Primary Process Emotion Approaches. *Proceedings of the Royal Society B*, **277**, 2905–2907.

Paul, E. S., Harding, E. J. & Mendl, M. (2005). Measuring Emotional Processes in Animals: The Utility of a Cognitive Approach. *Neuroscience and Biobehavioral Reviews*, **29**, 469–491.

Pavlov, I. P. (1927). *Conditioned Reflexes*. Oxford: Oxford University Press.

Payne, R. B., Thompson, W. L., Fiala, K. L. & Sweany, L. L. (1981). Local Song Traditions in Indigo Buntings: Cultural Transmission of Behaviour Patterns across Generations. *Behaviour*, **77**, 199–221.

Pennisi, E. (2016a). Tracking how Humans Evolve in Real Time. *Science*, **352**, 876–877.

Pennisi, E. (2016b). Fossil Fishes Challenge "Urban Legend" of Evolution. *Science*, **353**, 1483–1484.

Petherick, J. C., Seawright, E., Waddington, D., Duncan, I. J. H. & Murphy, L. B. (1995). The Role of Perception in the Causation of Dustbathing Behaviour in Domestic Fowl. *Animal Behaviour*, **49**, 1521–1530.

Petitto, L. A. & Marentette, P. F. (1991). Babbling in the Manual Mode: Evidence for the Ontogeny of lLnguage. *Science*, **251**, 1493-1496.

Pfaffmann, C. (1960). The Pleasures of Sensation. *Psychological Review*, **67**, 253-268.

Phelps, S. M., Ryan, M. J. & Rand, A. S. (2001). Vestigial Preference Functions in Neural Networks and Túngara Frogs. *Proceedings of the National Academy of Sciences*, **98**, 13161-134166.

Piersma, T. & van Gils, J. A. (2011). *The Flexible Phenotype*. Oxford: Oxford University Press.

Pievani, T. & Serrelli, E. (2011). Exaptation in Human Evolution: How to Test Adaptive vs Exaptive Evolutionary Hypotheses. *Journal of Anthropological Sciences*, **89**, 9-23.

Pinker, S. (1994). *The Language Instinct*. New York: William Morrow.

Pinker, S. (2007). *The Stuff of Thought*. New York: Viking Penguin.

Pittendrigh, C. S. (1958). Adaptation, Natural Selection, and Behavior. In A. Roe & G. G. Simpson (Eds.), *Behavior and Evolution*. New Haven, CT: Yale University Press, pp. 390-416.

Polsky, R. H. (1975). Hunger, Prey Feeding, and Predatory Aggression. *Behavioral Biology*, **13**, 81-93.

Pulvermüller, F. & Fadiga, L. (2010). Active Perception: Sensorimotor Circuits as a Cortical Basis for Language. *Nature Reviews Neuroscience*, **11**, 351-360.

Raby, C. R., Alexisi, D. M., Dickinson, A. & Clayton, N. S. (2007). Planning for the Future by Western Scrub-Jays. *Nature*, **445**, 919-921.

Raff, R. A. (2000). Evo-Devo: The Evolution of a New Discipline. *Nature Reviews Genetics*, **1**, 74-79.

Rasa, O. A. E. (1971). Appetence for Aggression in Juvenile Damsel Fish. *Zeitschrift für Tierpsychologie, Beiheft*, 7.

Réale, D, Reader, S. M., Sol, D., McDougall, P. T. & Dingemanse, N. J. (2007). Integrating Animal Temperament within Ecology and Evolution. *Biological Reviews*, **82**, 291-318.

Redish, A. D. (2001). The Hippocampal Debate: Are We Asking the Right Questions? *Behavioral Brain Research*, **127**, 81-98.

Reisbick, S. H. (1973). Development of Food Preferences in Newborn Guinea Pigs. *Journal of Comparative and Physiological Psychology*, **85**, 427-442.

Rendell, L. et al. (2011). Cognitive Culture: Theoretical and Empirical Insights into Social Learning Strategies. *Trends in Cognitive Sciences*, **15**, 68-76.

Rescorla, R. A. (1967). Pavlovian Conditioning and Its Proper Control Procedures. *Psychological Review*, **74**, 71-80.

Rescorla, R. A. (1988). Behavioral Studies of Pavlovian Conditioning. *Annual Review of Neuroscience*, **11**, 329-352.

Rescorla, R. A. & Wagner, A. R. (1972). A Theory of Pavlovian Conditioning. In A. H. Black & W.F. Prokasy (Eds.), *Classical Conditioning II*. New York: Appleton-Century-Crofts, pp. 64-99.

Resende, B. D., Nagy-Reis, M. B., Lacerda, F. N., Pagnotta, M. & Savalli, C. (2014). Tufted Capuchin Monkeys (*Sapajus* sp.) Learning how to Crack Nuts: Does Variability Decline throughout Development? *Behavioural Processes*, **109**, 89-94.

Rhoad, K. D., Kalat, J. W. & Klopfer, P. (1975). Aggression and Avoidance by *Betta splendens* toward Natural and Artificial Stimuli. *Animal Learning and Behavior*, **3**, 271-276.

Rial, R. V. et al. (2010). Evolution of Wakefulness, Sleep and Hibernation: From Reptiles to Mammals. *Neuroscience and Biobehavioral Reviews*, **34**, 1144-1160.

Rice, J. C. (1978). Effects of Learning Constraints and Behavioural Organization on the Association of Vocalizations and Hunger in Burmese Red Junglefowl Chicks. *Behaviour*, **67**, 259-298.

Richards, R. J. (1992). *The Meaning of Evolution*. Chicago, University of Chicago Press.

Richerson, P. J. & Boyd, R. (2005). *Not by Genes Alone: How Culture Transformed Human Evolution*. Chicago: University of Chicago Press.

Richter, C. P. (1942). Total Self Regulatory Functions in Animals and Human Beings. *The Harvey Lectures Series*, **38**, 63-103.

Riebel K. (2003). The "Mute" Sex Revisited: Vocal Production and Perception Learning in Female Songbirds. *Advances in the Study of Behavior*, **33**, 49-86.

Rivkees, S. A. (2003). Developing Circadian Rhythmicity in Infants. *Pediatrics*, **112**, 373-381.

Rizzolatti, G. & Craighero, L. (2004). The Mirror-neuron System. *Annual Review of Neuroscience*, **27**, 169-192.

Roberts, E. P. & Weigl, P. D. (1984). Habitat Preference in the Dark-eyed Junco (Junco hyemalis): The Role of Photoperiod and Dominance. *Animal Behaviour*, 32, 709-714.

Rodenburg, T. B. *et al.* (2004). Feather Pecking in Laying Hens: New Insights and Directions for Research? *Applied Animal Behaviour Science*, **86**, 291-298.

Roeder, K. D. (1967). *Nerve Cells and Insect Behavior*, revised edn. Cambridge, MA: Harvard University Press.

Roesch, M. R., Esber, G. R., Li, J., Daw, N. D. & Schoenbaum, G. (2013). Surprise! Neural Correlates of Pearce-Hall and Rescorla-Wagner Coexist within the Brain. *European Journal of Neuroscience*, **35**, pp. 1190-1200.

Romanes, G. J. R. (1883). *Mental Evolution in Animals*. London: Kegan Paul, Trench.

Roper, T. J. (1973). Nesting Material as a Reinforcer for Female Mice. *Animal Behaviour*, **21**, 733-740.

Roper, T. J. (1975). Nest Material and Food as Reinforcers for Fixed-ratio Responding in Mice. *Learning and Motivation*, **6**, 327-343.

Rozin, P. (1976). The Selection of Foods by Rats, Humans, and other Animals. *Advances in the Study of Behavior*, **6**, 21-76.

Rozin, P. (1984). The Acquisition of Food Habits and Preferences. In J. D. Matarazzo, S. M. Weiss, J. A. Herd, N. E. Miller & S. M. Weiss (Eds.), *Behavioral Health: A Handbook of Health Enhancement and Disease Prevention*. New York: Wiley.

Russell, J. A. (2003). Core Affect and the Psychological Construction of Emotion. *Psychological Review*, **110**, 145-172.

Rutter, M. (1991). A Fresh Look at "Maternal Deprivation." In P. Bateson (Ed.), *The Development and Integration of Behaviour*. Cambridge, UK: Cambridge University Press, pp. 331-374.

Rutter, M. (2002). Nature, Nurture, and Development: From Evangelism through Science toward Policy and Practice. *Child Development*, **73**, 1-21.

Ryan, M. J. (2009). The Evolution of Behavior, and Integrating It towards a Complete and Correct Understanding of Behavioral Biology. In J. J. Bolhuis & S. Verhulst (Eds.), *Tinbergen's Legacy*. Cambridge, UK: Cambridge University Press, pp. 127-146.

Ryan, M. J. (2005). Evolution of Behavior. In Bolhuis & Giraldeau, pp. 294-314.

Ryan, M. J. & Cummings, M. E. (2003). Perceptual Biases and Mate Choice. *Annual Review of Ecology, Evolution, and Systematics*, **44**, 437-459.

Ryan, M. J. & Rand, A. S. (1999). Phylogenetic Influences on Mating Call Preferences in Female Túngara Frogs (*Physalaemus pustulosus*). *Animal Behaviour*, **57**, 945-956.

Sachs, B. D. & Barfield, R. J. (1976). Functional Analysis of Masculine Copulatory Behavior in the Rat. *Advances in the Study of Behavior*, **7**, 92–154.

Sackett, G. P. (1970). Unlearned Responses, Differential Rearing Experiences, and the Development of Social Attachments by Rhesus Monkeys. In L. A. Rosenblum (Ed.), *Primate Behavior*. New York: Academic Press.

Saffran, J. R. (2002). Constraints on Statistical Language Learning. *Journal of Memory and Language*, **47**, 172–196.

Sahay, A., Wilson, D. A. & Hen, R. (2011). Pattern Separation: A Common Function for New Neurons in Hippocampus and Olfactory Bulb. *Neuron*, **70**, 582–588.

Sanotra, G. S., Vestergaard, K. S., Agger, J. F. & Lawson, L. G. (1995). The Relative Preferences for Feathers, Straw, Wood-shavings and Sand for Dustbathing, Pecking and Scratching in Domestic Chicks. *Applied Animal Behaviour Science*, **43**, 263–277.

Savory, C. J. (1979). Changes in Food Intake and Body Weight of Bantam Hens during Breeding. *Applied Animal Ethology*, **5**, 283–288.

Schachter, S. & Singer, J. E. (1962). Cognitive, Social, and Physiological Determinants of Emotional State. *Psychological Review*, **69**, 379–399.

Schacter, D. L. (2009). Memory Systems, Neural Basis Of. In T. Bayne, A. Clemens & P. Wilken (Eds.), *The Oxford Companion to Consciousness*. Oxford: Oxford University Press, pp. 432–436.

Schacter, D. L. & Tulving, E. (Eds.). (1994). *Memory Systems*. Cambridge, MA: MIT Press.

Schiller, P. H. (1957). Manipulative Patterns in the Chimpanzee (1949). In C. H. Schiller (Ed.), *Instinctive Behavior*. New York: International Universities Press, pp. 264–287.

Schjelderup-Ebbe, Th. (1922). *Beiträge zur Social-psychologie des Haushuhns. Zeitschrift für Psychologie*, **92**, 225–252.

Schluter, D. (1996). Ecological Causes of Adaptive Radiation. *American Naturalist*, **148** (Suppl.), S40–S64.

Schmid-Hempel, P., Kacelnik, A. & Houston, A. I. (1985). Honeybees Maximise Efficiency by Not Filling Their Crop. *Behavioral Ecology and Sociobiology*, **17**, 61–66.

Schneirla, T. C. (1949). Levels in the Psychological Capacities of Animals. In R. W. Sellars, V. J. McGill & M. Farber (Eds.), *Philosophy for the Future: The Quest for Modern Materialism*. New York: Macmillan.

Schneirla, T. C. (1956). Interrelationships of the "Innate" and the "Acquired" in Instinctive Behavior. In *L'instinct dans le comportement des animaux et de l'homme*. Paris: Masson.

Schneirla, T. C. (1965). Aspects of Stimulation and Organization in Approach/ Withdrawal Processes Underlying Vertebrate Behavioral Development. *Advances in the Study of Behavior*, **1**, 1–74.

Schneirla, T. C. & Rosenblatt, J. S. (1961). Behavioral Organization and Genesis of the Social Bond in Insects and Mammals. *American Journal of Orthopsychiatry*, **31**, 223–253.

Schneirla, T. C., Rosenblatt, J. S. & Tobach, E. (1963). Maternal Behavior in the Cat. In H. Rheingold (Ed.), *Maternal Behavior in Mammals*. New York: Wiley.

Schultz, W. (2006). Behavioral Theories and the Neurophysiology of Reward. *Annual Review of Psychology*, **57**, 87–115.

Sevenster, P. (1961). A Causal Analysis of a Displacement Activity (Fanning in *Gasterosteus aculeatus* L.). *Behaviour*, Suppl. **9**, 1–170.

Sevenster, P. (1968). Motivation and Learning in Sticklebacks. In D. Ingle (Ed.), *Central Nervous System and Fish Behaviour*. Chicago: University of Chicago Press, pp. 233–245.

Sevenster, P. (1973). Incompatibility of Response and Reward. In Hinde & Stevenson-Hinde, pp. 265–283.

Sevenster, P. & van Roosmalen, M. E. (1985). Cognition in Sticklebacks: Some Experiments on Operant Conditioning. *Behaviour*, **93**, 170–183.

Sevenster-Bol, A. C. A. (1962). On the Causation of Drive Reduction after a Consummatory Act. *Archive Néerlandaises de Zoologie*, **15**, 175–236.

Seyfarth, R. M., Cheney, D. L. & Marler, P. (1980). Vervet Monkey Alarm Calls: Semantic Communication in a Free-ranging Primate. *Animal Behaviour*, **28**, 1070–1094.

Shatz, C. J. (1992). The Developing Brain. *Scientific American*, **267**, 60–67. (Reprinted in Bolhuis & Hogan, 1999.)

Sheffield, F. D. (1966). A Drive Induction Theory of Reinforcement. In R. N. Haber (Ed.), *Current Research in Motivation*. New York: Holt, pp. 98–111.

Shelton, J. R. & Caramazza, A. (1999). Deficits in Lexical and Semantic Processing. Implications for Models of Normal Language. *Psychonomic Bulletin and Review*, **6**, 5–27.

Sherman, P. W. (1988). The Levels of Analysis. *Animal Behaviour*, **36**, 616–619.

Sherry, D. F. (1977). Parental Food-calling and the Role of the Young in the Burmese Red Junglefowl (*Gallus g. spadiceus*). *Animal Behaviour*, **25**, 594–601.

Sherry, D. F. (1981). Parental Care and Development of Thermoregulation in Red Junglefowl. *Behaviour*, **76**, 250–279.

Sherry, D. F. (1985). Food Storage by Birds and Mammals. *Advances in the Study of Behavior*, **15**, 153–188.

Sherry, D. F. (2009). Do Ideas about Function Help in the Study of Causation? In J. J. Bolhuis & S. Verhulst (Eds.), *Tinbergen's Legacy*. Cambridge, UK: Cambridge University Press, pp. 147–162.

Sherry, D. F. & Duff, S. J. (1996). Behavioural and Neural Bases of Orientation in Food-storing Birds. *Journal of Experimental Biology*, **199**, 165–171.

Sherry, D. F. & Schacter, D. L. (1987). The Evolution of Multiple Memory Systems. *Psychological Review*, **94**, 439–454.

Sherry, D. F. & Strang, C. G. (2015). Contrasting Styles in Cognition and Behaviour in Bumblebees and Honeybees. *Behavioural Processes*, **117**, 59–69.

Sherry, D. F. & Vaccarino, A. L. (1989). Hippocampus and Memory for Food Caches in Black-capped Chickadees. *Behavioral Neuroscience*, **103**, 308–318.

Sherry, D. F., Mrosovsky, N. & Hogan, J. A. (1980). Weight Loss and Anorexia during Incubation in Birds. *Journal of Comparative and Physiological Psychology*, **94**, 89–98.

Shettleworth, S. J. (1972). Constraints on Learning. *Advances in the Study of Behavior*, **4**, 1–68.

Shettleworth, S. J. (1975). Reinforcement and the Organization of Behavior in Golden Hamsters. *Journal of Experimental Psychology: Animal Behavior Processes*, **1**, 56–87.

Shettleworth, S. J. (2010). *Cognition, Evolution, and Behavior*. New York: Oxford University Press.

Shubin, N., Tabin, C. & Carroll, S. (2009). Deep Homology and the Origins of Evolutionary Novelty. *Nature*, **457**, 818–823.

Sih, A., Bell, A. & Johnson, J. C. (2004). Behavioral Syndromes: An Ecological and Evolutionary Overview. *Trends in Ecology and Evolution*, **19**, 372–378.

Silver, R. (1990). Biological Timing Mechanisms with Special Emphasis on the Parental Behavior of Doves. In D. A. Dewsbury (Ed.), *Contemporary Issues in Comparative Psychology*. Sunderland, MA: Sinauer, pp. 252–277.

Skinner, B. F. (1938). *The Behavior of Organisms*. New York: Appleton-Century-Crofts.

Skinner, B. F. (1950). Are Theories of Learning Necessary? *Psychological Review*, **57**, 193–216.

Skinner, B. F. (1953). *Science and Human Behavior*. New York: Macmillan.

Slack, J. M. W. (1991). *From Egg to Embryo*. Cambridge, UK: Cambridge University Press.

Slater, P. J. B. (1981). Individual Differences in Animal Behavior. In P. P. G. Bateson & P. H. Klopfer (Eds.), *Perspectives in Ethology*, vol. **4**. New York: Plenum, pp. 35–49.

Slater, P. J. B. (1986). The Cultural Transmission of Bird Song. *Trends in Ecology and Evolution*, **1**, 94–97.

Smith, K. S. & Graybiel, A. M. (2013). A Dual Operator View of Habitual Behavior Reflecting Cortical and Striatal Dynamics. *Neuron*, **79**, 361–374.

Snell-Rood, E. C. (2013). An Overview of the Evolutionary Causes and Consequences of Behavioural Plasticity. *Animal Behaviour*, **85**, 1044–1011.

Snowdon, C. T. & Hausberger, M. (Eds.). (1997). *Social Influences on Vocal Development*. Cambridge, UK: Cambridge University Press.

Sober, E. (2001). The Two Faces of Fitness. In R. S. Singh *et al.* (Eds.), *Thinking about Evolution*, vol. **2**. Cambridge, UK: Cambridge University Press, pp. 309–321.

Sokolowski, M. B. (2001). *Drosophila*: Genetics Meets Behaviour. *Nature Reviews Genetics*, **2**, 879–890.

Sokolowski, M. B. (2010). Social Interactions in "Simple" Model Systems. *Neuron*, **65**, 780–794.

Sol, D., Bacher, S., Reader, S. M. & Lefebvre, L. (2008). Brain Size Predicts the Success of Mammal Species Introduced into Novel Environments. *American Naturalist*, **172**, S63–S71.

Spalding, D. A. (1873/1954). Instinct with Original Observations on Young Animals. *British Journal of Animal Behaviour*, **2**, 2–11.

Speirs, A. L. (1962). Thalidomide and Congenital Abnormalities. *Lancet*, **279**, 303–305.

Spencer, H. (1864). *Principles of Biology*. London.

Staddon, J. E. R. & Ayres, S. L. (1976). Sequential and Temporal Properties of Behavior Induced by a Schedule of Periodic Food Delivery. *Behaviour*, **54**, 26–49.

Staddon, J. E. R. & Cerutti, D. T. (2003). Operant Conditioning. *Annual Review of Psychology*, **54**, 115–144.

Staddon, J. E. R. & Simmelhag, V. L. (1971). The "Superstition" Experiment: A Reexamination of Its Implication for the Principles of Adaptive Behavior. *Psychological Review*, **78**, 3–43.

Stamps, J. A. (2016). Individual Differences in Behavioural Plasticities. *Biological Reviews*, **91**, 534–567.

Stamps, J. A. & Groothuis, T. G. G. (2010). Developmental Perspectives on Personality: Implications for Ecological and Evolutionary Studies of Individual Differences. *Philosophical Transactions of the Royal Society B*, **365**, 4029–4041.

Steiner, A. P. & Redish, A. (2014). Behavioral and Neurophysiological Correlates of Regret in Rat Decision-making on a Neuroeconomic Task. *Nature Neuroscience*, **17**, 995–1002.

Steiner, J. E. (1979). Human Facial Expressions in Response to Taste and Smell Stimulation. *Advances in Child Development and Behavior*, **13**, 257–295.

Stellar, E. (1960). Drive and Motivation. In J. Field, H. W. Magoun & V. E. Hall (Eds.), *Handbook of Physiology* (Section 1, Vol. 3). Washington, DC: American Psychological Association, pp. 1501–1528.

Stephan, F. K. (2001). Food-entrainable Oscillators in Mammals. In J. S. Takahashi, F. W. Turek & R. Y. Moore (Eds.), *Handbook of Behavioral Neurobiology, vol.* **12**: Circadian Clocks. New York: Kluwer Academic/Plenum, pp. 223–246.

Stevens, M. (2005). The Role of Eyespots as Anti-predator Mechanisms, Principally Demonstrated in the Lepidoptera. *Biological Reviews*, **80**, 573–588.

Stevens, M. (2013). *Sensory Ecology, Behaviour, and Evolution*. Oxford: Oxford University Press.

Stevens, M., Hardman, C. J. & Stubbins, C. L. (2008). Conspicuousness, Not Eye Mimicry, Makes 'Eyespots' Effective Antipredator Signals. *Behavioral Ecology*, **19**, 525–531.

Stevenson, J. G. (1967). Reinforcing Effects of Chaffinch Song. *Animal Behaviour*, **15**, 427–432.

Struhsaker, T. T. (1967). Social Structure among Vervet Monkeys (*Cercopithecus aethiops*). *Behaviour*, **29**, 6–121.

Suboski, M. D. & Bartashunas, C. (1984). Mechanisms for the Social Transmission of Pecking Preferences in Neonatal Chicks. *Journal of Experimental Psychology: Animal Behavior Processes*, **10**, 182–194.

Sutherland, N. S. (1964). The Learning of Discriminations by Animals. *Endeavour*, **23**, 148–152.

Tarsitano, M. & Andrew, R. (1999). Scanning and Route Selection in the Jumping Spider Portia Labiata. *Animal Behaviour*, **58**, 255–265.

ten Cate, C. (1984). The Influence of Social Relations on the Development of Species Recognition in Zebra Finch Males. *Behaviour*, **91**, 263–285.

ten Cate, C. (1986). Sexual Preferences in Zebra Finch Males Exposed to Two Species. I. A Case of Double Imprinting. *Journal of Comparative Psychology*, **100**, 248–252.

ten Cate, C. (1987). Sexual Preferences in Zebra Finch Males Exposed to Two Species. II. The Internal Representation Resulting from Double Imprinting. *Animal Behaviour*, **35**, 321–330.

ten Cate, C. (1994). Perceptual Mechanisms in Imprinting and Song Learning. In Hogan & Bolhuis, pp. 116–146.

ten Cate, C. (1991). Behaviour-contingent Exposure to Song and Zebra Finch Song Learning. *Animal Behaviour*, **42**, 857–859.

Terman, M. (1983). Behavioral Analysis and Circadian Rhythms. In M. D. Zeiler & P. Harzen (Eds.), *Advances in Analysis of Behavior, vol.* 3: *Biological Factors in Learning*. New York: Wiley, pp. 103–141.

Thelen, E. (1985). Expression as Action: A Motor Perspective on the Transition from Spontaneous to Instrumental Behaviors. In G. Zivin (Ed.), *The Development of Expressive Behavior*. Orlando, FL: Academic Press, pp. 221–267.

Thelen, E. (1995). Motor Development: A New Synthesis. *American Psychologist*, **50**, 79–95.

Thelen, E. & Fisher, C. M. (1992). Newborn Stepping: An Explanation for a "Disappearing Reflex." *Developmental Psychology*, **18**, 760–775.

358   References

Thompson, T. I. (1963). Visual Reinforcement in Siamese Fighting Fish. *Science*, **141**, 55–57.

Thorndike, E. L. (1898). Animal Intelligence. *Psychological Review*, Monograph suppl. 2, 1–113.

Thorndike, E. L. (1911). *Animal Intelligence*. New York: Macmillan.

Thorndike, E. L. (1931). *Human Learning*. New York: Century.

Thorpe, W. H. (1956). *Learning and Instinct in Animals*. London: Methuen.

Thorpe, W. H. (1958). The Learning of Song Patterns by Birds, with Especial Reference to the Song of the Chaffinch (*Fringilla coelebs*). *Ibis*, **100**, 535–570.

Thorpe, W. H. (1961). *Bird Song*. Cambridge, UK: Cambridge University Press.

Thorpe, W. H. & Jones, F. G. W. (1937). Olfactory Conditioning and Its Relation to the Problem of Host Selection. *Proceedings of the Royal Society, Series B*, **124**, 56–81.

Timberlake, W. (1994). Behavior Systems, Associationism, and Pavlovian Conditioning. *Psychonomic Bulletin and Review*, **1**, 405–420.

Timberlake, W. (2004). Is Operant Contingency Enough for a Science of Purposive Behavior? *Behavior and Philosophy*, **32**, 197–229.

Tinbergen, J. M. (1981). Foraging Decisions in Starlings (*Sturnus vulgaris*). *Ardea*, **69**, 1–67.

Tinbergen, J. M. & Daan, S. (1990). Family Planning in the Great Tit (*Parus major*): Optimal Clutch Size as Integration of Parent and Offspring Fitness. *Behaviour*, **114**, 161–190.

Tinbergen, N. (1932/1972). On the Orientation of the Digger Wasp *Philanthus triangulum* Fabr. In N. Tinbergen (Ed.), *The Animal in Its World*, vol. 1. London: George Allen & Unwin, pp. 103–127.

Tinbergen, N. (1940). Die Übersprungbewegung. *Zeitschrift für Tierpsychologie*, **4**, 1–40.

Tinbergen, N. (1942). An Objectivistic Study of the Innate Behaviour of Animals. *Bibliotheca Biotheoretica*, **1**, 39–98.

Tinbergen, N. (1951). *The Study of Instinct*. Oxford: Oxford University Press.

Tinbergen, N. (1952). Derived Activities: Their Causation, Biological Significance, Origin and Emancipation during Evolution. *Quarterly Review of Biology*, **27**, 1–32.

Tinbergen, N. (1959). Comparative Studies of the Behaviour of Gulls (Laridae): A Progress Report. *Behaviour*, **15**, 1–70.

Tinbergen, N. (1963). On Aims and Methods of Ethology. *Zeitschrift für Tierpsychologie*, **20**, 410–433.

Tinbergen, N. (1965). *Animal Behavior*. New York: Time, Inc.

Tinbergen, N. (1972). Ethology (1969). In *The Animal in Its World*, Vol. 2. London: Allen and Unwin, pp. 130–160.

Tinbergen, N., Broekhuysen, G. J., Feekes, F., Houghton, J. C. W., Kruuk, H. & Szulc, E. (1963). Egg Shell Removal by the Black-headed Gull, *Larus ridibundus* L.; A Behaviour Component of Camouflage. *Behaviour*, **19**, 74–117.

Toates, F. M. (1983). *Animal Behaviour – A Systems Approach*. Chichester: Wiley.

Toates, F. M. & Jensen, P. (1991). Ethological and Psychological Models of Motivation: Towards a Synthesis. In J.-A. Meyer & S. Wilson (Eds.), *From Animals to Animats*, pp. 194–205. Cambridge, MA: MIT Press.

Tolman, E. C. (1948). Cognitive Maps in Rats and Men. *Psychological Review*, **55**, 189–208.

Tomasello, M. (1995). Language Is Not an Instinct. *Cognitive Development*, **10**, 131–156.

Tracy, J. L. & Randles, D. (2011). Four Models of Basic Emotions: A Review of Ekman and Cordaro, Izard, Levenson, and Panksepp and Watt. *Emotion Review*, **3**, 397–405.

Trivers, R. (1974). Parent-Offspring Conflict. *American Zoologist*, **14**, 249–264.

Trivers, R. (1972). Parental Investment and Sexual Selection. In B. Campbell (Ed.), *Sexual Selection and the Descent of Man*. Chicago: Aldine, pp. 139–179.

Trut, L. N. (1999). Early Canid Domestication: The Farm-Fox Experiment. *American Scientist*, **87**, 160–169.

Trut, L. N., Oskins, I. & Khariamova, A. (2009). Animal Evolution during Domestication: The Domesticated Fox as a Model. *BioEssays*, **31**, 349–360.

Underwood, H. (2001). Circadian Organization in Nonmammalian Vertebrates. In J. S. Takahashi, F. W. Turek & R. Y. Moore (Eds.), *Handbook of Behavioral Neurobiology*, vol. **12**: *Circadian Clocks*. New York: Kluwer Academic/Plenum, pp. 247–290.

Van der Kooy, D. & Hogan, J. A. (1978). Priming Effects with Food and Water Reinforcers in Hamsters. *Learning and Motivation*, **9**, 332–346.

van der Meer, M. A. A. & Redish, A. D. (2010). Expectancies in Decision Making, Reinforcement Learning, and Ventral Striatum. *Frontiers in Neuroscience*, **4**, 29–37.

van der Zee, E. A., *et al.* (2008). Circadian Time-Place Learning in Mice Depends on Cry Genes. *Current Biology*, **18**, 844–848.

van Gils, J. A., Schenk, I. W., Bos, O. & Piersma, T. (2003). Incompletely Informed Shorebirds that Face a Digestive Constraint Maximize Net Energy Gain when Exploiting Patches. *American Naturalist*, **161**, 777–793.

van Iersel, J. J. A. (1953). An Analysis of the Parental Behaviour of the Three-spined Stickleback (*Gasterosteus aculeatus* L.). *Behaviour, Suppl.* **3**, 1–159.

van Iersel, J. J. A. & Bol, A. C. A. (1958). Preening of Two Tern Species. A study on Displacement Activities. *Behaviour*, **13**, 1–88.

van Kampen, H. S. (1996). A Framework for the Study of Filial Imprinting and the Development of Attachment. *Psychonomic Bulletin and Review*, **3**, 3–20.

van Kampen, H. S. (1997). Courtship Food-calling in Burmese Red Junglefowl: II. Sexual Conditioning and the Role of the Female. *Behaviour*, **134**, 775–787.

van Kampen, H. S. (2015). Violated Expectancies: Cause and Function of Exploration, Fear, and Aggression. *Behavioural Processes*, **117**, 12–28.

van Kampen, H. S. & Bolhuis, J. J. (1991). Auditory Learning and Filial Imprinting in the Chick. *Behaviour*, **117**, 303–339.

van Lawick-Goodall, J. (1970). Tool Using in Primates and Other Vertebrates. *Advances in the Study of Behavior*, **3**, 195–249.

van Liere, D. W. (1992). The Significance of Fowls' Bathing in Dust. *Animal Welfare*, **1**, 187–202.

van Liere, D. W. & Bokma, S. (1987). Short Term Feather Maintenance as a Function of Dustbathing in Laying Hens. *Applied Animal Behaviour Science*, **18**, 197–204.

van Rhijn, J. G. (1991). *The Ruff*. London: Poyser.

van Schaik, C. P. *et al.* (2003). Orangutan Cultures and the Evolution of Material Culture. *Science*, **299**, 102–105.

Verkuil, Y. I. (2010). *The Ephemeral Shorebird: Population History of Ruffs*. Ph.D. thesis, *Rijksuniversiteit* Groningen.

Verkuil, Y. I. et al. (2012). Losing a Staging Area: Eastward Redistribution of Afro-Eurasian Ruffs Is Associated with Deteriorating Fuelling Conditions along the Western Flyway. *Biological Conservation*, **149**, 51–59.

Vestergaard, K. S. (1982). Dust-bathing in the Domestic Fowl: Diurnal Rhythm and Dust Deprivation. *Applied Animal Ethology*, **8**, 487–495.

Vestergaard, K. S. & Baranyiova, E. (1996). Pecking and Scratching in the Development of Dust Perception in Young Chicks. *Acta Veterinaria Brno*, **65**, 133–142.

Vestergaard, K. S. & Hogan, J. A. (1992). The Development of a Behavior System: Dustbathing in the Burmese Red Junglefowl. III. Effects of Experience on Stimulus Preference. *Behaviour*, **121**, 215–230.

Vestergaard, K. S., Damm, B. I., Abbot, U. K. & Bildsøe, M. (1999). Regulation of Dustbathing in Feathered and Featherless Domestic Chicks: The Lorenzian Model Revisited. *Animal Behaviour*, **58**, 1017–1025.

Vestergaard, K. S., Hogan, J. A. & Kruijt, J. P. (1990). The Development of a Behavior System: Dustbathing in the Burmese Red Junglefowl: I. The Influence of the Rearing Environment on the Organization of Dustbathing. *Behaviour*, **112**, 99–116.

Vestergaard, K. S., Kruijt, J. P. & Hogan, J. A. (1993). Feather Pecking and Chronic Fear in Groups of Red Junglefowl: Its Relations to Dustbathing, Rearing Environment and Social Status. *Animal Behaviour*, **45**, 1127–1140.

von Frisch, K. (1955). *The Dancing Bees*. New York: Harcourt, Brace.

von Helmholtz, H. (1878). Lecture: The Facts of Perception. *Selected Writings of Hermann Helmholtz*. Middletown, CT: Wesleyan University Press.

von Holst, E. (1935). *Über den Prozess der zentralnervösen Koordination. Pflügers Archiv für die gesamte Physiologie*, **236**, 149–158.

von Holst, E. (1937). *Vom Wesen der Ordnung im Zentralnervensystem. Naturwissenschaften*, **25**, 625–631, 641–647. [Tr. as: On the Nature of Order in the Central Nervous System. In von Holst, 1973, pp. 3–32.]

von Holst, E. (1954). Relations between the Central Nervous System and the Peripheral Organs. *British Journal of Animal Behaviour*, **2**, 89–94.

von Holst, E. (1973). *The Selected Papers of Erich von Holst*, vol. **1**. (Tr. by R. Martin). London: Methuen.

von Holst, E. & Mittelstaedt, H. (1950). *Das Reafferenzprinzip. Naturwissenschaften*, **37**, 464–476. [Tr. as: The Reafference Principle. In von Holst, 1973, pp. 139–173.]

von Holst, E. & Saint Paul, U. von. (1960). *Vom Wirkungsgefüge der Triebe. Naturwissenschaften*, 47, 409–422. [Tr. as: On the Functional Organisation of Drives. *Animal Behaviour*, 1963, 11, 1–20; also in von Holst, 1973, pp. 220–258.]

von Uexküll, J. (1921). *Umwelt und Innenwelt der Tiere*. Berlin.

von Uexküll, J. (1934). *Streifzüge durch die Umwelten von Tieren und Menschen*. Berlin: Springer. [Tr. as: *A Stroll through the World of Animals and Men*. In C. H. Schiller (Ed.), *Instinctive Behavior*. New York: International Universities Press, 1957, pp. 5–80.]

Vygotsky, L. (1934/1988). *Thought and Language*. (Tr. by A. Kozulin). Cambridge, MA: MIT Press.

Waddington, C. H. (1942). Canalization of Development and the Inheritance of Acquired Characters. *Nature*, **150**, pp. 563–565.

Waddington, C. H. (1953). Genetic Assimilation of an Acquired Character. *Evolution*, **7**, 118–126.

Waddington, C. H. (1959a). Evolutionary Systems – Animal and Human. *Nature*, **183**, 1634–1638.

Waddington, C. H. (1959b). Canalization of Development and Genetic Assimilation of Acquired Characters. *Nature*, **183**, 1654–1655.

Waddington, C. H. (1966). *Principles of Development and Differentiation*. New York: Macmillan. (Excerpt reprinted in Bolhuis & Hogan, 1999.)

Walker, S. (1983). *Animal Thought*. London: Routledge & Kegan Paul.

Warden, C. J., Warner, H. C. & Jenkins, T. N. (1931). *Animal Motivation: Experimental Studies on the Albino Rat*. New York: Columbia University Press.

Wasserman, E. A. (1973). Pavlovian Conditioning with Heat Reinforcement Produces Stimulus-directed Pecking in Chicks. *Science*, **181**, 875–877.

Watkins, C. (1969). The Indo-European Origin of English. In W. Morris (Ed.), *The American Heritage Dictionary of the English Language*. Boston, MA: Houghton Mifflin, pp. xix–xx.

Watson, A. (1973). Moults of Wild Scottish Ptarmigan, *Lagopus mutus*, in Relation to Sex, Climate and Status. *Journal of Zoology* (London) **171**, 207–223.

Watson, J. B. (1908). The behavior of Noddy and Sooty Terns. *Papers from the Tortugas Laboratory of the Carnegie Institution of Washington*, **2**, 187–255.

Watson, J. B. (1919). *Psychology from the Standpoint of a Behaviorist*. Philadelphia: Lippincott.

Wauters, A. M., Richard-Yris, M. A. & Talec, N. (2002). Maternal Influences on Feeding and General Activity in Domestic Chicks. *Ethology*, **108**, 529–540.

Wayne, R. K. & vonHoldt, B. (2012). Evolutionary Genomics of Dog Domestication. *Mammalian Genome*, **23**, 3–18.

Webb, W. B. (1998). Sleep. In G. Greenberg & M. M. Haraway (Eds.), *Comparative Psychology: A Handbook*. New York: Garland, pp. 327–331.

Wehner, R. & Menzel, R. (1990). Do Insects Have Cognitive Maps? *Annual Review of Neuroscience*, **13**, 403–414.

Werker, J. F. & Tees, R. C. (1992). The Organization and Reorganization of Human Speech Perception. *Annual Review of Neuroscience*, **15**, 377–402.

West-Eberhard, M. J. (2003). *Development Plasticity and Evolution*. Oxford: Oxford University Press.

West, M. J. & King, A. P. (1988). Female Visual Display Affects the Development of Males Song in the Cowbird. *Nature*, **334**, 244–246.

Whiten, A. *et al.* (2001). Charting Cultural Variation in Chimpanzees. *Behaviour*, **138**, 1481–1516.

Whiten, A., Goodall, J., McGrew, W. C., Nishida, T., Reynolds, V., Sugiyama, Y., Tutin, C. E. G., Wrangham, R. W. & Boesch, C. (1999). Cultures in Chimpanzees. *Nature*, **399**, 682–685.

Whitman, C. O. (1911). *The Behavior of Pigeons*. Washington, DC: Carnegie Institute, Publication, **257**, pp. 1–161.

Wiesel, T. N. (1982). Postnatal Development of the Visual Cortex and the Influence of the Environment. *Nature*, **299**, 583–591.

Wilczynski, W. Rand, A. S. & Ryan, M. J. (2001). Evolution of Calls and Auditory Tuning in the *Physalaemus pustulosus* Species Group. *Brain, Behavior and Evolution*, **58**, 137–151.

Williams, C. L. (1991). Development of a Sexually Dimorphic Behavior: Hormonal and Neural Controls. In H. N. Shair, G. A. Barr & M. A. Hofer (Eds.), *Developmental Psychobiology: New Methods and Changing Concepts*. New York: Oxford University Press, pp. 206–222.

Williams, G. C. (1966). *Adaptation and Natural Selection: A Critique of Some Current Evolutionary Thought*. Princeton: Princeton University Press.

Wilson, D. M. (1966). Central Nervous Mechanisms for the Generation of Rhythmic Behaviour in Arthropods. *Symposia of the Society for Experimental Biology*, **20**, 199–228.

Wilson, D. S. & Wilson, E. O. (2007). Rethinking the Theoretical Foundation of Sociobiology. *Quarterly Review of Biology*, **82**, 327–348.

Wilson, E. O. (1971). *The Insect Societies*. Cambridge, MA: Harvard University Press.

Wilson, E. O. (1975). *Sociobiology: The New Synthesis*. Cambridge, MA: Harvard University Press.

Wiltschko, R. & Wiltschko, W. (2003). Avian Navigation: From Historical to Modern Concepts. *Animal Behaviour*, **65**, 257–272.

Wise R.A. (2004). Dopamine, Learning and Motivation. *Nature Reviews Neuroscience*, **5**, 483–494.

Wolpert, D. M., Diedrichsen, J. & Flanagan, J. R. (2011). Principles of Sensorimotor Learning. *Nature Reviews Neuroscience*, **12**, 739–751.

Wood-Gush, D. G. M. (1971). *The Behaviour of the Domestic Fowl*. London: Heinemann.

Wouters, A. G. (2003). Four Notions of Biological Function. *Studies in History and Philosophy of Biological and Biomedical Sciences*, **34**, 633–668.

Wray, G. A. *et al.* (2014). Does Evolutionary Theory Need a Rethink? No, All Is Well. *Nature*, **514**, 161–164.

Wynne, C. D. L. (2004). *Do Animals Think?* Princeton: Princeton University Press.

Xie, L. *et al.* (2013). Sleep Drives Metabolite Clearance from the Adult Brain. *Science*, **342**, 373–377.

Young, P. T. (1961). *Motivation and Emotion*. New York: Wiley.

Zahavi, A. (1975). Mate Selection – A Selection for a Handicap. *Journal of Theoretical Biology*, **53**, 205–214.

Zeder, M. A. (2012). The Domestication of Animals. *Journal of Anthropological Research*, **68**, 161–190.

Zeiler, M. D. (1992). On Immediate Function. *Journal of the Experimental Analysis of Behavior*, **57**, 417–427.

Zuberbühler, K. (2000). Referential Labeling in Diana Monkeys. *Animal Behaviour*, **59**, 917–927.

Zuberbühler, K. (2009). Survivor Signals: The Biology and Psychology of Animal Alarm Calling. *Advances in the Study of Behavior*, **40**, 277–322.

Name Index

Subject Index

Printed in the United States
By Bookmasters